Pancham Shukla, Rajanikanth Aluvalu, Shilpa Gite and Uma Maheswari V (eds.)
**Computer Vision**

# De Gruyter Frontiers in Computational Intelligence

Volume 15

# Computer Vision

Applications of Visual AI and Image Processing

Edited by
Pancham Shukla, Rajanikanth Aluvalu, Shilpa Gite
and Uma Maheswari V

DE GRUYTER

**Editors**
Dr. Pancham Shukla
Principal Teaching Fellow
Department of Computing, Faculty of Engineering
Imperial College London
South Kensington Campus
London SW7 2AZ
E-mail: pancham.shukla@imperial.ac.uk

Dr. Rajanikanth Aluvalu
Department of IT,
Chaitanya Bharathi Institute of Technology
Gandipet, Hyderabad-500075
Telangana, India
Rajanikanth.aluvalu@ieee.org

Dr. Shilpa Gite
Computer Science Department
Symbiosis International (Deemed University)
Lavale, Mulshi 412115
Maharashtra, India
shilpa.gite@sitpune.edu.in

Prof. Uma Maheswari V
Department of Computer Science and Engineering
KG Reddy College of Engineering and Technology,
Moinabad, Hyderabad, Telangana, India
umamaheswari@ieee.org

ISBN 978-3-11-075667-8
e-ISBN (PDF) 978-3-11-075672-2
e-ISBN (EPUB) 978-3-11-075682-1
ISSN 2512-8868

**Library of Congress Control Number: 2022941864**

**Bibliographic information published by the Deutsche Nationalbibliothek**
The Deutsche Nationalbibliothek lists this publication in the Deutsche Nationalbibliografie; detailed bibliographic data are available on the internet at http://dnb.dnb.de.


Cover image: shulz/E+/getty images
Typesetting: Integra Software Services Pvt. Ltd.
Printing and binding: CPI books GmbH, Leck

MIX
Papier | Fördert gute Waldnutzung
FSC® C083411

www.degruyter.com

# Preface

Artificial intelligence (AI) has occupied almost all the fields of our lives nowadays by making machines work and think like humans. To build the ability to "see" the objects, identify scenes and act accordingly, machines should have "vision" ability. Computer vision (CV) is considered as one of the most powerful yet challenging fields of AI. To provide visual interpretation and contextual understanding of the digital images and videos by the machines, CV and its applications are well adopted by all the domains. The state of the art of CV models are most useful for image and video analysis and prediction of various applications in real life. CV tackles all images and visual representation/interpretation-related problems using the machine's cognitive abilities that are possible with the help of deep learning approaches. Hence, to make the readers aware about recent advances in CV image processing, this book was planned.

This book focuses on the details of such techniques and algorithms along with the interesting case studies in the industry and also in our day-to-day lives. Review articles are a part of this book which enlightens about agriculture and natural resources management using CV, image processing-related topics such as color constancy, hyperspectral imaging and also face recognition for dictionary reading. This book discusses latest developments in the fields of visual AI, image processing and CV. It provides state-of-the-art applications that highlight CV algorithms, processes, novel architectures and fascinating results underlying machine intelligence with detailed execution flow of models. It also demonstrates research in basic processes like image preprocessing, feature extraction, enhancement along with interesting applications in biometrics, healthcare and others.

Readers of this book will gain a thorough understanding of CV, image processing and aligned application domains. Not only UG/PG/PhD students but also CV researchers and industry professionals will be benefited by its contents. Anyone with a basic background in image processing will find this book easily readable and adaptable. Happy reading!

https://doi.org/10.1515/9783110756722-202

# Contents

# About the Editors

**Dr. Pancham Shukla** holds PhD in electrical and electronic engineering from Imperial College London, MPhil in electronic and electrical engineering from the University of Strathclyde (Glasgow, UK), ME in electronic and computer engineering and BE in electronic engineering from Sardar Patel University (India).

Dr. Pancham Shukla is a principal teaching fellow in the Faculty of Engineering at Imperial College London. Dr. Shukla previously worked as a senior lecturer in the School of Computing and Digital Media at London Metropolitan University; as a research fellow in Applied Digital Signal and Image Processing Research Centre (ADSIP) at the University of Central Lancashire; and as a research associate in Communications and Signal Processing (CSP) group at Imperial College London. Dr. Shukla holds more than 25 years of teaching and research experience in various universities in the UK and India.

Dr. Pancham Shukla has supervised a broad range of projects and published papers in signal/image processing, and electronic and communications systems. He has organized conferences and has served as a technical session chair along with reviewing many international conference and journal papers. Dr. Shukla was a deputy director of the Centre for Communications Technology, vice chair of the Research Student Progress Group and PhD Degrees in the School of Computing and Digital Media at London Metropolitan University.

Dr. Shukla received Vice Chancellor's award for his outstanding contribution to learning and teaching as well as Student Union's award as an outstanding member of academic staff while working at London Metropolitan University.

**Dr. Rajanikanth Aluvalu** is working as Professor, Department of IT, Chaitanya Bharathi Institute of Technology, Hyderabad, India. He worked as professor and Head, Department of Computer Science and Engineering, Vardhaman College of Engineering, Hyderabad, India; vice-chair, entrepreneurship and startup committee, past treasurer and secretary at IEEE Computer Society, Hyderabad Section. He is senior IEEE member, life member of ISTE and member of MIR Labs, USA. He is having more than 17 years of teaching experience. He obtained his PhD in cloud computing as specialization. He is the academic editor for *Peerj Computer Science* journal. He has published more than **90 Research articles** in various peer-reviewed journals and conferences. He is the editor for *The Fusion of Internet of Things, Artificial Intelligence, and Cloud Computing in Health Care* book published by Springer. He is serving as an academic editor for *Peerj Computer Science* journal, editor for *IJDMMM* journal published by Inderscience Publisher and reviewer for many Scopus-indexed and SCI journals.

He is recipient of IUCEE Faculty Fellow award in the year 2018 and Best Advisor from IEEE Hyderabad Section in the year 2020.

https://doi.org/10.1515/9783110756722-204

**Dr. Shilpa Gite** received PhD for deep learning for assistive driving in semiautonomous vehicles from Symbiosis International (Deemed University), Pune, India, in 2019. Currently, she is working as an associate professor at the Computer Science Department of Symbiosis Institute of Technology, Pune. She is also working as an associate faculty at Symbiosis Centre of Applied A.I. (SCAAI). She has around 13 years of teaching experience. Her research areas include deep learning, machine learning, medical imaging and computer vision. She is currently guiding six PhD students in biomedical imaging, self-driving cars and natural language processing areas. She is also the recipient of Best Paper award at 11th IEMERA Conference held virtually at Imperial College, London, in October 2020. She has published more than 50 research articles in various peer-reviewed journals with Scopus and Web of Science indexed and many international conferences. She is also working as a reviewer for many Scopus-indexed and SCI journals.

**Dr. Uma Maheswari V** is working as Associate Professor and Head, Department of CSE, KG Reddy College of Engineering and Technology, Hyderabad, India. She worked earlier in Vardhaman College of Engineering, Hyderabad, India. She is volunteering as treasurer at IEEE-RAS, Hyderabad Section. She completed her PhD from Visveswaraya Technological University, Belgaum, in the area of image analytics and data science. She has published 20+ research articles in SCI, ESCI, WoS, DBLP, and SCOPUS-indexed journals and conferences. She has published four Indian patents on facial expression analysis in the field of medical, E-commerce,

education, and security. She is a member of IEEE. Dr. Uma Maheswari has done an enormous study and has given contributions in the field of facial expression analysis and applications. She has constructed feature vector for a given image based on the directions and introduced dynamic threshold values while comparing the images, which helps to analyze any image. She has done research on the similarity of images in a given database to retrieve the relevant images. She also worked with convolutional neural networks by giving the preprocessed input image to improve the accuracy. It has been proved that the maximum edge intensity values are enough to retrieve the required feature from the image instead of working on total image data. Prof. Uma Maheswari is the coordinator for TEDxVCE, organized various technical programs and served as a technical committee member and reviewer for various conferences. She has delivered sessions in various capacities.

Prof. Uma Maheswari had received the Best Faculty award under the innovation category for the year 2019 by the CSI Mumbai Chapter.

# List of Contributors

**Suresh Kumar B.**
Vardhaman College of Engineering,
Hyderabad, India
E-mail: sureshkumargoud2006@gmail.com

**Viswanadha Raju S.**
JNTUCEJ, Jagityal, India
E-mail: svraju.jntu@gmail.com

**Uma Maheswari V.**
KG Reddy College of Engineering and
Technology, Hyderabad, India
E-mail: umamaheswari@ieee.org

**Mehul S. Raval**
Ahmedabad University, Ahmedabad, India
E-mail: mehul.raval@gmail.com

**G. P. Obi Reddy**
ICAR-National Bureau of Soil Survey and Land
Use Planning, Nagpur, India
E-mail: obireddygp@gmail.com

**Sanjay Chaudhary**
Ahmedabad University, Ahmedabad, India
E-mail: sanjay.chaudhary@ahduni.edu.in

**J. Adinarayana**
Indian Institute of Technology Bombay,
Mumbai, India
E-mail: adi@csre.iitb.ac.in

**Umamaheswari P.**
Department of CSE, SRC, SASTRA Deemed
University, Tamil Nadu, India,
E-mail: pum@it.sastra.edu

**Vaishnavi D.**
Department of CSE, SRC, SASTRA Deemed
University, Tamil Nadu, India,
E-mail: vaishnavi@src.sastra.edu

**Indira Linginani**
Department of CSE, GITAM
E-mail: indiralingineni@gmail.com

**Akkalakshmi Muddana**
Department of CSE, GITAM Hyderabad, India
E-mail: amuddana@gitam.edu

**Gagan Dubey**
IT, Manipal University, Jaipur, Rajasthan,
India,
E-mail: gagan.dubey23@gmail.com

**G. Malini Devi**
Dept. of CSE, G. Narayanamma Institute of
Technology and Science (For Women),
E-mail: gmalini12@gnits.ac.in

**Kiran P. Kamble**
Shivaji University, Kolhapur
E-mail: kirankamble5065@gmail.com

**Sonali Patil**
Walchand College of Engineering, Sangli,
E-mail: sonalipatil25oct@gmail.com

**Pratibha Garje**
Walchand College of Engineering, Sangli
E-mail: pratibhagarje123@gmail.com

**Pooja D. Mundada**
Walchand College of Engineering, Sangli
Email:

**Vaibhav Tasgaonkar**
Symbiosis Institute of Technology, Pune,
Maharashtra, India,
E-mail: tasgaonkarvaibhav@gmail.com

**Ujwala Kshirsagar**
Symbiosis Institute of Technology, Pune,
Maharashtra, India,
E-mail: kshirsagarujwala9@gmail.com

**Soham Hudnurkar**
Symbiosis Institute of Technology, Pune,
Maharashtra, India
E-mail: sohamhudnurkar@gmail.com

https://doi.org/10.1515/9783110756722-205

**Pravin Sridhar**
Symbiosis Institute of Technology, Pune, Maharashtra, India
E-mail: pravinsrdhr@gmail.com

**J. Srinivas**
Matrusri College of Engineering, Hyderabad, India
E-mail: jagirdar.srinivas@gmail.com

**Rajesham Gajula**
CSED, Muffakam Jah College of Engineering and Technology, Hyderabad,
E-mail: rajeshamgajula@gmail.com

**Rutuja D. Bhandari**
MIT ADT University,MIT SOE,Pune,India
E-mail: rutuja2199@gmail.com

**Shraddha Phansalkar**
MIT ADT University, MIT SOE, Pune, India
E-mail: shradda.phansalkar@mituniversity.edu.in

**Megharani Patil**
Thakur college of Engineering and Technology, Mumbai, India
E-mail: megharani.patil@thakureducation.org

**Yash Gupta**
Thakur college of Engineering and Technology, Mumbai, India
E-mail: gupta.yash211l@gmail.com

**Vatsal Kadakia**
Thakur college of Engineering and Technology, Mumbai, India
E-mail: vatsalkadakia03@gmail.com

**Harsh Doshi**
Thakur college of Engineering and Technology, Mumbai, India
E-mail: harshgautamdoshi16@gmail.com

**Kiran L. N. Eranki**
CSE and AI, SR University, Warangal,
E-mail: eranki.kiran@sru.edu.in

**Abdulnasser Alwaheed**
Leeds Beckett University, School of Built Environment, Engineering and Computing, Caedmon Hall, Leeds, LS6 3QS, United Kingdom,
E-mail: a.alwaheed6446@student.leedsbeckett.ac.uk

**Md Akmol Hussain**
Leeds Beckett University, School of Built Environment, Engineering and Computing, Caedmon Hall, Leeds, LS6 3QS, United Kingdom

**Akbar Sheikh-Akbari**
Leeds Beckett University, School of Built Environment, Engineering and Computing, Caedmon Hall, Leeds, LS6 3QS, United Kingdom; School of Built Environment, Engineering and Computing, Leeds Beckett University

**Henry Ndu**
School of Built Environment, Engineering and Computing, Leeds Beckett University

**Iosif Mporas**
School of Engineering and Computer Science, University of Hertfordshire

**Sadu Chiranjeevi**
Rajiv Gandhi University of Knowledge Technologies Nuzvid, India
E-mail: c.sadu@iitg.ac.in

**Pradip Kumar Das**
Indian Institute of Technology Guwahati, India-781039
E-mail: pkdas@iitg.ac.in

**Pavan Kumar Asundi**
Central Manufacturing Technology Institute, Bengaluru, India

**Deepa Radhakrishna**
Central Manufacturing Technology Institute, Bengaluru, India

**Kavitha Veeraswamy**
Central Manufacturing Technology Institute, Bengaluru, India

Suresh Kumar B., Viswanadha Raju S., Uma Maheswari V.

# 1 OpenCV libraries for computer vision

**Abstract:** From the last few decades, researchers are enthusiastic to develop machines that work and act as human beings. Computer vision (CV) is the most excellent field for researchers to fulfill their dreams. CV is a subfield of artificial intelligence. CV works with visual inputs like digital images and videos. Mainly it focuses on creating systems, which process and analyze the visual input data, to get meaningful information from those inputs. It aims to perform different types of operations on supplied visual inputs. Object classification, object identification, object tracking, edge detection, feature extraction and human identification are few applications of CV. To perform different types of operations, special algorithms with minimum computation time are required. Various libraries are developed to perform CV operations. Among all open sources, computer vision library (OpenCV) is the most popular and widely used open-source library for developing CV applications. OpenCV is designed to develop CV applications using inbuilt methods with less computational time. The major goal of OpenCV library is to provide a classy view to researchers with an optimized code mechanism on basic infrastructure. In OpenCV, optimized code is developed for a lot of applications like filters and edge detection. Using this optimal code, researchers can develop their applications. OpenCV supports various operating systems, including mobile operating systems. This chapter discusses developing computer visual applications using OpenCV. This will be useful for researchers to understand and use OpenCV for their applications.

**Keywords:** computer vision, OpenCV, image processing, image manipulations, classification, edge detection, machine learning, face recognition

## 1.1 Introduction

### 1.1.1 What is the computer vision

Computer vision (CV) is the field where system can get the prominent perceptive from image and video streams. It includes various methods like analyzing, processing and extracting meaningful information from the images, which in turn leads to take decisions for image analysis [1]. Currently, lot of algorithms in CV are designed

**Suresh Kumar B.,** Vardhaman College of Engineering, Hyderabad, India
**Viswanadha Raju S.,** JNTUCEJ, Jagityal, India
**Uma Maheswari V.,** KG Reddy College of Engineering and Technology

https://doi.org/10.1515/9783110756722-001

for pattern analysis. In this process, systems will be trained for enormous visual data. In training, the system will process the visual data and recognize the patterns [2]. Let us consider an example to identify dogs and cats in the set of images. To do this, create a model which will train on a set of images and finds patterns of the images and as a result it will detect whether a particular image belongs to a dog or a cat.

To develop this type of method, CV requires a strong library. OpenCV is one of the beautiful and most popular library which supports in lot of CV applications. Let us see the brief introduction to OpenCV [3, 4]. Gary Bradsky introduced OpenCV in 1990 and it is released in 2000. To run this OpenCV on Intel platform, Vadim Pisarevsky joined with Gary Bradsky. OpenCV has won DARPA grand challenge in 2005. At present, OpenCV supports in fields such as machine learning and artificial intelligence, and is going up step by step [5]. OpenCV supports lot of programming languages like C++, Java and Python, and OpenCV also supports a wide variety of operating systems like Ubuntu and Windows.

## 1.1.2 Installation of OpenCV

This chapter mainly focuses on practical part of the CV application with OpenCV with Python programming, and this chapter covers mostly used methods in OpenCV [6]. There are few prerequisites for CV applications; those are like knowledge on python and few packages like numpy and matplot and jupyter notebook. Jupyterlab is a web-based interface for a user. It helps to work with Jupyter notebook.

### 1.1.2.1 Installation

With the help of pip, JupyterLab can be installed in any operating systems. To install JupyterLab using pip, use the following command:

```
pip install jupyterlab
```

Note: pip3 will support Python 3.x versions.

## 1.1.3 Working with images

We need to install few packages for the development of CV applications. To enhance the OpenCV, we are using Jupyter notebook. After installing Jupyter notebook, open the terminal in your system and type “Jupyter notebook.”

**Figure 1.1:** Opening Jupyter notebook.

It will open the web browser with Jupyter notebook and click new button and select python3 version to start. Refer Figure 1.1. To install specified package, use pip to install any sort of libraries. Let us check how to install numpy package. Refer Figure 1.2.

```
pip install numpy
```

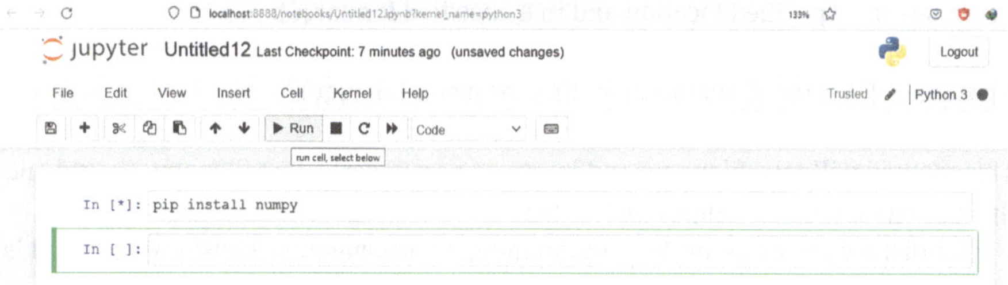


**Figure 1.2:** Installation of numpy package.

Use the same method to install other packages like matplotlib and pandas. Refer Figure 1.3.

```
pip install matplotlib
```

**Figure 1.3:** Installation of matplotlib package.

pip install opencv-python

### 1.1.4 Reading an image

The initial step to work for CV application is reading an image or video frame. Initially, we focus on reading of an image and in the later stage we will work on how to deal with video frames.

Firstly, use import command to apply OpenCV package in our programming:

syntax: **import cv2**

To read any image, cv2.imread() method is used [7, 8]. This method works to import an image from the desired location. It will return Empty matrix if it is not a valid path. Let us see the syntax:

Syntax: cv2.imread(path, flag)

This method will take two parameters: image path file and flag bit. Flag represents how image is supposed to be read, and by default it is cv2.IMREAD_COL. Basically, this method supports different types of file formats like .jpeg and .png. To save an image into the local drive, cv2.imwrite() method is used. This method helps to save the image in a specified location and in a specified format [6]:

**Syntax:** cv2.imwrite(filename(including extension), image)

This method will take filename and image as parameters. Here filename should end .jpeg, .png or any other supported format.

To display an image in the programming environment, cv2.imshow() method is used.

This method presents the image in window and automatically the window fits to the size of the image:

**Syntax:** cv2.imshow(windowname,image)

This method is having parameters window_name and the image which is to be displayed. Refer Figure 1.4.

Let us see an example of program for reading, writing and showing an image:

```
# import the cv2 library
import cv2 as cv
# read an image.
```

```
img_originalimage = cv.imread('image1.png',0)
# Display the originalimage image in a window.
cv.imshow('graycsale image',img_ originalimage)
#To close all the created windows use cv.destroyAllWindows()
cv.destroyAllWindows()
# Save the image
cv.imwrite('grayscale.jpg',img_grayscale)
```

**Figure 1.4:** Read, write and save the image using OpenCV.

To change Color space of any of an image, cv2.cvtColor() method is used. Nearly 150 color space conversion methods are present in OpenCV:

**Syntax: cv2.cvtColor(source,color space conversion)**

This method accepts parameters like source, color space conversion and few optional parameters.

Let us see few basic operations of OpenCV, to know the pixel values and to provide row and column coordinates. This process will return the values of blue, green and red in an array of the BGR image. Firstly, load the BGR image:

```
#importing numpy for array creation
import numpy
#importing Opencv
import cv2  as cv
#read original image
image = cv.imread("image1",1)
#accessing the pixel values
pixel1 = image[150,150]
#print pixel values
print(pixel1)
Output:
[255 255 255]
```

To access image properties like size, shape and dimensions, use ndarry.shape method, and by providing the index position we will get the height, width and channels.

Let us see an example:

```
import cv2
# read image
image= cv2.imread(r'image1.png',1)
 # height, width, number of channels in image
image_ height = image.shape[0]
image_width = image.shape[1]
image_channels = image.shape[2]
image_size = image.size
print('Height of the image : ', image_height)
print('Width of the image : ', image_width)
print('Number of Channels in an image: ', image_channels)
print('Size of the image:', image_ size)
```

OpenCV is having lot of methods for image resize, resizing nothing but scaling.

The syntax for image resize is:

**cv2.resize(source, dsize[, dst[, fx[, fy[, interpolation]]]]).** This method is having source of the input image, desired size for the output image and few optional parameters. Resizing is one of the important factors in many real-time applications. The main intention behind the resizing is to reduce the number of pixels in the images without disturbing the information of the image. To resize the image, **INTER_AREA, INTER_CUBIC, INTER_LINEAR** are available for resizing the image

```
import cv2
 Original _image = cv2.imread('image1.png',1)
 print('Original Dimensions : ', Original _image.shape)
 width1 = 450
height1 = 350
dimension = (width1, height1)
 # resize image
resize_image1 = cv2.resize(Original _image, dimension, interpolation = cv2.INTER_AREA)
 print('Resized Dimensions : ',resize_image1.shape)
 cv2.imshow("Resized image", resize_image1)
cv2.waitKey(0)
cv2.destroyAllWindows()
```

## 1.2 Image smoothing with a variety of low-pass filters

Image blurring is mainly used for noise removal. It is accomplished by a low-pass filter. It eliminates high-frequency information from the image. OpenCV provides few most important smoothing techniques.

### 1.2.1 Averaging

In this method, average kernel is applied on the image pixel, and it will calculate the average value and replace the average in the center of the kernel area. This is done by applying the following functions: cv2.blur() or cv2.boxFilter(). We provide the detail of the kernel whenever it is used in programming. Refer figure 1.5.

Let us see one example where the kernel size is 3 × 3:

```
import cv2 as cv
import numpy as np
from matplotlib import pyplot as plt
original_image = cv.imread('image1.png')
converted_blur1 = cv.blur(original_image,(3,3))
cv.imshow(original_image)
cv.imshow(converted_blur1)
```

**Result:**

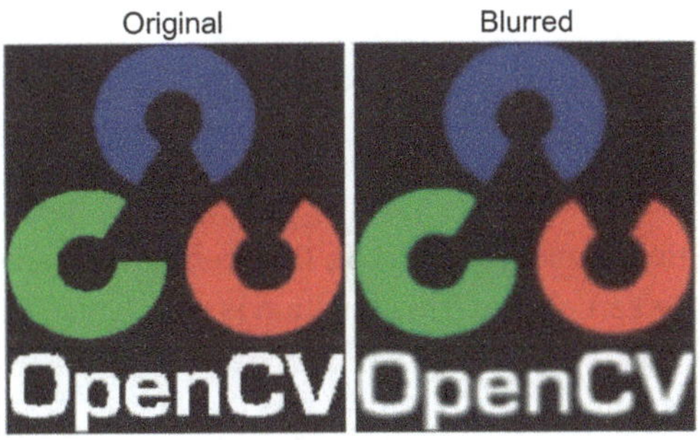


**Figure 1.5:** Image blurring using OpenCV.

### 1.2.2 Gaussian blurring

In this method, Gaussian kernel is used. It is done by applying the function cv.GaussianBlur().

The syntax for Gaussian blurring method is GaussianBlur(src, dst, ksize, sigmaX). In this, kernel should be positive and odd. Let us see the example:

```
# importing cv2 and numpy libraries
import cv2
Original_image1 = cv2.imread('image1.png',1)
cv2.imshow('Original', Original_image1)
Gaussian_image = cv2.GaussianBlur(image, (7, 7), 0)
cv2.imshow('Blurred', Gaussian_image)
cv2.waitKey(0)
```

**Kindly refer Figure 1.6 for Result:**

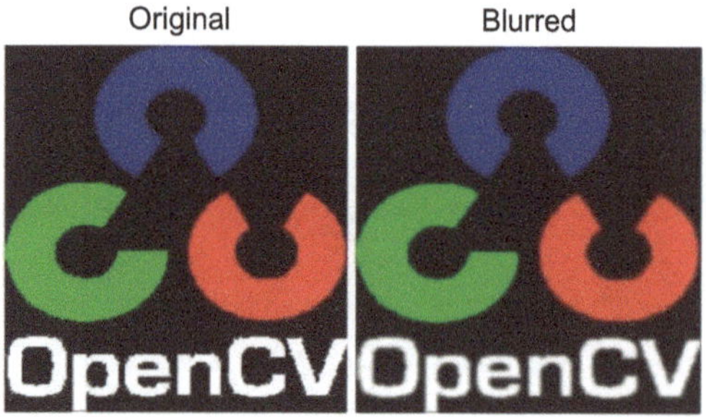


**Figure 1.6:** Image blurring using GaussianBlur.

### 1.2.3 Median blurring

Median filter is applied using the function cv.medianBlur(). In this, central element is replaced with median value under the kernel area. This function is mainly useful in removing the graying noise in an image.

The syntax for median blurring is as follows:

medianBlur(source, distination, kernal size)

Let us see one example here:

```
# importing cv2 and numpy libraries
import cv2 as cv
Original_image1= cv.imread('image1.png',1)
```

**cv.imshow('Original', Original_image1)**
**# Gaussian Blur**
**median_image = cv.medianBlur(Original_image1,5)**
**cv.imshow('Median', median_image)**
**cv.waitKey(0)**

**Kindly Refer Figure 1.7 for Result:**

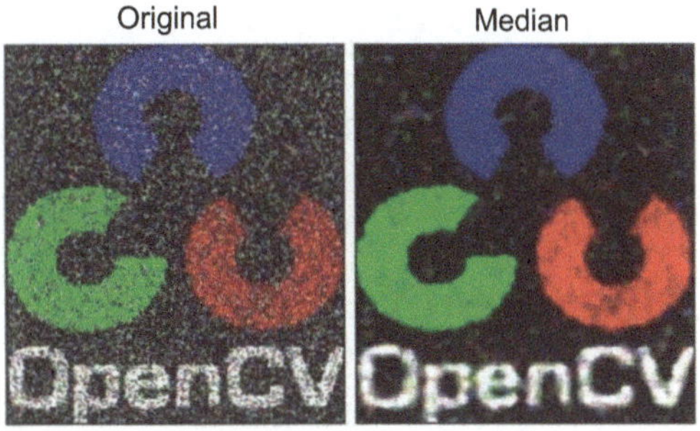


**Figure 1.7:** Image blurring using median filter.

## 1.2.4 Bilateral filtering

This function is used effectively to remove noise as maintaining edges sharp [9, 10]. The function used here is cv.bilateralFilter(). The main drawback of this function is that when compared with other functions, the bilateral filtering operations are very slow:

import cv2 as cv
# Read the **Original image.**
**Original_image1** = cv.imread('image1.png')
# Apply bilateral filter with d = 20, and sigmaColor =50 sigmaSpace = 50.
bilateral_image = cv.bilateralFilter(**Original_image1**, 15, 75, 75)
cv.imwrite('bilateral.jpg', bilateral_image)

**Kindly Refer Figure 1.8 for Result:**

**Figure 1.8:** Image blurring using bilateral filter.

## 1.3 Cropping an image using OpenCV

Selecting a region which we are interested is nothing but cropping. The main intention of cropping is removing unnecessary information like object or any areas from specified image and also it is helpful for extracting particular features from the images. In OpenCV, we do not have any specific function to crop an image. To do this particular operation, we will convert the image into numpy nd-array and apply the slicing operations. Let us see the basic syntax:

```
cropped_image = img[starting_row:ending_row,starting_col:ending_col]
Example:
# Import packages
import cv2 as cv
import numpy as np
from matplotlib import pyplot as plt
Original_image1 = cv.imread('image1.png')
print(Original_image1.shape) # Print image shape
# Cropping an image
cropped_image = Original_image1 [150:300, 150:250]
# Display cropped image
cv.imshow("cropped", cropped_image)
cv.destroyAllWindows()
```

**Kindly Refer Figure 1.9 for Result:**

```
(249, 202, 3)
```

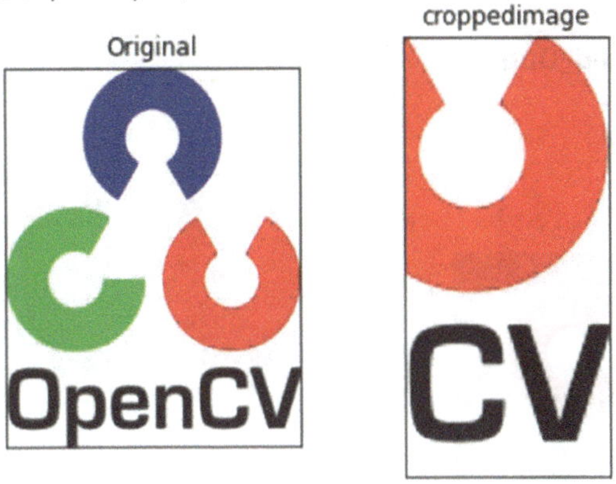


**Figure 1.9:** Cropping an image using OpenCV.

## 1.4 Image rotation

Many of the deep learning applications are developed based on the quality and quantity of the data. Many researchers believe that to develop deep learning models, it needs more amount of data. But, in some cases, it is not possible to get huge amount of data and it is not always possible to manually collect and label the data; so, in order to overcome this problem, data augmentation is introduced [11]. Let us take a dive into data augmentation. With the help of this method, one can generate more data for different purposes. Data augmentation utilizes the available data to produce the novel data, by appropriate image functions like rotation, scaling, and translation.

In data augmentation, rotation is one of the popularly used techniques. Rotation will perform by rotating the image at a subjective position and given that it is the same label as the original image. Let us see the implementation part of the rotation:

```
#importing the required libraries for data augmentation
import numpy as np
import cv2 as cv
import matplotlib.pyplot as plt
original_image = cv.imread('image1.png')
rows,cols = image.shape[:2]
#(col/4,rows/4) is the one fourth of center rotation for the image
```

```
# one fourth is the coordinates of the center
Rotated_image1 = cv2.getRotationMatrix2D((cols/4,rows/4),90,1)
Rotated_Output1 = cv2.warpAffine(image,Middle,(cols,rows))
cv.imshow('Input',original_image)
cv.imshow('rotatedOutput', Rotated_Output)
```

**Kindly refer figure 1.10 for Result:**

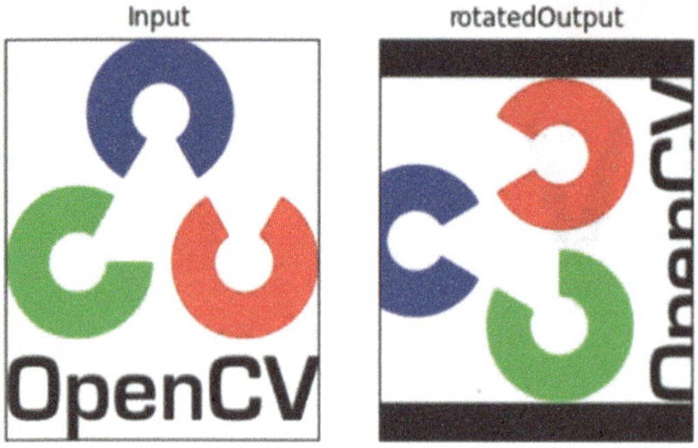


**Figure 1.10:** Image rotation using OpenCV.

## 1.5 Image translation

Image translation is also a mostly used data augmentation. The name suggests that using this technique, the original image is translated from one place to another place [12]. By applying translation different from one pixel, it will generate a new position. Let us consider x1,y1 location points are present and dx,dy are translations that are different to produce new positions X1,Y1:

$$X1 = x1 + dx$$

$$Y1 = y1 + dy$$

This translation is normally used in the image preprocessing stage. Let us see the implementation part of translation:

```
#importing the required libraries
import numpy as np
import cv2 as cv
import matplotlib.pyplot as plt
#reading the original_image
original_image = cv.imread('image1.png')
#translating the original image in 200 pixels in both dimensions
```

```
center_image = np.float32([[1,0,-200],[0,1,-200]])
Translation = cv.warpAffine(image,center,(cols,rows))
plt.subplot(121),plt.imshow(image),plt.title('Input')
plt.xticks([]), plt.yticks([])
plt.subplot(122),plt.imshow(Translation),plt.title('Translation')
plt.xticks([]), plt.yticks([])
cv.destroyAllWindows()
```

**Kindly refer Figure 1.11 for Result:**

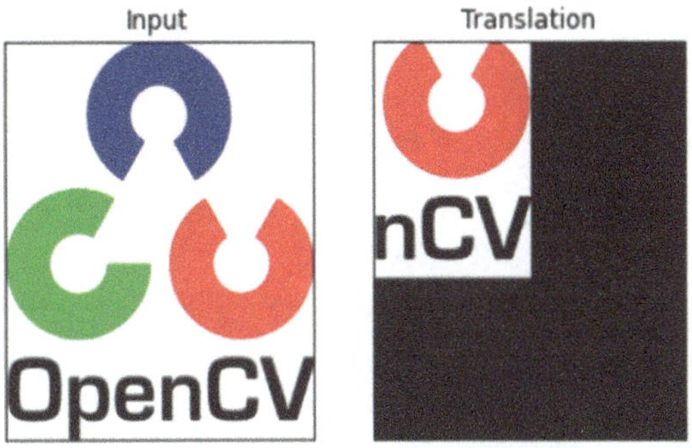


**Figure 1.11:** Image translation using OpenCV.

## 1.6 Simple image threshold

Threshold is one type of segmentation method. It updates the pixel values by comparing them with a threshold value. Multiple threshold methods are presented in OpenCV [13]. It is possible to apply thresholding only if the image is a grayscale image. Let us see a simple implementation code:

```
#importing the required libraries
import numpy as np
import cv2 as cv
import matplotlib.pyplot as plt
original_image1 = cv.imread('image1.png',0)
image1,Binary_image = cv2.threshold(gray_image1,127,255,cv2.THRESH_BINARY)
image2,binary_invimage = cv2.threshold(gray_image1,127,255,cv2.THRESH_BI-
NARY_INV)
image3,trunc1_image = cv2.threshold(gray_image1,127,255,cv2.THRESH_TRUNC)
image4,tozero1_image = cv2.threshold(gray_image1,127,255,cv2.THRESH_TOZERO)
```

```
image5,tozero_invimage = cv2.threshold(gray_image1,127,255,cv2.THRESH_
TOZERO_INV)
#DISPLAYING THE DIFFERENT THRESHOLDING STYLES
names=
['OriginalImage','BINARY','BINARY_INV','TRUNC','TOZERO','TOZERO_INV']
images=original_image1,Binary_image,binary_invimage,trunc1_image,to-
zero1_image,tozero_invimage
for i in range(6):
plt.subplot(2,3,i+1),plt.imshow(images[i],'gray')
plt.title(names[i])
plt.xticks([]),plt.yticks([])
plt.show()
```

**Kindly Refer Figure 1.12 for Result:**

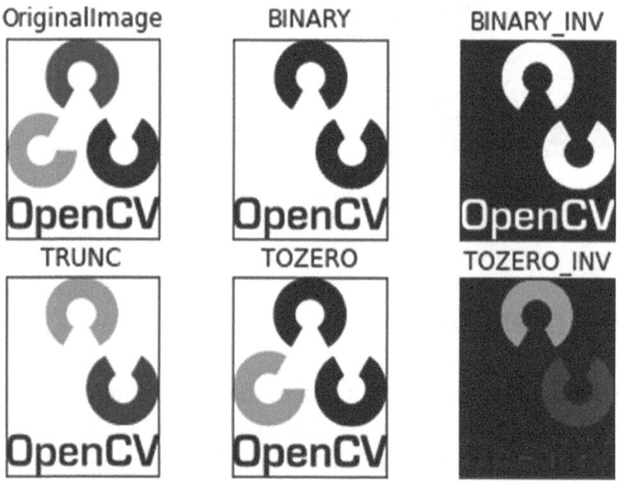


**Figure 1.12:** Image thresholding.

## 1.6.1 Adaptive thresholding

Adaptive thresholding is one of the most used thresholding techniques. It worked for different parts with different thresholds. This method provides better outcomes for images with unstable illumination situations:

```
#import the libraries
import numpy as np
import matplotlib.pyplot as plt
```

```
import cv2 as cv
#ADAPTIVE THRESHOLDING
original_image1 = cv.imread('image1.png',0)
image1,global_image = cv.threshold(original_image1,127,255,cv.THRESH_BINARY)
mean_image=cv.adaptiveThreshold(original_image1,255,cv.ADAPTIVE_TH-
RESH_MEAN_C,cv.THRESH_BINARY,11,2)
gaussian_image=cv.adaptiveThreshold(original_image1,255,cv.ADAPTIVE_TH-
RESH_GAUSSIAN_C,cv.THRESH_BINARY,11,2)
names=['Original_Image','Global_image','Adaptive_Mean_Threshold','Adapti-
ve_Gaussian_Thresholding']
images = [original_image1,global_image,mean_image,gaussian_image]
for i in range(4):
   plt.subplot(2,3,i+1),plt.imshow(images[i],'gray')
   plt.title(names[i])
   plt.xticks([]),plt.yticks([])
plt.show()
```

**Kindly Refer Figure 1.13 for Result:**

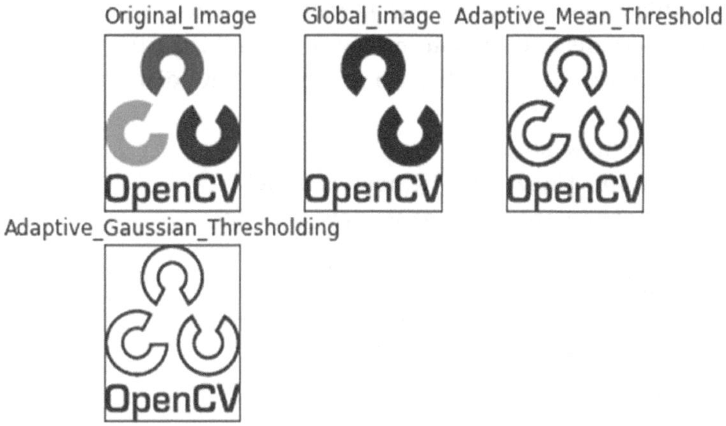


**Figure 1.13:** Adaptive thresholding using OpenCV.

# 1.7 Image segmentation (watershed algorithm)

Classifying every pixel of the image into some different classes is called image segmentation. Classification of each pixel as background or foreground is one of the best examples for segmentation [14, 15]. Image segmentation is mostly used to take out the relevant parts from a given image.

The watershed algorithm is one of the best image segmentation algorithms. Let us implement this algorithm:

```
#importing required libraries
import numpy as np
import cv2 as cv
import matplotlib.pyplot as plt
#reading the image
original_image = cv.imread('image1.png')
#converting image to grayscale format
gray_image = cv.cvtColor(original_image,cv2.COLOR_BGR2GRAY)
#apply thresholding
image1,thresh_image = cv2.threshold(gray_image,0,255,cv2.THRESH_BINAR-
Y_INV+cv2.THRESH_OTSU)
#get a kernel
new_kernel1 = np.ones((3,3),np.uint8)
new_opening1 = cv.morphologyEx(thresh_image,cv2.MORPH_OPEN,new_ker-
nel1,iterations = 2)
#extract the background from image
new_sure_bg = cv.dilate(new_opening1,new_kernel1,iterations = 5)
new_dist_transform = cv.distanceTransform(new_opening1,cv2.DIST_L2,5)
ret,new_sure_fg = cv.threshold(new_dist_transform,0.7*new_dist_transform.
max(),255,0)
new_sure_fg = np.uint8(new_sure_fg)
new_unknown = cv.subtract(new_sure_bg,new_sure_bg)
ret,new_markers = cv.connectedComponents(new_sure_fg)
new_markers = new_markers+1
new_markers[new_unknown==255] = 0
new_markers = cv.watershed(original_image,new_markers)
original_image[new_markers==-1] = [255,0,0]
plt.imshow(new_sure_fg)
```

**Kindly Refer Figure 1.14 for Result:**

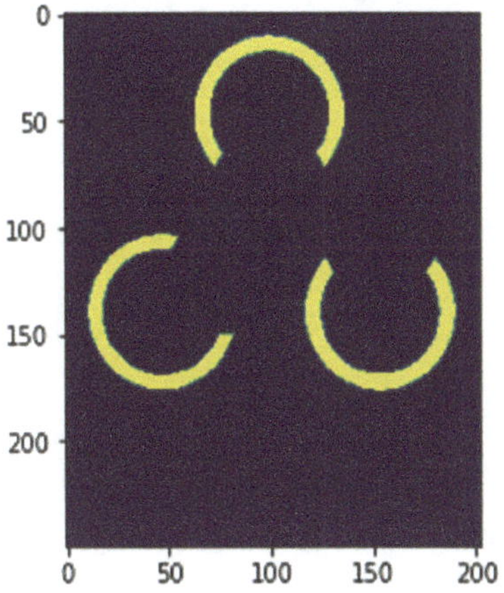


**Figure 1.14:** Image segmentation using OpenCV.

## 1.8 Bitwise operations

Bitwise operations include AND, OR, NOT and XOR [16, 17]. In applications of CV, if we want to apply any mask to the image or to select any region, bitwise operations are very helpful. Let us implement the bitwise operations on image:

```
import numpy as np
import matplotlib.pyplot as plt
import cv2 as cv
#read the image
original_image = cv.imread('image1.png')
#apply thresholding
image1,image_masking=      cv.threshold(sure_fg,0,255,cv2.THRESH_BINAR-
Y_INV+cv2.THRESH_OTSU)
#apply AND operation on image and mask generated by thrresholding bitwiseand
image_bitwise = cv.bitwise_and(original_image,original_image,image_mask =
image_mask)
#plot the result
plt.imshow(image_bitwise)
```

**Kindly Refer Figure 1.15 for Result:**

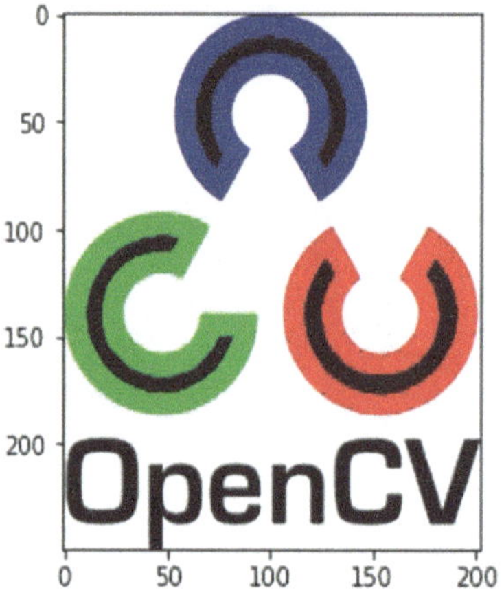


**Figure 1.15:** Applying bitwise operations on image using OpenCV.

## 1.9 Edge detection

Edges in the images are the points where brightness of the image pixels altered piercingly or discontinuities. In many CV applications, edge detections play the major role. Edge detection features are helpful in classification of objects. Even lot of deep learning models use edge detection techniques. Let us check the implementation of edge detection for an image:

```
import numpy as np
import cv2
import matplotlib.pyplot as plt
#read the image
original_image = cv.imread('image1.png')
#calculate the edges using Canny edge algorithm
edgedetection_image = cv.Canny(image,250,250)
#plot the edges
plt.imshow(edgedetection_image)
```

**Kindly Refer Figure 1.16 for Result:**

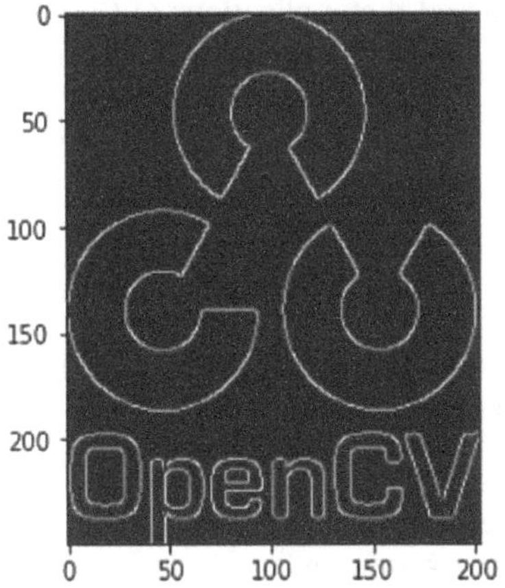


**Figure 1.16:** Edge detection using OpenCV.

# 1.10 Application of computer vision with OpenCV

1. Face recognition with Python and OpenCV
2. Lot of research is going on in face detection algorithms but still it is an open problem. Many researchers are working on face detection in various application purposes like surveillance systems and automatic attendance in various sectors. OpenCV has provided different types of algorithms to do this task like Haar cascading; once the face detection is successfully done, researchers worked on facial expression. Facial expression detection is one of the finest fields of research. Directional-based edge detection helps to retrieve the texture feature to identify the expressions as well [18–20].
3. Medical field: OpenCV library is also helpful in the medical field. Using OpenCV.
4. Number plate recognition: Number plate recognition is also one of the finest applications developed using OpenCV.
5. Not only there are lot of applications like building autonomous vehicles developed using OpenCV.

# 1.11 Conclusion

Computer is the trending term to retrieve the information from images and videos to classify the objects. As many researchers have done tremendous work in this field, still there is a gap to fulfill in terms of accuracy and efficiency. An efficient and convenient platform like OpenCV is required to develop the mechanisms to retrieve the features from the visionary points. This chapter presented all libraries which help to extract the features from image such as edge detection, thresholding, cropping and filtering. Finally, this chapter discussed the applications of CV with OpenCV and suggested suitable libraries to newcomers in the field of computer vision.

## References


[1] Borgefors, G. (1988). Hierarchical chamfer matching: A parametric edge matching algorithm. IEEE Transactions on Pattern Analysis and Machine Intelligence, 10(6), 849–865.

[2] Ansari, S. (2020). Building Computer Vision Applications Using Artificial Neural Networks. Springer Science and Business Media LLC.

[3] Pulli, K., Baksheev, A., Kornyakov, K., Eruhimov, V. (1 April 2012). Realtime computer vision with OpenCV. Queue, 10(4), 40:40–40:56. 10.1145/2181796.2206309.

[4] Bay, H., Ess, A., Tuytelaars, T., Van Gool, L. (2008). Speeded Up Robust Features (SURF). Computer Vision and Image Understanding, 110(3), 346–359.

[5] Kaehler, A., Bradski, G. (14 December 2016). Learning OpenCV 3: Computer Vision in C++ with the OpenCV Library. O'Reilly Media, 26ff. ISBN 978-1-4919-3800-3.

[6] Baggio, D.L. (2012). Mastering OpenCV with Practical Computer Vision Projects. PACKT Publishing.

[7] https://www.pyimagesearch.com

[8] https://www.geeksforgeeks.org

[9] Brunelli, R. (2009). Template Matching Techniques in Computer Vision: Theory and Practice. Wiley & Sons, Inc.

[10] Canny, J. (1986). A computational approach to edge detection. IEEE Transactions on Pattern Analysis and Machine Intelligence, 8(6), 679–698.

[11] Cyganek, B. (2013). Object Detection and Recognition in Digital Images: Theory and Practice. Wiley & Sons, Inc. Dalal, H., Triggs, B. (2005). Histograms of oriented gradients for human detection. CVPR: 886–893.

[12] Doermann, D., Mihalcik, D. (2000). Tools and techniques for video performance evaluation. 15th International Conference on Pattern Recognition, 4, 167–170.

[13] Farneback, G. (2003). Two-frame motion estimation based on polynomial expansion. Lecture Notes in Computer Science, 2749, 363–370.

[14] Flusser, J., Suk, T., Zitov´a, B. (2009). Moments and Moment Invariants in Pattern Recognition. Wiley & Sons, Inc. Freeman, H. (1961). On the encoding of arbitrary geometric configurations. IRE Transactions on Electronic Computers, 10, 260–268.

[15] Gonz´alez, R.C., Woods, R.E. (2007). Digital Image Processing. Prentice Hall.

[16] Harris, C., Stephen, M. (1988). A combined corner and edge detector. Proceedings of the 4th Alvey Vision Conference, 147–151.

[17] Illingworth, J., Kittler, J. (1988). A survey of the Hough transform. Computer Vision, Graphics, and Image Processing, 44(1), 87–116.

[18] Maheswari, V.U., Raju, S.V., Reddy, K.S., (2019 November). Local directional weighted threshold patterns (LDWTP) for facial expression recognition. In *2019 Fifth International Conference on Image Information Processing (ICIIP)*. IEEE, 167–170.

[19] Maheswari, V.U., Varaprasad, G., Raju, S.V. (2021). Local directional maximum edge patterns for facial expression recognition. Journal of Ambient Intelligence and Humanized Computing, 12(5), 4775–4783.

[20] Maheswari, V.U., Prasad, G.V., Raju, S.V. (2021). Facial expression analysis using local directional stigma mean patterns and convolutional neural networks. International Journal of Knowledge-based and Intelligent Engineering Systems, 25(1), 119–128.

Mehul S. Raval, G. P. Obi Reddy, Sanjay Chaudhary,
J. Adinarayana

# 2 Data science, computer vision and machine learning for agriculture and natural resource management: an overview

**Abstract:** The progress of humanity is dependent on agriculture and natural resources. A developed society with modern amenities and infrastructure will be useless without access to food or natural resources. As population and living standards rise, so is the demand for food. On the other hand, the Earth has finite natural resources. Therefore, researchers must develop innovative solutions to improve the yield and effectively manage natural resources. Data science, computer vision and machine learning are spearheading key innovations in agriculture in the present times. In the near future, the farms will be a great source of data. The data obtained from farms will drive digital agriculture. It will help farmers or customers to make optimal decisions in real time. The chapter shows how to best use available voluminous, dynamic and real-time data. In turn, it will drive the effective development of agriculture and proper utilization of natural resources.

**Keywords:** agriculture, computer vision, data science, machine learning, natural resource management, remote sensing

## 2.1 Introduction

The areas of concern for agricultural development are lack of natural resources, rise in population, urbanization, climate change, energy and food waste. The entire agriculture supply chain has to gear up for the challenges as the food consumption pattern changes. One of the crucial questions arising is: "Who will farm?" [1]. Urbanization is

---

**Acknowledgment:** The authors would like to thank the editors for accepting the chapter proposal for the book. Special thanks to Dr. Pancham Shukla for his valuable feedback, which motivated structural changes in the chapter and improved the overall quality. The authors express their gratitude to Ahmedabad University, ICAR-National Bureau of Soil Survey & Land Use Planning, Indian Institute of Technology Bombay (IITB), for providing time and environment during the penning of the review chapter. The authors would also like to thank Mr. Bhavesh Oza, Ph.D. scholar at Ahmedabad University, for proofreading and formatting.

---

**Mehul S. Raval, Sanjay Chaudhary,** Ahmedabad University, Ahmedabad, India
**G. P. Obi Reddy,** ICAR-National Bureau of Soil Survey and Land Use Planning, Nagpur, India
**J. Adinarayana,** Indian Institute of Technology Bombay, Mumbai, India

https://doi.org/10.1515/9783110756722-002

spreading rapidly with a shrinking rural population. The spiraling population growth and the rise in living standards have placed unprecedented pressure on agriculture and natural resources. Agriculture is a domain with many uncertainties aggravated by climate change, land pollution, water pollution and modifications in demands due to socioeconomic upliftment [2].

These challenges demand improvement in productivity from agriculture while reducing the overuse/misuse of natural resources. The traditional approach of farming has benefited from technology interventions [3]. It has to accelerate further using science and technology-driven agriculture 4.0. Precision agriculture (PA) is turning out to be the fulcrum of agriculture 4.0 [1]. The PA or intelligent agriculture domain has shown promise to handle uncertainties in agriculture and natural resource management (NRM). Emerging technologies like remote sensing – satellites and unmanned aerial vehicles (UAVs), Internet of things (IoT), global positioning system (GPS) and cloud computing are driving onfield and off-field activities in PA. The information and communication (ICT) technology helps at every stage of the typical crop cycle – soil preparation, sowing, growth management, irrigation, harvesting and storage.

IoT combines physical objects or "things" and creates their network. Each of these objects is equipped with sensors, and they connect to exchange information. As more advanced machines and sensors are involved in farming, the knowledge builds in amount and breadth. This enables penetration of digital technologies in the farms. The digital technologies monitor and measure various in-field parameters, for example, weather and off-field parameters producing "big data" quickly. The sensor produces data with massive *volume*, *variety* and *velocity*. It necessitates large-scale data collection from hybrid sources, preprocessing, storage, modeling, and analysis. The agri-data needs real-time processing and integration to extract economic benefits. Mark [144] shows five major "V" dimensions of big data as follows:

1. Volume: It shows the amount of data available for processing and analysis.
2. Velocity: The rate at which data arrives and its processing for deriving timely and relevant outcomes.
3. Variety: The data arrives from heterogeneous sources with different dimensions, formats, resolutions and applications.
4. Veracity: It deals with data accuracy, reliability, and confidence in the data.
5. Value: It is about extracting a value from the data after all the data processing.

Later, many other V's have been added [2]; for example, the veracity dealing with the reliability of data, valorization for knowledge propagation and visualization for presenting the information. The typical applications, which handle the prominent "5 V's," are shown in Table 2.1.

Big data sets contain both structured and unstructured records. The survey paper [2] shows different data sources as follows: ground sensors – chemical sensors, biological sensors, weather devices; governmental records – statistics handbooks, reports, regulatory body reports; remote sensors – UAVs, satellites, robotic vehicles; online

**Table 2.1:** Agriculture applications using big data.

| V's | Sample agriculture application |
|---|---|
| Volume | Earth observation, yield estimation, land use/land cover, weather forecasting |
| Velocity | Pest and disease identification, weed identification, financial transactions in agri domain, weather forecasting |
| Variety | Crop estimation, quality monitoring, plant phenotyping, plant species identification |
| Veracity | Yield estimation, agriculture econometrics, weather forecasting |
| Value | Crop prediction, plant phenotyping, land use/land cover, disease identification |

web services, public repositories, archives, live feeds; crowdsourcing – images, social media mining and cell phone data. The sources are heterogeneous and generate data at a different rate and in various formats. The dataset is also governed by the application for which it is used, for example, plant phenotyping would often use ground sensors or platforms. At the same time, remote sensing is required for applications related to land mapping and monitoring. Kamilaris et al. [2] also suggested that images and videos are popular sensing modalities that computer vision (CV) algorithms can process. The output by CV is most often used with machine Learning (ML) for prediction, clustering, and classification problems. The importance and integration of CV and ML in solving problems in agriculture are described in the next section.

## 2.2 Vision and machine learning in agriculture

Technology-driven PA is dependent on sensors, signal-processing devices, computation and ICT. PA is expected to balance the supply and demand side of the equation, making the agriculture cycle more efficient, safe and green. It is expected that PA will penetrate and increase the connectivity among the farms. The growing number of devices based on vision will bring both challenges and opportunities. CV uses visual cues and helps the computer to understand the surrounding world. Broadly, it is an interdisciplinary domain that tries to mimic human visual systems to automate the understanding of the surrounding environment. A classical CV task has four main parts as follows:

1. Data acquisition through spatial or spatiotemporal sensors like cameras or radar.
2. Data preprocessing to improve the visual signal quality, for example, noise reduction and contrast enhancement.
3. Image analysis extracts practical information from the signal that breaks the given image or video into different segments.
4. Image understanding provides semantic meaning to the constituent parts and meaning to the surrounding world.

CV offers many advantages for agriculture domain processing. Some of them are summarized as follows:
- The high-resolution sensors with many modalities, for example, visible spectrum and near infrared (NIR) bands, are available at low cost.
- Present-day electronics allow high-performance computations. Cloud computing provides resources on an as-you-need basis.
- The CV, when combined with ML, creates a robust mechanism for nondestructive testing. It provides state-of-the-art accuracy with a very high repeatability rate.
- The sensing technology allows data capturing at multiple resolutions and distances.

It is vital to use technology that helps to interact with the environment (sensing and actuating), learn, reason and draw inferences. As a sample, this section shows ML and agro-vision to solve some of the agriculture problems [4–6]:
- Crop growth monitoring
- Weed pest and disease identification
- Product quality testing
- Yield estimation

Common to all of these tasks is to detect the objects of interest and classify them. It requires a robust pipeline with high-quality sensing, high-performance computing and a generalized data analysis framework.

### 2.2.1 Crop growth monitoring

Plant phenology study changes in the plant in reaction to the variations of environmental conditions – temperature, light and humidity. The automation in quantitative measurements (plant phenotyping) of plant properties offers an efficient and productive alternative. Seventeen micronutrients, macronutrients and secondary nutrient elements are essential for plant growth [7]. Crop growth monitoring is one of the essential aspects of PA, and it helps in understanding the growth environment. A minor adjustment in the environment can significantly increase production [8]. Faragó et al. [9] suggested a CV-based method to measure morphological and physiological parameters of in vitro plants. Using a Canon camera, the size, shape and color of the plants were repeatedly captured. The image analysis measured several parameters – rosette size, convex area and ratio, chlorophyll and anthocyanin. The imaging proof of concept revealed several significant differences between wild-type and transgenic *Arabidopsis* plants. Rico-Fernández et al. [10] performed a foliage segmentation for measuring plant growth. The paper presented an approach consistent across different crop species – tomato, maize and carrot. The approach is also robust to changes in environmental conditions. It contextualized the information and color space transformation to segment the foliage accurately using a support vector machine (SVM). On average, it achieved a segmentation quality of 0.9 which is showing very good accuracy.

Naik et al. [11] showed an end-to-end phenotyping pipeline for automated analysis of the iron-deficiency chlorosis (IDC) in soybean plants. The paper shows experiments with ten classifiers and selects the best for use in a smartphone. Several preprocessing steps were carried out for effective segmentation. Calculating the proportion of yellow regions (chlorosis) and brown regions (necrosis) is essential. The accuracy and interpretability – the ability to interpret data by the user – are chosen as metrics for evaluation. It was observed that the hierarchical model performed well.

It is essential to develop an automatic measurement system for grain [12]. The authors proposed an automated wheat-head growth monitoring system. The images are captured directly from the field under different light conditions. The two-stage detection mechanism has been employed to detect the spike – the wheat head. In the first step, wheat ear detection occurs, and in the second step, non-ear regions are eliminated. The sift-invariant feature transform forms a low-level visual descriptor encoded by the Fisher vector to form a mid-level feature. Then SVM is employed to classify the ear. The results match reasonably with the human observations. Table 2.2 summarizes the application of ML and CV to growth monitoring of the crop. It shows the successful use of conventional ML approaches on color images. Figure 2.1 shows a semantic segmentation for sample images in the GrassClover dataset [13], which contains images collected in outdoor agriculture scenarios.

**Table 2.2:** Summary of crop growth monitoring.

| Author | Application | Sensor/ image size | Classifier/results |
|---|---|---|---|
| [9] | Measuring morphological and physiological parameters of in vitro *Arabidopsis* plants | RGB camera/ 3000 × 4000 pixels | Rosette size and plant size are highly correlated ($R^2 > 0.9$) |
| [10] | Foliage segmentation for growth monitoring – tomato, maize and carrot | Robot mounted RGB camera | SVM/segmentation accuracy – 0.9 |
| [11] | IDC – soybean | RGB DSLR camera | 10 classifiers – decision trees, random forest, naïve Bayes, SVM. Mean per class accuracy – 95.9% |
| [12] | Wheat development in the field | RGB camera | SVM/accuracy – 66.9% |

Semantic segmentation is further helpful for growth monitoring. In one such challenge, wheat ear heads were counted using CV and ML. The global wheat head dataset (https://www.aicrowd.com/challenges/global-wheat-challenge-2021) contains 6000 images collected from 11 countries. It aims at finding the size and density of wheat ears from different wheat types. Sample images for successful detection is shown in

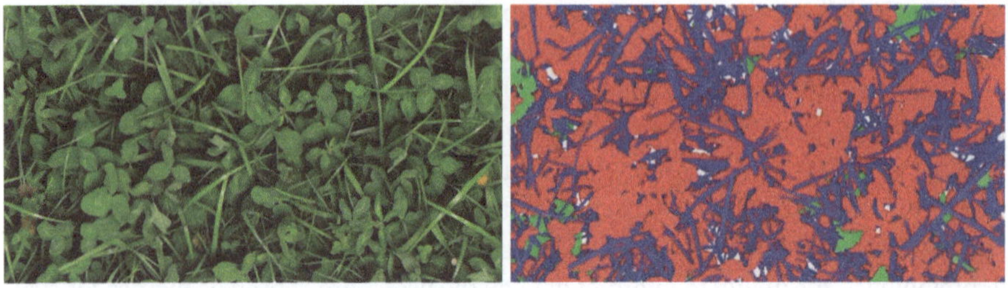

**Figure 2.1:** Semantic segmentation for yield estimation. Red is clover, blue is grass, green is weeds and light gray is soil [13]. Grassclover dataset: https://vision.eng.au.dk/grass-clover-dataset/.

Figure 2.2. It can be seen that the digital detection is challenging due to overlapping wheat heads, variation in wheat type, head orientation, presence of barbs and wind. Figure 2.3 shows the cases where CV and ML approaches fail to detect the wheat head.

**Figure 2.2:** Yellow bounding box shows detection of the wheat head. Top row – difficulty level of detection is easy. Bottom row – difficulty level of detection is hard (source: https://www.aicrowd.com/challenges/global-wheat-challenge-2021).

**Figure 2.3:** The cases of failure for wheat head detection (source: https://www.aicrowd.com/challenges/global-wheat-challenge-2021).

## 2.2.2 Weed, pest and disease identification

The crop quality is controlled by protecting against weed, pest attacks and disease. A quick and full diagnostic measure to automatically estimate the severity can limit the loss in the yield [14, 15]. Ma et al. [15] presented a segmentation of the foliar disease spots on images captured in the greenhouse using a color information and region growing method for segmentation. The method was evaluated on the cucumber downy mildew images captured under challenging conditions. The proposed algorithm achieved a good precision compared to the state-of-the-art method.

Ramcharan et al. [16] train a convolutional neural network (CNN) to detect disease symptoms appearing on the foliar of cassava plants. The CNN identified symptoms of three diseases (cassava mosaic disease, cassava brown streak disease and brown leaf spot), two types of pest damage (green mite damage and red mite damage) and nutrient deficiency. The model used single-shot multibox [142] model with the MobileNet classifier pretrained on the COCO dataset. The developed algorithm is deployed in the mobile app and tested in the fields of Tanzania. The model predicted two levels of severity of the disease – mild and heavy. It was observed that the performance of the trained model drops when tested on the images and videos captured in the open field.

Aphids are the pests that heavily impact the wheat crop [143]. The insects feed on the sap of the wheat, severely affecting its development. The paper proposed identifying the red and green species and monitoring the density of aphids in the field. Preprocessing is used for image enhancement and noise reduction on the image captured in the field. A histogram of oriented gradients (HoGs) and SVM were used for classification. The HoG feature is used to train a binary classifier that detects the presence or absence of the pest.

Sabzi et al. [17] proposed a neural network-based system for identifying three weeds and potato plant types. Weed removal is necessary as they compete with plants for nutrients, minerals and even water. It affects the growth and quality of the main crop. The paper used color features and texture features to classify three weeds and potato plants. The images are captured using a camera mounted on a moving platform, and 3459 objects were extracted for training. Table 2.3 summarizes ML and vision's application to detect weed, pest and disease in the crop. Figure 2.4 shows a sample weed image and its corresponding segmentation mask. The tuple trains the ML algorithm to perform semantic segmentation of the images.

**Figure 2.4:** Weed and its corresponding mask [18]. CWFI dataset https://github.com/cwfid/dataset.

**Table 2.3:** Summary of weed, pest and disease identification.

| Author | Application | Sensor/image size | Classifier/results |
|---|---|---|---|
| [15] | Detecting cucumber downy mildew | Color CCD camera/ 4320 × 3240 pixels | Region growing/precision – 97.2% |
| [16] | Disease, pest damage and nutrient deficiency in Casava | 20 MP camera/2415 images | MobileNet/mean average precision 75% on image and 81.2% on video |
| [143] | Pest detection – aphids | Digital camera/ 1200 × 1800 pixels | SVM/average detection rate – 86.8% Error rate – 8.91% |
| [17] | Weed detection | Video camera on moving platform/ | Neural network/accuracy – 98.3% |

### 2.2.3 Product quality monitoring

The quality of the agricultural product is achieved by automatic grading and quality inspection. It helps promote commercialization and meet the standards required with product exports [19]. Quality testing has been automated to lower the cost of the traditional system [20]. The automation also overcomes the inconsistencies of manual quality checks and increases the satisfaction levels of the customers. Deng et al. [21] developed an automated carrot grading system. It improved the grading efficiency through the automated carrot sorting system using CV. The overall system has image acquisition, a conveyor belt to carry crops and a control system for sorting. The quality is determined by checking the surface defects. The system detects three types of defects such as unnatural shape, fibrous root and surface crack. After removing the carrots with defects, they are graded based on their length. A convex polygon is used to detect the shape of the carrot. The system achieved good accuracy in grading and sorting.

Nondestructive testing and measurements are the obvious mechanisms in post-harvesting for checking the damage to the crop. The mechanical damage to the citrus fruit is not easily detectable, specifically through deformation in appearance. The paper [22] proposes using image processing and UV radiation to detect the mechanical damage to sweet lemon. First, the fruits were dropped from the height (2–3 m) on the ground, and then UV light of different wavelength images are captured. The images are analyzed using the green spot index, which shows mechanical damage. It significantly increases the detection level, and all 135 fruits were classified as damaged and undamaged with 100% accuracy.

It is vital to classify fruits into different grades considering their popularity. Iraji et al. [23] proposed many methods to classify tomatoes in different grades. The paper proposes adaptive neuro-fuzzy inference system with input features, regression and an extreme learning machine. In another approach, deep-stacked sparse autoencoders on images are used for tomato classification. The method produces excellent accuracy in grading the tomatoes without extracting features.

Wang et al. [24] used hyperspectral imaging to measure the internal mechanical damage of the blueberries. The method uses two CNN architectures, ResNet and its newer variant, ResNeXt. The fine-tuned networks achieved the accuracy of 0.88 and 0.89, respectively. For comparison, the author also used five other ML algorithms with k-fold cross-validation. The accuracy for CNN is better than five classical ML techniques. Table 2.4 summarizes the use of machine learning for quality monitoring of agriculture products. The approaches show the successful use of deep learning to solve the problem.

**Table 2.4:** Summary of product quality monitoring.

| Author | Application | Sensor | Classifier/best results |
|---|---|---|---|
| [21] | Carrot grading and sorting | Industrial-grade camera | Convex polygon for curvature – 95.5%, concave point method for fibrous root – 98%, Hough transform for surface crack – 88.3% |
| [22] | Sweet lemon mechanical damage | UV camera | Binary classifier, accuracy – 100% |
| [23] | Tomato-quality grading | RGB camera | Deep-stacked autoencoder/95.5% |
| [24] | Internal damage in blueberry | Hyperspectral camera | CNN ResNet Accuracy – 0.89 |

## 2.2.4 Yield prediction

Real-time information from farms can help estimate the yield in PA. It is one of the critical mechanisms in production planning. UAV provides a platform for acquiring images with low cost and high resolution for quick implementation [25]. The details in images can help with better planning in agriculture economics and production outputs. Maldonado et al. [26] proposed a method to estimate orange crop yield on three varieties. It exploits a known correlation between the number of fruits detected in an image and fruits present in an orange tree. All images are taken by a camera placed at a distance of 2 m from the tree canopy. The method uses series of image processing techniques to extract the green fruit feature. The fruits were classified using SVM, and a lower distance between plant canopy and camera resulted in significant false positives. The algorithm could not detect all the fruits in images. However, under good lighting conditions, it has a false-positive rate of less than 3%.

The quality of the field scanned has significantly improved with the use of UAVs. Drone technologies have grown in leaps and bounds, facilitating the treatment and analysis of sugarcane farms [27]. The paper evaluates the degree of canopy for different planting approaches and row spacing to assess the potential application in predicting the yield of the sugarcane field. The images are taken considering the biomass accumulation curve. The leaf area index (LAI) and green–red vegetation index (GRVI) are computed by field sensor and UAV. The use of GRVI resulted in an $R^2$ value of 0.69 between the onfield sensor and UAV image. When combined with LAI, the yield estimates improved to $R^2 = 0.79$. It shows that UAV is an effective mechanism in estimating the yield.

Han et al. [28] used UAVs to capture structural and spectral information and use ML algorithms to estimate the maize biomass. The feature elimination obtained essential features. Four ML regression algorithms were evaluated: multiple linear regression, SVM, artificial neural network (ANN) and random forest (RF) regressor. In this paper, the authors used three methods to estimate the height of the plants and compared them with the manual ground-based measurement. The results indicated that RFR produced the best results with the lowest estimation error. The proposed method increased the ratio of explained variance with the least error rates.

Multispectral and hyperspectral sensing also gain prominence in the PA for yield estimation [29]. The paper used an unmanned ground vehicle to capture line-scan hyperspectral images in a mango orchard. The best model was obtained after the preprocessing of tree delimitation and identifying pixels belonging to mango in the image. The count obtained through RGB techniques is compared with that obtained after manually counting fruits on the tree. The model was then mapped to find field count – number of trees and mango count per tree. The model achieved a high degree of determination coefficient value. Table 2.5 summarizes the approaches for yield prediction using ML and CV. Figure 2.5 shows the use of CV and ML for fruit detection and yield prediction. It shows the field image and corresponding

**Table 2.5:** Summary of yield prediction.

| Author | Application | Sensor | Regression best results |
|---|---|---|---|
| [26] | Orange yield estimation | RGB – camera/ 2592 × 1944 pixels | Coefficient of determination ($R^2$) manual – automatic yield prediction – $R^2 = 0.48$<br>SVM for correct fruit identification. Accuracy – 49% |
| [29] | Sugarcane field yield estimation | Onfield and UAV images | Correlation GRVI – $R^2 = 0.69$ |
| [28] | Maize biomass estimation | Digital camera, multispectral camera | RFR $R^2 = 0.944$ |
| [27] | Mango yield estimation | Ground-based hyperspectral sensing | Field count $R^2 = 0.75$ (18 trees)<br>Mango count $R^2 = 0.83$ (216 trees) |

The section reviewed the applications of CV and ML in four essential areas of PA. It shows that how agriculture automation can be achieved with low cost and high efficiency. The penetration of technology will continue to grow in PA with the use of AI. It will improve the efficiency and quality of the agroproducts.

## 2.3 Data science and ML for natural resource management

With the advancements in remote sensing, GPS and real-time kinematics, it is easier to monitor crop health, identify nutrient deficiencies and reduce yield. Google Earth Engine (GEE) is an emerging big data platform. The United States Geological Survey,

**Figure 2.5:** Mango yield estimation with field image and corresponding ground truth image [30] (dataset: https://github.com/avadesh02/MangoNet-Semantic-Dataset).

in partnership with several US universities and international institutes, produced the world's first Landsat satellite-derived, global cropland extent product at 30 m resolution (GCEP30) for the nominal year 2015 by using the GEE platform [31]. The emerging Web-based geoportal platforms help visualize, access, query and disseminate the information to the users for efficient planning, monitoring and sustainable management of natural resources. The section shows the following application for DS and ML in NRM:

- Drought monitoring
- Land-use changes monitoring
- Crop types mapping and monitoring
- Prediction of soil properties
- Land degradation assessment
- Irrigation water management

## 2.3.1 Drought monitoring

Drought is considered one of the most widespread natural disasters in almost all ecosystems [32, 33]. It severely affects natural ecosystems, global food production and human livelihoods [34]. The period, frequency and degree of droughts varied from region to region [35]. Various input parameters are needed to analyze the nature, extent and severity of droughts, usually derived from climate, remote sensing, hydrological and atmospheric systems. The advance Earth Observation (EO) and climate detection systems provide massive EO data with high spatial, temporal and radiometric resolutions [36]. The probabilistic methods help predict drought due to quantifying uncertainties from drought-causing hydroclimatic variables [37].

ML algorithms like RF and decision tree (DT) were also used in drought monitoring. They increase the decision-making capacity and reduce the subjectivity in obtaining the results [38]. RF algorithm that works based on the combination of several classifications and regression tree (CART) models. Kuswanto and Naufal [39] used CART and RF algorithms to classify the droughts and reported that both methods had been proved to be computationally efficient in predicting droughts. Dimension reduction algorithms like principal component analysis eliminate redundancy by converting the original feature space axes. Drought information can be shown in a new model space without correlation [40]. Alizadeh and Nikoo [41] studied the drought events using new fusion approaches from high-resolution satellite data at the feature level. They reported that ML approaches of RF, support vector regressor (SVR) and ANN showed excellent performance in obtaining significant results.

In recent times, data mining techniques were also applied to build drought monitoring and forecasting models by using ML algorithms such as SVR and ANN [42]. Chiang and Tsai [43] found SVM superior in predicting hydrological drought compared to the classical model. ANN models were used to predict quantitative values of drought indices to measure the degree of dryness for any period [44]. The drought forecast model was proposed based on the RF method to predict the monthly standardized precipitation index (SPI) [45]. Park et al. [46] developed drought prediction models for a short period using remote-sensing data and climate variability indices over East Asia through the RF model. A regional-level seasonal drought analysis was carried out based on SPI and DT algorithm at watershed level in Turkey [47]. The hybrid models like ANN and RT with the fuzzy segmentation approach using satellite images were used to identify and predict the drought in Ethiopia [145].

## 2.3.2 Land-use changes monitoring

In recent years, the application of ML algorithms in the classification and analysis of remotely sensed imageries in land-use/land-cover (LULC) mapping has been increased considerably [146]. ML algorithms are popular in remote-sensing and EO studies, especially LULC classification and vegetation monitoring [48]. The information generated by time-series satellites data is widely used in various EO studies, including vegetation monitoring [49] and land-use changes [31]. Zhang et al. [50] used MODIS data to analyze the driving factors of space observations of plant greenness and phenology. RF is one of the most widely used machine learning algorithms [51]. It can be used for classification and regression purposes with categorical and continuous variables [52]. RF showed its potential in land-cover classification due to its straightforward and understandable decision-making process and excellent classification results [53] from geo-big data computations.

SVM, neural networks and classification and regression tree algorithms were performed in land-cover classification with limited training data. It was reported that SVM outperforms other classifiers due to its overall high capacity to generalize complex features [54]. DL algorithms were widely used in remote-sensing data analysis, including scene classification, LULC analysis and object detection [55, 56]. Srivastava et al. [57] compared the performance of the ANN and SVM as classification algorithms for LULC change studies. Based on the performance analysis, they reported that the ANN was better than the SVM. Keshtkar et al. [58] advocated that the SVM classifier was performed better than the RF to classify land-cover types of a heterogeneous landscape.

However, the selection of suitable algorithms in vegetation monitoring and assessment depends upon the study's objectives. Suitable ML algorithm, input datasets and parameters are needed to be considered to obtain accurate results. Noi and Kappas [59] compared the performances of RF, k-nearest neighbor (kNN) and SVM classifiers for LULC classification using Sentinel-2 image data of a river delta in Vietnam. They reported that all three classifications provided high overall accuracy (OA) ranged from 90% to 95%. However, SVM produced the highest accuracy among the three classifiers with the least sensitivity to the training sample sizes, followed by RF and kNN. However, few limitations are also associated with ML algorithms, as DT is too sensitive to small changes in the training dataset. It is occasionally unstable and tends to overfit in the model [60].

### 2.3.3 Crop-type mapping and monitoring

In PA, accurate and detailed mapping of crop types are crucial for crop monitoring and accurate crop yield estimations. Since there is limited scope for increasing agricultural land horizontally, more food is needed from the existing land and water resources [61, 62]. To ensure food and nutritional security, sustainable intensification in the crop fallow areas through cereal-based systems needs an hour [63]. Hence, crop-type mapping and monitoring at higher resolution assume greater importance in agriculture and ensure global food security. The spectral response of cropland areas varies significantly among the crop's phenological stage and the cropping calendars [64]. Remote sensing has been widely used for rice fallows mapping [65, 66] to provide policy inputs in decision-making.

Pelletier et al. [67] reported that the RF algorithm helps classify large datasets with high spatial and temporal resolutions for extensive area land-cover mapping. GEE platform is widely employed to analyze multitemporal satellite data in crop-type mapping using ML algorithms [68]. Landsat data on the GEE platform was used to produce cropland at 30-m resolution and reported that the RF algorithm could map croplands rapidly and accurately at various scales [69]. Gumma et al. [70] used Landsat 30-m TM (Thematic Mapper) data on the GEE platform by adopting the RF algorithm in cropland mapping in South Asia. They reported that the RF algorithm gives better OA in producing agricultural croplands. Moumni and Lahrouni [71] used high-resolution Sentinel-1 and Sentinel-2 fused satellite data by adopting ANN, SVM and maximum likelihood in mapping crop-type irrigated areas. They reported that fused images of optical and microwave improved the accuracy in crop-type classification compared to the classification results of optical or SAR data alone.

The SVM algorithm can classify high-dimensional data with small training samples [54]. Inglada et al. [72] applied RF and SVM on high-resolution multitemporal optical imagery of SPOT4 and Landsat 8 to produce crop-type maps at the global scale. SVM classifier using Gaussian kernel density function was superior in crop-type mapping [73]. In general, several studies reported that RF and SVM were more suitable for classification in agricultural regions using high-resolution satellite imageries [74, 75]. However, further research is to be carried out to develop hybrid algorithms to precisely analyze and discriminate crop types, especially in the small agriculture fields of heterogeneous landscapes.

## 2.3.4 Prediction of soil properties

Accurate prediction of soil properties immensely helps in the optimum management of soil resources. Digital soil mapping (DSM) technique showed its great potential in producing spatial information on soil resources [76]. In soil mapping, the information on environmental variables such as climate, lithology, vegetation and landforms play an essential role in soil properties mapping and soil landscape analysis [77]. DSM techniques offer to generate soil property distribution maps. It is done through numerical models that take soil environmental covariates into account in inferring spatial and temporal variations of soil properties [78]. Tree-based ML algorithms and the simplest version of regression tree and their variants have been used in the DSM [79].

RF algorithms were widely used in DSM [80, 81]. Vaysse and Lagacherie [82] introduced a quantile RF to map the uncertainty associated with predicting the soil properties. Forkuor et al. [83] used SVM to map soil properties. It can handle large datasets, learn complex data classes and make better decisions regarding the separation of classes. Soil organic carbon (SOC) plays a significant role in soil health management, water holding capacity, nutrient availability and plant growth. The spatial variability of SOC at the field to the regional scale is highly related to the soil-forming factors, including the climate, organisms, relief, parent materials and time [84]. In the context of the availability of huge remotely sensed imageries in the public domain and easy accessibility of climatic and digital elevation model data at different resolutions, the application of ML algorithms in the prediction of SOC is significantly increased [85].

Many ML algorithms have been successfully applied for quantitative mapping of various soil properties like the prediction of SOC [86], map soil textural classes like clay, silt and sand content [81, 87], soil pH [88] and cation exchange capacity [83]. ML algorithms were also applied to make maps of soil nutrients such as nitrogen [83], phosphorus [89], potassium, calcium or magnesium [90]. Emadi et al. [91] reported that the deep neural network (DNN) algorithm outperformed other ML algorithms regarding the power of the prediction uncertainty at the province scale. It demonstrated that DNN is suitable for use as a robust estimator for SOC mapping. In SOC assessment, ML algorithms were proven their merits than the geostatistical methods due to their higher ability to obtain more information for unsampled points by investigating nonlinear relationships between SOC and environmental auxiliary variables.

### 2.3.5 Land degradation assessment

On a global scale, soil erosion by water is recognized as one of the world's most serious environmental problems of the twenty-first century [92] and the leading cause of land degradation [93]. More than 83% of the global land degradation in all climatic regions is caused by soil erosion alone [94]. Using advanced tools assumed greater importance in the context of changing climate, land-use practices and meeting the global commitments in climate change and land degradation neutrality. Soil erosion caused by water is the major challenge of India, which is contributing to land degradation when it exceeds the natural soil formation rates [95–97]. By using the six parameters of the empirical universal soil loss equation model, that is, rainfall erosivity (R) factor, soil erodibility (K) factor, slope length (LS) factor, cover (C) factor and support practice (P) factor in GIS, estimated the extent and spatial distribution of soil erosion by many authors [77, 98–100].

Current trends in automating the identification and assessment of land degradation are based on vegetation indices [101, 102]. The analysis of time-series vegetation indices derived from satellite data can effectively distinguish and map soil degradation [103]. The selection of satellite images can be performed using deep ML algorithms [104] to perform the data mining procedures [105]. Yousefi et al. [106] analyzed 12 independent conditional factors for their relationships to range quality by applying RF, CART and SVM. They reported that RF was determined to be the most robust. Various ML algorithms were used in erosion susceptibility prediction on heterogeneous landscapes [107, 108] such as ANN [109], SVM [110], logistic regression [111] and RF [112, 113].

Haghighi et al. [114] employed SVM, multivariate adaptive regression splines (MARS), generalized linear model (GLM) and dragonfly algorithm (DA) ML algorithms to generate high-quality and accurate land degradation risk maps in a watershed of central Iran. They reported that DA had the highest accuracy and efficiency with the most outstanding learning and prediction power in land degradation risk mapping. Since the GEE platform offers time-series satellite data at various resolutions, ground observations can map more extensive areas in a short period, along with ground surveys.

### 2.3.6 Irrigation water management

As water resources are limited, smart irrigation is more important in PA and enhancing crop productivity. Big data coupled with sensor systems showing promising ways to plan water systems optimally, schedule irrigation plans, mitigate climate change and detect the water-induced changes in the ecosystem. ICT technologies facilitate disseminating extensive data analysis and modeling to the farmers to access weather forecasts and market information. It allows to make better decisions, improve the livelihoods, optimum utilization of water resources and ensure food security. Al-Ghobari and Mohammad [115] tested the evapotranspiration (ET), ICT and Internet Information Services (IIS)-based technologies in Saudi Arabia in wheat and tomato fields. They concluded that IIS was more feasible in water usage than ICT and ET-based systems.

Cai et al. [116] showed strategies for solving extensive nonlinear models for water resources management. They combine genetic algorithms (GAs) with linear programming. Hinnell et al. [117] developed neurodrip irrigation systems by using ANNs to predict the spatial distribution of water in the subsurface. Kim et al. [118] used a distributed wireless network for sensing and control of irrigation systems from a remote location. Wall and King [119] developed an intelligent system that controlled valves of sprinklers using temperature and moisture sensors deployed in the field. Wardlaw and Bhaktikul [120] developed GAs to solve irrigation water scheduling. They optimized the utilization of water resources in irrigation systems operating on a rotational basis and field soil moisture constraints. Twarakavi et al. [121] developed SVMs for estimating the hydraulic parameters describing the soil water retention and hydraulic conductivity. Cruz et al. [147] exploited the ANN feed-forward and back-propagation technologies to optimize the water resources in an innovative farm. Different ML algorithms are used for water needs estimations in automatic irrigation systems. Yamaç and Todorovic [122] compared kNN and adaptive boosting (AdaBoost) algorithms to estimate water needs for the potato crop.

Tang et al. [123] used AI models for actual crop ET modeling, and water needs estimation in maize croplands. Kisi [124] compared SVR with MARS and M5 model tree (M5Tree) in modeling reference ET. Davis et al. [125] conducted a study in Florida. They found that ET-based watering scheduling controllers are more useful in irrigation cost, size and labor requirement. Viani et al. [126] reported that fuzzy logic-based decision support system (DSS) provides more water-saving over single-threshold and multithreshold-based irrigation scheduling. Gutiérrez et al. [127] proposed an automated irrigation system using a wireless sensor network (WSN) and GPRS module to save water in irrigation through real-time monitoring and irrigation control. Roopaei et al. [128] developed an intelligent irrigation monitoring system using thermal imaging techniques to identify crop water requirements and irrigation monitoring.

Albeit tremendous progress using DS in agriculture and NRM, there are still many gaps and challenges to be addressed. The following section delves into some crucial issues.

## 2.4 Gaps and challenges

WSN, IoT and cloud computing are the primary driving forces in PA. The sensor measures the physical quantity and transmits the collected information to the cloud or the device. The physical layer contains all the sensory data. The network layer takes care of data transmission, and the application layer analyzes and processes the data. The sensing may retrieve real-time information about soil, crop, fertilizer, weather, water requirements, market and government policies. PA may use in-field or remote-sensing techniques to collect data at multiple stages. This needs collection of big data from various sources at different geographic locations. The collected information is processed and analyzed to generate agro-recommendations for enhancing the yields. Therefore, sensing, communication and computing combine to satisfy the precise needs of the crop. This also shows that the application of ML to PA involves a cross-domain paradigm and requires an interdisciplinary research. There are several challenges in implementing technology or data-driven PA, and issues are listed further and explained:
- Data acquisition for in-field and remote sensing
- Data transmission
- Data management
- Data security
- Nontechnical factors

### 2.4.1 Challenges in data acquisition for in-field and remote sensing

The use of the GPS for civilian use initiated the PA era. The next decade allowed yield monitoring and generate recommendations for soil reaction rectification or fertilizer. Typical sensors for in-field measurements are: the location sensors use GPS satellites to find latitude, longitude and altitude. An optical sensor measures the clay properties, organic matter, soil temperature and soil moisture. Electrochemical sensors sense pH and soil nutrients. A mechanical sensor measures soil compaction through a strain gauge, and an airflow sensor measures the soil permeability.

The above sensor types show the heterogeneity in captured data, their formats, the devices and specifications, the frequency of data collection, the distance between the sensor and the object, and the transmission protocol. The environment also impacts the accuracy of data collection by the sensor. The in-field sensors are susceptible to variations in environmental conditions – wind, temperature changes and rain.

A remotely placed sensor senses the data for PA. Remote sensing measures the physical phenomenon by looking at reflections from a distance. There are many types of cameras to capture images in its field of view. Imaging platforms are also a critical factor when capturing an image. RGB camera appears to be the most prevalent sensor for image acquisition [129] in PA. However, they are sensitive to illumination variations in the field impacting object detection. Integrating visible spectrum cameras with NIR modality can improve the performance against changes in illumination. The cameras are mounted on different airborne platforms – UAV, airplane or satellite [130]. Spectral-range remote sensing is mounted on airborne platforms, which has its characteristics. Table 2.6 shows the typical features of such platforms and sensors [131].

The collection and distribution of images in real time are also hindrances for open solutions. It can be seen that data acquisition is impacted by the spatial resolution, data collection frequency, the spectral range of the sensor and distance. The images captured by remote sensing, especially the airborne and satellite, are impacted by the cloud formation and haze formed due to aerosols. It needs the development of advanced techniques for atmospheric corrections, cloud detection and noise cancellation. The images should also target specific applications using appropriate imaging platforms [129]; for example, the ground-based platform with a top-down camera will be more appropriate than a UAV for capturing images of the weed and observing the dynamics of crop characteristics/traits. At the same time, UAVs are well suited for large field-of-view applications, for example, field scouting.

**Table 2.6:** Characteristics of platforms and sensors for remote sensing – multispectral (MS), hyperspectral (HS), panchromatic (PAN), synthetic aperture radar (SAR), short-wave infrared (SWIR), light detection and ranging (LiDAR).

| Platform | Spatial resolution | Coverage | Difficulty in use | Sensor | Height |
|---|---|---|---|---|---|
| UAV | <1 m | Tens of meters | Easy | Camera, LiDAR | Few meters |
| Airborne | <1 m | Few km | Medium | Camera, LiDAR, SAR | Up to 10 km |
| Satellite | <1 m to km | Swath width – few tens of km | High | MS, HS, SWIR, PAN, X-band, C-band | Hundreds of km |

The type of agricultural task governs the data acquisition in PA; for example, growth monitoring may span over a whole season as it would require image collection over multiple growth stages. On the other hand, some tasks can be completed quickly, for example, fruit picking within a week [129]. The existing datasets do not address the large-scale diversity in the species. It, in turn, limits the generalization ability of the machine learning algorithm.

### 2.4.2 Challenges in data transmission

The main challenge for any IoT-based sensing system is latency in data transfer, low power devices, low bandwidth availability, intermittent Internet connectivity, cloud congestion due to transmission by a large number of devices and line of sight. Many WSN protocols are cellular, 6LoWPAN, ZigBee, Bluetooth, RFID, LoRaWAN and WiFi. The protocols have their specifications, including bandwidth, the number of channels, power, price and availability; for example, the deployment of 4G technology provides high bandwidth with reliable connectivity but requires higher battery power consumption, infrastructure deployment and high operational costs.

Table 2.7 summarizes the technologies available in communication that differ highly in terms of power, data rate, range and cost. Environmental conditions can hinder communication between the in-field sensor and the central processing server; for example, noise in the wireless communication channel may increase during the rainy season due to disturbances. The development of new technology, for example, 5G, can provide better security at a higher speed.

**Table 2.7:** Summary of fundamental networking and communication technologies in PA.

| Protocol | Power | Data rate | Range | Standard |
|---|---|---|---|---|
| ZigBee [132] | Low | Few hundreds of kbps | ~100 m | IEEE 802.15.4 |
| Bluetooth [133] | Low | Few Mbps | ~ 30 m | IEEE 802.15.1 |
| RFID [134] | Ultra low | Few tags per second | ~ cm | RFID |
| LoRaWAN [135] | Very low | Tens of kbps | ~10 km | IEEE 802.11ah |
| WiFi [136] | Medium | Tens of Mbps | ~50 m | IEEE 802.11 |
| 4 G [137] | High | Tens of Mbps | ~Tens of km | GSM |

### 2.4.3 Challenges in data management

The in-field and/or remote sensor generate the data. The measurement from sensors has to be real time and precise for decision making. They would need continuous calibration for generating accurate data. In the PA domain, the information influx is heavy as the in-field sensors create a time series at different frequencies, and the airborne platforms generate high-resolution MS or HS images. It would need high-performance computing for subsequent data processing and analysis. It also increases the storage requirements. It demands newer scalable facilities in terms of software and hardware platforms for effective management of big data. The cloud computing platforms can facilitate software as a service, focusing on merging sensor data and data management functions.

The use of different digital standards in PA causes interoperability and better data/knowledge issues, severely limiting the adoption of the newer technologies in PA. It impacts the growth rate and efficiency of digital agriculture applications. Protocols for the machine-to-machine (M2M) communication and development of better data sharing protocols between machine and information management systems handle the interoperability issue.

### 2.4.4 Challenges in data security

Data-driven farming has raised privacy issues and rights to data usage. The information about farming activities can be used by rivals and create speculations in the commodity markets [2]. There are issues related to security and data access within the ambit of the law of the land. The development of technology in PA has facilitated autonomy and the fusion of intelligence [138]. It needs a peer-to-peer system that can securely verify, monitor and analyze agricultural data. The penetration of IoT devices into agriculture will bring in various IoT attacks by hackers. Due to a lack of standardization in IoT, security attackers can execute forgery routines, data blocks and encryption [138]. IoT also works with M2M communication. The attacker can use industrial control system to launch an attack and encrypt the data or launch ransomware across many operating systems [139]. Data security can be handled by creating blockchain-based information systems in PA. Blockchain will allow reliable sharing of data in a decentralized environment while transparently manipulating data.

### 2.4.5 Challenges due to non-technical factors

Big data analytics in agriculture demands vast investments required for data collection, storage and processing. There is a considerable gap in available hardware, software resources and even baseline Internet connectivity. There exists a digital divide between developed and developing countries. The lack of financial resources restricts technology implementation, resulting in lower use of PA in the present context of sustainable development and climate change aberrations. The real-time information availability demands high-quality sensing elements and sensor procurement, deployment and maintenance that involve substantial costs. Therefore, digitization seems to be benefiting technology-savvy and wealthy farmers. It concludes that PA was found to be cost-effective for sizeable arable land. However, low-cost solutions for small arable land are also emerging on the horizon [140]. Literacy also impacts the adoption of PA in developing countries [141]. Farmers use traditional practices for the crop cycle, which is based on intuition and experience. Many times, lack of education may limit the benefit arising out of using the technology. Technology concentration in the hands of few corporations may lead to farmers' monopolistic practices and dependence on these big corporations [2].

## 2.5 Discussions and summary

The content in Sections 2.2 and 2.3 highlighted applications in agriculture and NRM, handled using data science. The sections analyze images as they emerge from sensors on the ground or sensors placed on UAVs or satellites. The CV-based non-destructive approaches gain popularity as data can be systematically obtained over a desired geographic area – large or small. The coarse-resolution satellite images from popular sensors – AVHRR, Landsat and MODIS – have become a primary source for data analysis. The high spatial resolution from Landsat or SPOT satellites is used for decision-making and NRM.

The farmers can learn from cooperatives' success, for example, in India's dairy industry, and leverage their strength to create *the agri-data ecosystem.* It will bring in parity in data-driven agricultural practices. The government should promote and legislate laws for copyright protection and data ownership. A robust policy framework for data management and security is required to further adopt DS into agriculture and NRM.

It is essential to provide large-scale access to the hardware and software resources. It requires investments in cloud infrastructure for storage, analysis and data visualization. It is crucial to develop easy accessibility to the data through user-friendly online platforms [2]. The use of well-accepted open-source libraries could help develop tools added to existing platforms to support large-scale data analysis. The emergence of publicly available datasets is also a healthy sign for data-driven approaches [2]. The large-scale challenges using CV and ML provide complex and large annotated datasets.

With the increasing availability of big data, well-known data analytic techniques through open source have tremendous potential to adopt digital technologies in agriculture and NRM further. It will accelerate producing a large quantity of quality food while protecting the environment and existing natural resources. Government support in making open data policy and making data readily available and affordable will enhance the quality and productivity and, in turn, improve the data/informatics culture. Federated systems will augment the research capabilities and innovations, thereby having a social impact.

# References


[1] De Clercq, M., Vats, A., Biel, A. (2018). Agriculture 4.0: The future of farming technology. Proceedings of the World Government Summit, Dubai, UAE, 11–13.

[2] Kamilaris, A., Kartakoullis, A., Prenafeta-Boldú, F.X. (2017). A review on the practice of big data analysis in agriculture. Computers and Electronics in Agriculture, 143, 23–37.

[3] World Bank. (2007). World development report 2008: Agriculture for development. Washington, DC. https://openknowledge.worldbank.org/handle/10986/5990 [Last accessed on 23.06.2021]

[4] Bulanon, D.M., Hestand, T., Nogales, C., Allen, B., Colwell, J. (2020). Machine vision system for orchard management. In: Machine Vision and Navigation. Cham, Springer, 197–240.

[5] Mavridou, E., Vrochidou, E., Papakostas, G.A., Pachidis, T., Kaburlasos, V.G. (2019). Machine vision systems in precision agriculture for crop farming. Journal of Imaging, 5(12), 89.

[6] Patrício, D.I., Rieder, R. (2018). Computer vision and artificial intelligence in precision agriculture for grain crops: A systematic review. Computers and Electronics in Agriculture, 153, 69–81.

[7] Culman, M.A., Gomez, J.A., Talavera, J., Quiroz, L.A., Tobon, L.E., Aranda, J.M., . . . Bayona, C.J. (2017 April). A novel application for identification of nutrient deficiencies in oil palm using the internet of things. In 2017 5th IEEE International Conference on Mobile Cloud Computing, Services, and Engineering (MobileCloud), 169–172. IEEE.

[8] Li, K., Lian, H., Van Deun, R., Brik, M.G. (2019). A far-red-emitting NaMgLaTeO6: Mn4+ phosphor with perovskite structure for indoor plant growth. Dyes and Pigments, 162, 214–221.

[9] Faragó, D., Sass, L., Valkai, I., Andrási, N., Szabados, L. (2018). PlantSize offers an affordable, nondestructive method to measure plant size and color in vitro. Frontiers in Plant Science, 9, 219.

[10] Rico-Fernández, M.P., Rios-Cabrera, R., Castelan, M., Guerrero-Reyes, H.I., Juarez-Maldonado, A. (2019). A contextualized approach for segmentation of foliage in different crop species. Computers and Electronics in Agriculture, 156, 378–386.

[11] Naik, H.S., Zhang, J., Lofquist, A., Assefa, T., Sarkar, S., Ackerman, D., . . . Ganapathysubramanian, B. (2017). A real-time phenotyping framework using machine learning for plant stress severity rating in soybean. Plant Methods, 13(1), 1–12.

[12] Zhu, Y., Cao, Z., Lu, H., Li, Y., Xiao, Y. (2016). In-field automatic observation of wheat heading stage using computer vision. Biosystems Engineering, 143, 28–41.

[13] Skovsen, S., Dyrmann, M., Mortensen, A.K., Laursen, M.S., Gislum, R., Eriksen, J., . . . Jorgensen, R.N. (2019). The grassclover image dataset for semantic and hierarchical species understanding in agriculture. In Proceedings of the IEEE/CVF Conference on Computer Vision and Pattern Recognition Workshops.

[14] Ip, R.H., Ang, L.M., Seng, K.P., Broster, J.C., Pratley, J.E. (2018). Big data and machine learning for crop protection. Computers and Electronics in Agriculture, 151, 376–383.

[15] Ma, J., Du, K., Zhang, L., Zheng, F., Chu, J., Sun, Z. (2017). A segmentation method for greenhouse vegetable foliar disease spots images using color information and region growing. Computers and Electronics in Agriculture, 142, 110–117.

[16] Ramcharan, A., McCloskey, P., Baranowski, K., Mbilinyi, N., Mrisho, L., Ndalahwa, M., . . . Hughes, D.P. (2019). A mobile-based deep learning model for cassava disease diagnosis. Frontiers in Plant Science, 10, 272.

[17] Sabzi, S., Abbaspour-Gilandeh, Y., García-Mateos, G. (2018). A fast and accurate expert system for weed identification in potato crops using metaheuristic algorithms. Computers in Industry, 98, 80–89.

[18] Haug, S., Ostermann, J. (2014 September). A crop/weed field image dataset for the evaluation of computer vision based precision agriculture tasks. In European conference on computer vision, 105–116. Springer, Cham.
[19] Gongal, A., Amatya, S., Karkee, M., Zhang, Q., Lewis, K. (2015). Sensors and systems for fruit detection and localization: A review. Computers and Electronics in Agriculture, 116, 8–19.
[20] Hossain, M.S., Al-Hammadi, M., Muhammad, G. (2018). Automatic fruit classification using deep learning for industrial applications. IEEE Transactions on Industrial Informatics, 15(2), 1027–1034.
[21] Deng, L., Du, H., Han, Z. (2017). A carrot sorting system using machine vision technique. Applied Engineering in Agriculture, 33(2), 149–156.
[22] Firouzjaei, R.A., Minaei, S., Beheshti, B. (2018). Sweet lemon mechanical damage detection using image processing technique and UV radiation. Journal of Food Measurement and Characterization, 12(3), 1513–1518.
[23] Iraji, M.S. (2019). Comparison between soft computing methods for tomato quality grading using machine vision. Journal of Food Measurement and Characterization, 13(1), 1–15.
[24] Wang, Z., Hu, M., Zhai, G. (2018). Application of deep learning architectures for accurate and rapid detection of internal mechanical damage of blueberry using hyperspectral transmittance data. Sensors, 18(4), 1126.
[25] Yao, H., Qin, R., Chen, X. (2019). Unmanned aerial vehicle for remote sensing applications – A review. Remote Sensing, 11(12), 1443.
[26] Maldonado, W. Jr, Barbosa, J.C. (2016). Automatic green fruit counting in orange trees using digital images. Computers and Electronics in Agriculture, 127, 572–581.
[27] Sanches, G.M., Duft, D.G., Kölln, O.T., Luciano, A.C.D.S., De Castro, S.G.Q., Okuno, F.M., Franco, H.C.J. (2018). The potential for RGB images obtained using unmanned aerial vehicle to assess and predict yield in sugarcane fields. International Journal of Remote Sensing, 39(15–16), 5402–5414.
[28] Han, L., Yang, G., Dai, H., Xu, B., Yang, H., Feng, H., . . . Yang, X. (2019). Modeling maize above-ground biomass based on machine learning approaches using UAV remote-sensing data. Plant Methods, 15(1), 1–19.
[29] Gutiérrez, S., Wendel, A., Underwood, J. (2019). Ground-based hyperspectral imaging for extensive mango yield estimation. Computers and Electronics in Agriculture, 157, 126–135.
[30] Kestur, R., Meduri, A., Narasipura, O. (2019). MangoNet: A deep semantic segmentation architecture for a method to detect and count mangoes in an open orchard. Engineering Applications of Artificial Intelligence, 77, 59–69.
[31] Kar, R., Reddy, G.P.O., Kumar, N., Singh, S.K. (2018). Monitoring spatio-temporal dynamics of urban and peri-urban landscape using remote sensing and GIS – A case study from Central India. Egyptian Journal of Remote Sensing & Space Sciences, 21(3), 401–411.
[32] Dai, A. (2011). Drought under global warming: A review. Wiley Interdisciplinary Reviews: Climate Change, 2, 45–65.
[33] Hao, Z., AghaKouchak, A., Nakhjiri, N., Farahmand, A. (2014). Global integrated drought monitoring and prediction system. Scientific Data, 1, 140001.
[34] Hu, X., Ren, H., Tansey, K., Zheng, Y., Ghent, D., Liu, X., Yan, L. (2019). Agricultural drought monitoring using European Space Agency Sentinel 3A land surface temperature and normalized difference vegetation index imageries. Agriculture For Meteorology, 279, 107707.
[35] Sandeep, P., Reddy, G.P.O., Jegankumar, R., Kumar, K.C.A. (2021). Monitoring of agricultural drought in semi-arid ecosystem of Peninsular India through indices derived from time series CHIRPS and MODIS data sets. Ecological Indicators, 21(Article), 107033.

[36] Xia, J., Yang, C., Li, Q. (2018). Building a spatiotemporal index for earth observation big data. International Journal of Applied Earth Observation and Geoinformation: ITC Journal, 73, 245–252.
[37] Fung, K.F., Huang, Y.F., Koo, C.H., Soh, Y.W. (2019). Drought forecasting: A review of modelling approaches 2007–2017. Journal of Water and Climate Change.
[38] Tyralis, H., Papacharalampous, G., Langousis, A. (2019). A brief review of random forests for water scientists and practitioners and their recent history in water resources. Water, 11, 910.
[39] Kuswanto, H., Naufal, A. (2019). Evaluation of performance of drought prediction in Indonesia based on TRMM and merra-2 using machine learning methods. Methods X, 6, 1238–1251.
[40] Lasaponara, R. (2006). On the use of principal component analysis (PCA) for evaluating Interannual vegetation anomalies from SPOT/VEGETATION NDVI temporal series. Ecological Model, 194, 429–434.
[41] Alizadeh, M.R., Nikoo, M.R. (2018). A fusion-based methodology for meteorological drought estimation using remote sensing data. Remote Sensing of Environment, 211, 229–247.
[42] Borji, M., Malekian, A., Salajegheh, A., Ghadimi, M. (2016). Multi-time-scale analysis of hydrological drought forecasting using support vector regression (SVR) and artificial neural networks (ANN). Arabian Journal of Geosciences, 9, 725.
[43] Chiang, J.L., Tsai, Y.S. (2012). Reservoir drought prediction using support vector machines. Applied Mechanics and Materials, 145, 455–459.
[44] Morid, S., Smakhtin, V., Bagherzadeh, K. (2007). Drought forecasting using artificial neural networks and time series of drought indices. International Journal of Climatology, 27(15), 2103–2111.
[45] Chen, J., Li, M., Wang, W. (2012). Statistical uncertainty estimation using random forests and its application to drought forecast. Mathematical Problems in Engineering, Article ID 915053.
[46] Park, S., Im, J., Jang, E., Rhee, J. (2016). Drought assessment and monitoring through blending of multi-sensor indices using machine learning approaches for different climate regions. Agriculture For Meteorology, 216, 157–169.
[47] Yurekli, K., Sattari, M.T., Anli, A.S., Hinis, M.A. (2012). Seasonal and annual regional drought prediction by using data-mining approach. Atmósfera, 25(1), 85–105.
[48] Abdi, A. (2019). Land cover and land use classification performance of machine learning algorithms in a boreal landscape using Sentinel-2 data. Gucci Remote Sensing, 1–20.
[49] Reddy, G.P.O., Kumar, N., Sahu, N., Srivastava, R., Singh, S.K., Naidu, L.G.K., Chary, G.R., Biradar, C.M., Gumma, M.K., Reddy, B.S., Kumar, J.N. (2020). Assessment of Spatio-temporal vegetation dynamics in tropical arid ecosystem of India using MODIS time-series vegetation indices. Arabian Journal of Geosciences, 13(15), 1–13.
[50] Zhang, Y., Ma, R., Duan, H., Loiselle, S., Xu, J. (2016). Satellite analysis to identify changes and drivers of CyanoHABs dynamics in Lake Taihu. Water Supply, 16(5), 1451–1466.
[51] Breiman, L. (2001). Random Forests. Machine Learning, 45(1), 5–32.
[52] Woznicki, S.A., Baynes, J., Panlasigui, S., Mehaffey, M., Neale, A. (2019). Development of a spatially complete floodplain map of the conterminous United States using random forest. Science of the Total Environment, 647, 942–953.
[53] Belgium, M., Drˇagut, L. (2016). Random Forest in remote sensing: A review of applications and future directions. Official Publication of the International Society for Photogrammetry and Remote Sensing, 114, 24–31.
[54] Shao, Y., Lunetta, R.S. (2012). Comparison of support vector machine, neural network, and CART algorithms for the land-cover classification using limited training data points. ISPRS Journal of Photogrammetry and Remote Sensing, 70, 78–87.

[55] Chen, Y., Lin, Z., Zhao, X., Wang, G., Gu, Y. (2014). Deep learning-based classification of hyperspectral data. IEEE Journal of Selected Topics in Applied Earth Observations and Remote Sensing, 7, 2094–2107.

[56] Li, W., Fu, H., Yu, L., Cracknell, A. (2017). Deep learning-based oil palm tree detection and counting for high-resolution remote sensing images. Remote Sensing, 9, 22.

[57] Srivastava, P.K., Han, D., Rico-Ramirez, M.A., Bray, M., Islam, T. (2012). Selection of classification techniques for land use/land cover change investigation. Advances in Space Research, 50, 1250–1265.

[58] Keshtkar, H., Voigt, W., Alizadeh, E. (2017). Land-cover classification and analysis of change using machine-learning classifiers and multi-temporal remote sensing imagery. Arabian Journal of Geosciences, 10.

[59] Noi, P.T., Kappas, M. (2018). Comparison of random forest, k-nearest neighbor, and support vector machine classifiers for land cover classification using sentinel-2 Imagery. Sensors, 18, 18.

[60] Prasad, A., Iverson, L., Liaw, A. (2006). Newer classification and regression tree techniques: Bagging and random forests for ecological prediction. Ecosystems, 9, 181–199.

[61] Dias, L.C.P., Pimenta, F.M., Santos, A.B., Costa, M.H., Ladle, R.J. (2016). Patterns of land use, extensification, and intensification of Brazilian agriculture. Global Change Biology, 22(8), 2887–2903.

[62] Reddy, G.P.O., Singh, S.K. (2014). Achieving food security through land management, Geospatial Today, July, 2014, 14–20 p.

[63] Biradar, C., Krishna, G., Sarker, A., Swain, N., Nangia, V., Devkota, M., Govind, A., Swelam, A., Yigezu, Y., Sahoo, R., Wery, J. (2020). Digital augmentation for sustainable intensification of dryland farming systems. 16th ESA Congress, 1–4 September 2020, Sevilla, Spain.

[64] Whitcraft, A.K., Becker-Reshef, I., Justice, C.O. (2014). Agricultural growing season calendars derived from MODIS surface reflectance. International Journal of Digital Earth.

[65] Chandna, P.K., Mondal, S. (2020). Analyzing multi-year rice-fallow dynamics in Odisha using multi-temporal Landsat-8 OLI and Sentinel-1 Data. GIScience & Remote Sensing.

[66] Krupnik, T.J., Schulthess, U., Ahmed, Z.U., McDonald, A.J. (2017). Sustainable crop intensification through surface water irrigation in Bangladesh? A geospatial assessment of landscape-scale production potential. Land Use Policy, 60, 206–222.

[67] Pelletier, C., Valero, S., Inglada, J., Champion, N., Dedieu, G. (2016). Assessing the robustness of Random Forests to map land cover with high resolution satellite image time series over large areas. Remote Sensing of Environment, 187, 156–168.

[68] Wang, S., et al. (2019). Crop type mapping without field-level labels: Random Forest transfer and unsupervised clustering techniques. Remote Sensing of Environment, 222, 303–317.

[69] Teluguntla, P., Thenkabail, P.S., Oliphant, A., Xiong, J., Gumma, M.K., Congalton, R.G., Yadav, K., Huete, A. (2018). A 30-m Landsat-derived cropland extent product of Australia and China using random forest machine learning algorithm on Google Earth Engine cloud computing platform. ISPRS Journal of Photogrammetry, Remote Sensing, 144, 325–340.

[70] Gumma, M.K., Thenkabai, P.S., Teluguntla, P.G., Oliphant, A., Xiong, J., Giri, C., Pyla, V., Dixit, S., Whitbread, A.M. (2020). Agricultural cropland extent and areas of South Asia derived using Landsat satellite 30-m time-series big-data using random forest machine learning algorithms on the Google Earth Engine cloud. Gucci Remote Sensor, 57, 302–322.

[71] Moumni, A., Lahrouni, A. (2021). Machine learning-based classification for crop-type mapping using the fusion of high-resolution satellite imagery in a semiarid area. Scientifica, Article ID 8810279.

[72] Inglada, J., Arias, M., Tardy, B., Hagolle, O., Valero, S., Morin, D., Dedieu, G., Sepulcre, G., Bontemps, S., Defourny, P., Koetz, B. (2015). Assessment of an operational system for crop

type map production using high temporal and spatial resolution satellite optical imagery. Remote Sensing, 7, 12356–12379.
[73] Kumar, P., Prasad, R., Choudhary, A., Mishra, V.N., Gupta, D.K., Srivastava, P.K. (2017). A statistical significance of differences in classification accuracy of crop types using different classification algorithms. Geocarto International, 32, 206–224.
[74] Duro, D.C., Franklin, S.E., Duve, M.G. (2012). A comparison of pixel-based and object-based image analysis with selected machine learning algorithms for the classification of agricultural landscapes using SPOT-5 HRG imagery. Environmental Remote Sensing, 18, 259–272.
[75] Li, M.C., Ma, L., Blaschke, T., Cheng, L., Tiede, D. (2016). A systematic comparison of different object-based classification techniques using high spatial resolution imagery in agricultural environments. International Journal of Applied Earth Observation and Geoinformation, 49, 87–98.
[76] McBratney, A.B., Santos, M.L.M., Minasny, B. (2003). On digital soil mapping. Geoderma, 117, 3–52.
[77] Reddy, G.P.O., Nagaraju, M.S.S., Ramteke, I.K., Sarkar, D. (2013). Terrain characterization for soil resource mapping in part of semi-tract of Central India using high resolution satellite data and GIS. Journal of the Indian Society of Remote Sensing, 41(2), 331–343.
[78] Lagacherie, P., McBratney, A.B. (2007). Spatial soil information systems and spatial soil inference systems: Perspectives for digital soil mapping. In: Lagacherie, P., McBratney, A.B., Voltz, M., eds. Digital Soil Mapping: An Introductory Perspective. Amsterdam, Elsevier, vol. 31, 3–22.
[79] Taghizadeh-mehrjardi, R., Toomanian, N., Khavaninzadeh, A.R., Jafari, A., Triantafilis, J. (2016). Predicting and mapping of soil particle-size fractions with adaptive neuro-fuzzy inference and ant colony optimization in central Iran. European Journal of Soil Science, 67, 707–725.
[80] Tziachris, P., Aschonitis, V., Chatzistathis, T., Papadopoulou, M. (2019). Assessment of spatial hybrid methods for predicting soil organic matter using DEM derivatives and soil parameters. Catena, 174, 206–216.
[81] Vaysse, K., Lagacherie, P. (2015). Evaluating digital soil mapping approaches for mapping GlobalSoilMap soil properties from legacy data in Languedoc-Roussillon (France). Geoderma Regional, 4, 20–30.
[82] Vaysse, K., Lagacherie, P. (2017). Using quantile regression forest to estimate uncertainty of digital soil mapping products. Geoderma, 291, 55–64.
[83] Forkuor, G., Hounkpatin, O.K.L., Welp, G., Thiel, M. (2017). High resolution mapping of soil properties using remote sensing variables in south-western Burkina Faso: A comparison of machine learning and multiple linear regression models. PLOS ONE, 12, e0170478.
[84] Jenny, H. (1994). Factors of Soil Formation: A System of Quantitative Pedology. North Chelmsford, MA, USA, Courier Corporation.
[85] Padarian, J., Minasny, B., McBratney, A.B. (2019). Using deep learning for digital soil mapping: A review aided by machine learning tools. Soil, 5, 79–89.
[86] Were, K., Bui, D.T., Dick, Ø.B., Singh, B.R. (2015). A comparative assessment of support vector regression, artificial neural networks, and random forests for predicting and mapping soil organic carbon stocks across an Afromontane landscape. Ecological Indicators, 52, 394–403.
[87] Akpa, S.I.C., Odeh, I.O.A., Bishop, T.F.A., Hartemink, A.E. (2014). Digital mapping of soil particle-size fractions for Nigeria. Soil Science Society of America Journal. Soil Science Society of America, 78, 1953–1966.

[88] Dharumarajan, S., Hegde, R., Singh, S.K. (2017). Spatial prediction of major soil properties using random forest techniques-A case study in semi-arid tropics of South India. Geoderma Regional, 10, 154–162.
[89] Song, C., Sondergaard, M., Cao, X., Zhou, Y., Song, C., Sondergaard, M., et al. (2018). Nutrient utilization strategies of algae and bacteria after the termination of nutrient amendment with different phosphorus dosage: A mesocosm case. Geomicrobiology Journal, 35, 294–299.
[90] Hengl, T., Leenaars, J.G.B., Shepherd, K.D., Walsh, M.G., Heuvelink, G.B.M., Mamo, T., Tilahun, H., Berkhout, E., Cooper, M., Fegraus, E., et al. (2017). Soil nutrient maps of sub-Saharan Africa: Assessment of soil nutrient content at 250 m spatial resolution using machine learning. Nutrient Cycling in Agroecosystems, 109, 77–102.
[91] Emadi, M., Taghizadeh-Mehrjardi, R., Cherati, A., Danesh, M., Mosavi, A., Scholten, T. (2020). Predicting and mapping of soil organic carbon using machine learning algorithms in Northern Iran. Remote Sensing, 12, 2234.
[92] Shiferaw, B., Holden, S. (1999). Soil erosion and smallholders' conservation decisions in the highlands of Ethiopia. World Development, 27, 739–752.
[93] Borrelli, P., Robinson, D.A., Fleischer, L.R., Lugato, E., Ballabio, C., Alewell, C., Meusburger, K., Modugno, S., Schütt, B., Ferro, V., Bagarello, V., Oost, K.V., Montanarella, L., Panagos, P. (2017). An assessment of the global impact of twenty-first century land use change on soil erosion. Nature Communications, 8(1), 2013.
[94] Oldeman, L. (1994). The global extent of soil degradation; ISRIC: Wageningen. The Netherlands, 1994, 19–36.
[95] Maji, A.K., Reddy, G.P.O., Sarkar, D. (2010). Degraded and wastelands of India: Status and spatial distribution. Directorate of Information and Publications of Agriculture. New Delhi, ICAR, Krishi Anusandhan Bhavan I, Pusa, 158 p.
[96] Minhas, P.S., Reddy, G.P.O. (2017). Edaphic stresses and agricultural sustainability: An Indian perspective. Agricultural Research, 6(1), 8–21.
[97] Sharda, V.N., Juyal, G.P., Prakash, C., Joshi, B.P. (2007). Training manual soil conservation and watershed management, Vol-II, (soil water conservation engineering). Central Soil and Water Conservation Research and Training Institute (CSWCRTI), Dehradun, Uttaranchal, 1–410.
[98] Mahapatra, S.K., Yadav, R.P., Singh, S.P., Aggarwal, R.K., Sharma, J.P., Tiwari, A.K., Sidhu, G.S., Sarkar, D., Singh, S.K., Sharda, V.N., Mishra, P.K. (2014). Soil Erosion in Delhi, NBSS Publ.166, NBSS&LUP. Nagpur, India, 34.
[99] Sahoo, A.K., Sarkar, D., Sah, K.D., Maji, A.K., Dhyani, B.L. (2005). Soil Erosion of Orissa, NBSS Publ. 126, NBSS&LUP. Nagpur, 48.
[100] Sidhu, G.S., Pal, S., Tiwari, A.K., Sarkar, D., Sharda, V.N. (2013) Soil Erosion in Punjab, NBSS Publ. 151, NBSS&LUP, Nagpur, p 33.
[101] Ayalew, D.A., Deumlich, D., Šarapatka, B., Doktor, D. (2020). Quantifying the sensitivity of NDVI-based C factor estimation and potential soil erosion prediction using spaceborne earth observation data. Remote Sensing, 12, 1136.
[102] De Carvalho, D.F., Durigon, V.L., Antunes, M.A.H., De Almeida, W.S., Oliveira, P.T.S. (2014). Predicting soil erosion using RUSLE and NDVI time series from TM Landsat 5. Pesquisa Agropecuária Brasileira, 49, 215–224.
[103] Huang, Y., Chen, Z.-X., Yu, T., Huang, X.-Z., Gu, X.-F. (2018). Agricultural remote sensing big data: Management and applications. Journal of Integrative Agriculture, 17, 1915–1931.
[104] Shao, Z., Pan, Y., Diao, C., Cai, J. (2019). Cloud detection in remote sensing images based on multiscale features-convolutional neural network. IEEE Transactions on Geoscience and Remote Sensing, 57, 4062–4076.

[105] Hastie, T.J., Tibshirani, R., Friedman, J.H. (2008). The elements of statistical learning: Data mining, inference, and prediction, 2nd edn. Springer Series in Statistics, New York, NY, USA, Springer, 763.
[106] Yousefi, S., Pourghasemi, H.R., Avand, M., Janizadeh, S., Tavangar, S., Santosh, M. (2020). Assessment of land degradation using machine-learning techniques: A case of declining rangelands. Land Degradation & Development, 32(2).
[107] Azareh, A., Rahmati, O., Rafiei-Sardooi, E., Sankey, J.B., Lee, S., Shahabi, H., Ahmad, B.B. (2019). Modelling gully-erosion susceptibility in a semi-arid region, Iran: Investigation of applicability of certainty factor and maximum entropy models. Science of the Total Environment, 655, 684–696.
[108] Rotunno, A.F., Callari, C., Froiio, F. (2017). Computational modeling of backward erosion piping. In: Model Simulation, Exp Issues Struct Mech. Berlin, Springer, 225–234.
[109] Sarkar, T., Mishra, M. (2018). Soil erosion susceptibility mapping with the application of logistic regression and artificial neural network. Journal of Geovisualization and Spatial Analysis, 2(1), 8.
[110] Chen, S.H., Su, H.B., Tian, J., Zhang, R.H., Xia, J. (2011). Estimating soil erosion using MODIS and TM images based on support vector machine and à trous wavelet. International Journal of Applied Earth Observation and Geoinformation, 13(4), 626–635.
[111] Bhattacharya, R.K., Chatterjee, N.D., Das, K. (2021). Land use and land cover change and its resultant erosion susceptible level: An appraisal using RUSLE and logistic regression in a tropical plateau basin of West Bengal. Environment, Development and Sustainability, 23(2), 1411–1436.
[112] Moradi, H.R., Avand, M.T., Janizadeh, S. (2019). Landslide Susceptibility Survey using Modeling Methods. Amsterdam, Elsevier, 259–276.
[113] Paul, S.S., Li, J., Li, Y., Shen, L. (2019). Assessing land use–land cover change and soil erosion potential using a combined approach through remote sensing, RUSLE and random forest algorithm. Geocarto International, 34, 1–15.
[114] Haghighi, A.T., Darabi, H., Karimidastenaei, Z., Davudirad, A.A., Rouzbeh, S., Rahmati, O., Sajedi-Hosseini, F., Klöve, B. (2021). Land degradation risk mapping using topographic, human-induced, and geo-environmental variables and machine learning algorithms, for the Pole-Doab watershed, Iran. Environmental Earth Sciences, 80, 1.
[115] Al-Ghobari, H.M., Mohammad, F.S. (2011). Intelligent irrigation performance: Evaluation and quantifying its ability for conserving water in arid region. Applied Water Science, 1, 73–83.
[116] Cai, X., McKinney, D.C., Lasdon, L. (2001). Solving nonlinear water management models using a combined genetic algorithm and linear programming approach. Advances in Water Resources, 24(6), 667–676.
[117] Hinnell, A.C., Lazarovitch, N., Furman, A., Poulton, M., Warrick, A.W. (2010). Neuro-drip: Estimation of subsurface wetting patterns for drip irrigation using neural networks. Irrigation Science, 28, 535–544.
[118] Kim, Y.J., Evans, R.G., Iversen, W.M. (2008). Remote sensing and control of an irrigation system using a distributed wireless sensor network. IEEE Transactions on Instrumentation and Measurement, 57(7), 1379–1387.
[119] Wall, R.W., King, B.A. (2004) Incorporating Plug and Play Technology into Measurement and Control Systems for Irrigation. Management, 2004, Ottawa, Canada August 1–4.
[120] Wardlaw, R., Bhaktikul, K. (2004). Application of genetic algorithms for irrigation water scheduling. Irrigation and Drainage, 53, 397–414.
[121] Twarakavi, N.K.C., Simunek, J., Schaap, M.G. (2009). Development of pedotransfer functions for estimation of soil hydraulic parameters using support vector machines. Soil Science Society of America Journal, 73, 1443–1452.

[122] Yamaç, S.S., Todorovic, M. (2020). Estimation of daily potato crop evapotranspiration using three different machine learning algorithms and four scenarios of available meteorological data. Agricultural Water Management, 2020, 228.
[123] Tang, D., Feng, Y., Gong, D., Hao, W., Cui, N. (2018). Evaluation of artificial intelligence models for actual crop evapotranspiration modeling in mulched and non-mulched maize croplands. Computers and Electronics in Agriculture, 152, 375–384.
[124] Kisi, O. (2016). Modeling reference evapotranspiration using three different heuristic regression approaches. Agricultural Water Management, 169, 162–172.
[125] Davis, S.L., Dukes, M.D., Miller, G.L. (2009). Landscape irrigation by evapotranspiration-based irrigation controllers under dry conditions in Southwest Florida. Agricultural Water Management, 96, 1828–1836.
[126] Viani, F., Bertolli, M., Salucci, M., Polo, A. (2017). Low-cost wireless monitoring and decision support for water saving in agriculture. IEEE Sensors Journal, 17, 4299–4309.
[127] Gutiérrez, J., Villa-medina, J.F., Nieto-Garibay, A., Porta-gándara, M.Á., Gutierrez, J., Villa-medina, J.F., Nieto-Garibay, A., Porta-Gandara, M.A. (2014). Automated irrigation system using a wireless sensor network and GPRS module. IEEE Transactions on Instrumentation and Measurement, 63, 166–176.
[128] Roopaei, M., Rad, P., Choo, K.R., Choo, R. (2017). Cloud of things in smart agriculture: Intelligent irrigation monitoring by thermal imaging. IEEE Cloud Computing, 4, 10–15.
[129] Lu, Y., Young, S. (2020). A survey of public datasets for computer vision tasks in precision agriculture. Computers and Electronics in Agriculture, 178, 105760.
[130] Rudd, J.D., Roberson, G.T., Classen, J.J. (2017). Application of satellite, unmanned aircraft system, and ground-based sensor data for precision agriculture: A review. In 2017 ASABE Annual International Meeting, 1. American Society of Agricultural and Biological Engineers.
[131] Toth, C., Jóźków, G. (2016). Remote sensing platforms and sensors: A survey. ISPRS Journal of Photogrammetry and Remote Sensing, 115, 22–36.
[132] Anurag, D., Roy, S., Bandyopadhyay, S. (2008 May). Agro-sense: Precision agriculture using sensor-based wireless mesh networks. In 2008 first ITU-T kaleidoscope academic conference-innovations in NGN: Future network and services, 383–388. IEEE.
[133] Vuran, M.C., Salam, A., Wong, R., Irmak, S. (2018). Internet of underground things in precision agriculture: Architecture and technology aspects. Ad Hoc Networks, 81, 160–173.
[134] Ruiz-Garcia, L., Lunadei, L. (2011). The role of RFID in agriculture: Applications, limitations and challenges. Computers and Electronics in Agriculture, 79(1), 42–50.
[135] Haxhibeqiri, J., De Poorter, E., Moerman, I., Hoebeke, J. (2018). A survey of LoRaWAN for IoT: From technology to application. Sensors, 18(11), 3995.
[136] Afaqui, M.S., Garcia-Villegas, E., Lopez-Aguilera, E. (2016). IEEE 802.11 ax: Challenges and requirements for future high-efficiency WiFi. IEEE Wireless Communications, 24(3), 130–137.
[137] Khan, A.H., Qadeer, M.A., Ansari, J.A., Waheed, S. (2009 April). 4G as a next-generation wireless network. In 2009 International conference on future computer and communication, 334–338. IEEE.
[138] Torky, M., Hassanein, A.E. (2020). Integrating blockchain and the internet of things in precision agriculture: Analysis, opportunities, and challenges. Computers and Electronics in Agriculture, 105476.
[139] Makhdoom, I., Abolhasan, M., Lipman, J., Liu, R.P., Ni, W. (2018). Anatomy of threats to the internet of things. IEEE Communications Surveys & Tutorials, 21(2), 1636–1675.
[140] Pathak, H.S., Brown, P., Best, T. (2019). A systematic literature review of the factors affecting the precision agriculture adoption process. Precision Agriculture, 20(6), 1292–1316.

[141] Mittal, S., Mehar, M. (2016). Socio-economic factors affecting adoption of modern information and communication technology by farmers in India: Analysis using multivariate probit model. The Journal of Agricultural Education and Extension, 22(2), 199–212.
[142] Liu, W., Anguelov, D., Erhan, D., Szegedy, C., Reed, S., Fu, C.Y., Berg, A.C. (2016 October). Ssd: Single shot multibox detector. In European conference on computer vision, 21–37. Springer, Cham.
[143] Liu, T., Chen, W., Wu, W., Sun, C., Guo, W., Zhu, X. (2016). Detection of aphids in wheat fields using a computer vision technique. Biosystems Engineering, 141, 82–93.
[144] Mark, T. (2013). Big Data Meets Big Data Analytics. www.sas.com/resources/.../ WR46345.pdf. retrieved 10/03/22.
[145] Berhan, G., Hovav, A., Atnafu, S. (2010). Knowledge Discovery from Satellite Images for Drought Monitoring in Food Insecure Areas, AMCIS 2010 Proceedings.
[146] Maxwell, A. E., Warner, T. A., Fang, F. (2018). Implementation of machine-learning classification in remote sensing: an applied review. International Journal of Remote Sensing, 39(9), 2784–2817.
[147] Cruz, J.R.D., Baldovino, R.G., Bandala, A.A. and Dadios, E.P. (2017). Water usage optimization of smart farm automated irrigation system using artificial neural network, in Proceedings of the 2017 5th International Conference on Information and Communication Technology (ICoIC7), pp. 1–5, Melaka, Malaysia, May 2017.

Umamaheswari P., Vaishnavi D.*

# 3 Image caption generating system using convolutional neural networks and long short-term memory networks

**Abstract:** Human beings have the ability of easily determining the environment and the situations happening around them. Making a computer to automatically understand the surroundings can be done by training it with the images and ensuring that it describes the content of images. In this way, a computer will be able to learn the surroundings. Image caption generation is a process of creating appropriate captions for an image which is linguistically suitable and semantically precise for a scene and is also same to the understanding of humans. It is gaining attention because of its wide applications in scene recognition systems, human–robot communication systems, information retrieval systems and assistance systems for visually disabled persons. In this research, an image caption generating system is implemented using state-of-the-art CNN model InceptionResnetV2 for image feature extraction (transfer learning), and long short-term memory networks for image caption generation. The model is trained using Flickr8k dataset and it is evaluated using metrics like BLEU and METEOR scores.

**Keywords:** image caption generation, scene recognition, LSTM, transfer learning, CNN, InceptionResnetV2, BLEU score, METEOR score

## 3.1 Introduction

Human beings have the ability to easily determine the environment and the situations that happen around them. Making a computer to automatically understand the surroundings can be done by training it with the images and ensuring that it describes the content of images. In this way, a computer can able to learn the surroundings. Captioning an image is the process of generating captions which are linguistically and semantically correct for a particular scenario. It has large number of applications in scene recognition systems, human–robot communication systems, information retrieval systems, and assistance systems for visually disabled persons.

***Corresponding author: Vaishnavi D.,** Department of CSE, SRC, SASTRA Deemed University, Tamil Nadu, India, e-mail: vaishnavi@src.sastra.edu
**Umamaheswari P.,** Department of CSE, SRC, SASTRA Deemed University, Tamil Nadu, India, e-mail: pum@it.sastra.edu

https://doi.org/10.1515/9783110756722-003

### 3.1.1 Image captioning

The working method of image captioning is significantly harder than that of image classification, which has been the main motivation in the computer vision community because image classification deals only with identifying objects in an image, whereas image captioning must capture the relationship between the objects in an image. Recognizing the context of an image and describing it in the form of captions are highly challenging tasks. It requires high level of understanding of the contents of the image, and it expresses the information in a human speech.

### 3.1.2 Transfer learning

Transfer learning is one of the research-based problems present in machine learning that mainly focuses on storing the knowledge obtained during the problem solvation and applying it to the different problems. Considering an example, the knowledge gained in learning to find and recognize the objects can be applied when we are trying to recognize the relationship between the objects present in the new image. This area of research carried out some kind of relationship with the history on transfer learning. In this project, we will be using InceptionResnetV2 state-of-the-art CNN model for doing transfer learning, to obtain and store the feature vectors of all the dataset images. ResNet and Inception are used largely in image recognition systems in the current years. They can be assembled at minimum cost as well. InceptionResNetV2 joins the architecture of the Inception network with the architecture of residual network connections.

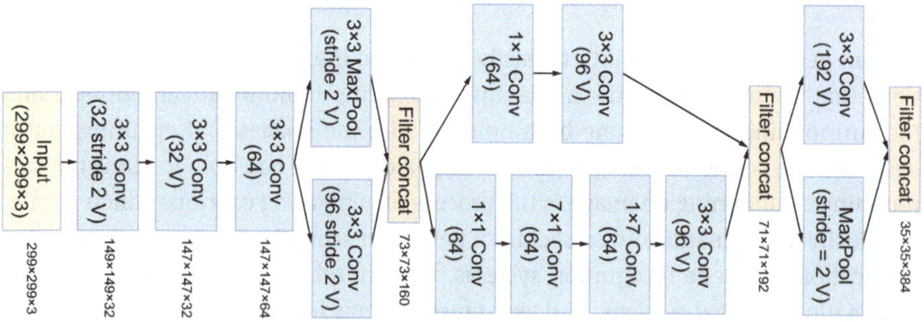


**Figure 3.1:** Inception ResNet architecture.

InceptionResNetV2 is a state-of-the-art CNN architecture model that is trained on ImageNet database which contains over lakhs of images of different number of classes. It has 164 layers in depth network and has the ability to classify the objects/ images into 1000 categories. Few of the categories are women, pen, lion, dog, hills and so on. This network is having feature representations which are learnt for a large number of images. The image size accepted by the network is of 299 × 299, and the output found is a list which is having the probabilities of all the two categories present. This model gives a feature vector which is of 1536 size for each and every image. It has been drawn up based on a fusion of residual connection and inception structure. The InceptionResNet block is made up of various sized convolutional filters that are combined with residual connections. The degradation problems are avoided by residual connections which are caused by deep structures, and training time is also minimized. Inception-Resnet-v2 network architecture is shown in Figure 3.1.

### 3.1.3 Long short-term memory networks

Long short-term memory (LSTM) is one of the recurrent neural networks (RNNs), which is able to remember long-distance temporal dependencies. This is mainly used in sequential generation, that is, in text generation, in time series data for stock market prediction and so on. The LSTM network is a long chain of a recurrent unit which is called as a memory cell and it is also referred to as LSTM cell. The memory cell or the LSTM cell states to the deep neural network architecture and it contains three important logic gates: the first is "Forget logic," second is "Input logic" and the final is "Output logic."

## 3.2 Related works

The image caption generation problem using natural language processing in the automatic generation of visual data has been studied by Gerber and Nagel [27]. Consider And–Or graphs or logic systems that are transformed to natural language by using rule-based systems. It was found that those works are weak and are used only on very restricted areas like sports and traffic scenes. Li et al. identified the object as a sentence by separating the items and matches the corresponding phrases; finally, all phrases are joined to form a sentence.

Kulkarni, Li and Baby suggested a system that makes a fitted connection between the particular image and the sentence generating method. The model uses visual detectors to detect specific objects, as well as attributes of a single object and relationships between multiple objects. Numerous detection graphs had been used in their work, beyond triplets with high complexity, and text generation is done by using templates.

- Language parsing was used by numerous language models. These methodologies are improper in terms of text generation.
- Ushiku and Yamaguchi et al. suggested various models for converting a given image. Those approaches are dealt with the text and images in the same vector space.

- Socher R and Karpathy et al. have suggested that bidirectional image sentence mapping and neural networks are used to embed images and sentences together along with image cropping and subsentence mapping techniques. Combination of deep convolutional neural network (CNN) will give sequence modeling. This method along with the RNNs to design a separate network generates image explanations. The disadvantages of RNNs lead to low-quality caption generation.

## 3.3 Methodology

### 3.3.1 Architecture diagram

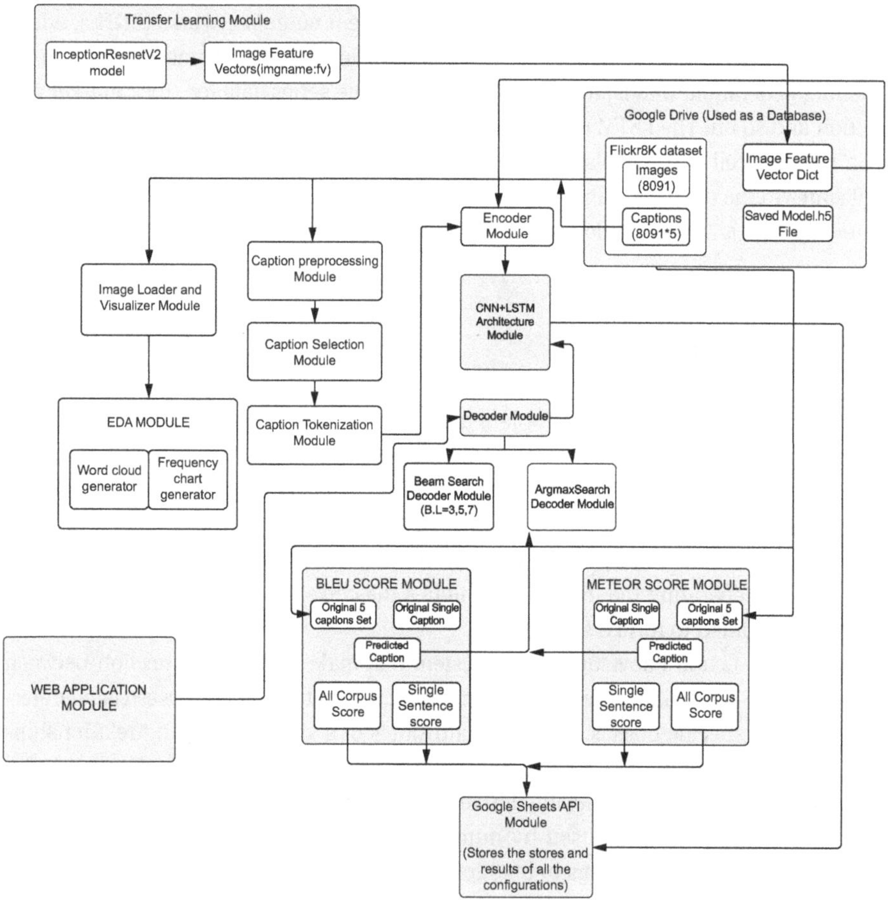


**Figure 3.2:** Framework of image caption generating system.

The main architecture of the project was designed by constructing various modules. Each module falls in one of the three stages. The main one is stage 2, which builds the main deep learning model that contains the EDA module as shown in Figure 3.2., captions preprocessing module, caption selection module, caption tokenization module, encoder, CNN + LSTM, decoder modules, BLEU and METEOR score evaluation modules and sheets API module. Stage 1 contains transfer learning module where we used InceptionResnetV2 model and stage 3 contains designing the Web application module.

### 3.3.2 Image feature extraction (transfer learning)

In the proposed model, we used CNNs to extract image features. CNN is a powerful algorithm which has the ability to find the underlying patterns in the images and produce an image feature vector. Since every image has different context and various objects/animals, it is very much needed to perform object detection using CNN. Since there may be more than 1000 different kinds of objects in an image, it is very difficult to implement a highly efficient CNN algorithm with the resources available with us. So, transfer learning was used. We used a state-of-the-art CNN architecture InceptionResnetV2, which was trained on ImageNet dataset and used to extract feature vectors from all the images. Other state-of-the-art CNN models like InceptionV3, Xception, VGG16, VGG19, NasNet and Mobile Net can also be used for transfer learning process, in order to improve the accuracy of the generated captions. Flickr8K is the dataset used in this project, which contains 8091 images and 5 captions per image. The dataset is obtained from Kaggle repository.

### 3.3.3 Text preprocessing

All the captions were preprocessed before being sent into the final model. It includes removal of unnecessary characters, numbers and lowercase conversions in order to maintain that the captions trained should be of good quality.

### 3.3.4 Exploratory dataset analysis

All the images in the dataset are displayed along with its five captions. The frequency of every word in the captions list is evaluated and displayed using histograms and Word clouds. (Most common words are man, woman, dog, etc.)

### 3.3.5 Caption selection and dataset splitting

Only a selected caption for every image is considered in order to reduce noise and for getting good and fluent captions which must be grammatically correct. This is inspired from the concept of evolution of learning. Then, the entire dataset is splitted into training and testing (20% for testing and 80% for training).

### 3.3.6 Caption tokenization

The selected caption is needed to be tokenized, because we cannot supply text into the model, so each and every word is mapped to a unique number. The total number of such words forms the vocabulary of the entire model. The longest length sentence is also obtained here, with which we can estimate the number of time steps taken by our LSTM model.

### 3.3.7 Preprocessing (encoding)

During training, the input given to the model should be formatted like [feature_vector, input_text (embedded)], the embedded form of input text will be a list whose length is max length caption in the dataset. The feature vector length is based on the type of model we choose during transfer learning (1536 for InceptionResnetV2). Each caption is subdivided further in terms of words, where the output to that input will be the next word (embedded form). All outputs will be one hot encoding on total vocab size. The sample of encoding is shown in Figure 3.3.

```
                                  ENCODING SAMPLE

        X(i/p)                                                     Y(o/p)
imageFV,startseq----------------------------------------------> a
[[]*1536, [0,0,0,0,0,0,0,0,0,0,0,0,0,0,0,0,0,1]]-------------->[0,0,1,0,0,0,0,0,0,0,..............]

imageFV,startseq a--------------------------------------------> child
[[]*1536, [0,0,0,0,0,0,0,0,0,0,0,0,0,0,0,0,0,1,2]]-------------->[0,0,0,1,0,0,0,0,0,0,..............]

imageFV,startseq a child----------------------->in
imageFV,startseq a child in ------------------->a
imageFV,startseq a child in a------------------>pink
.
.
```

Figure 3.3: Encoding sample.

### 3.3.8 Designing CNN-LSTM deep learning model

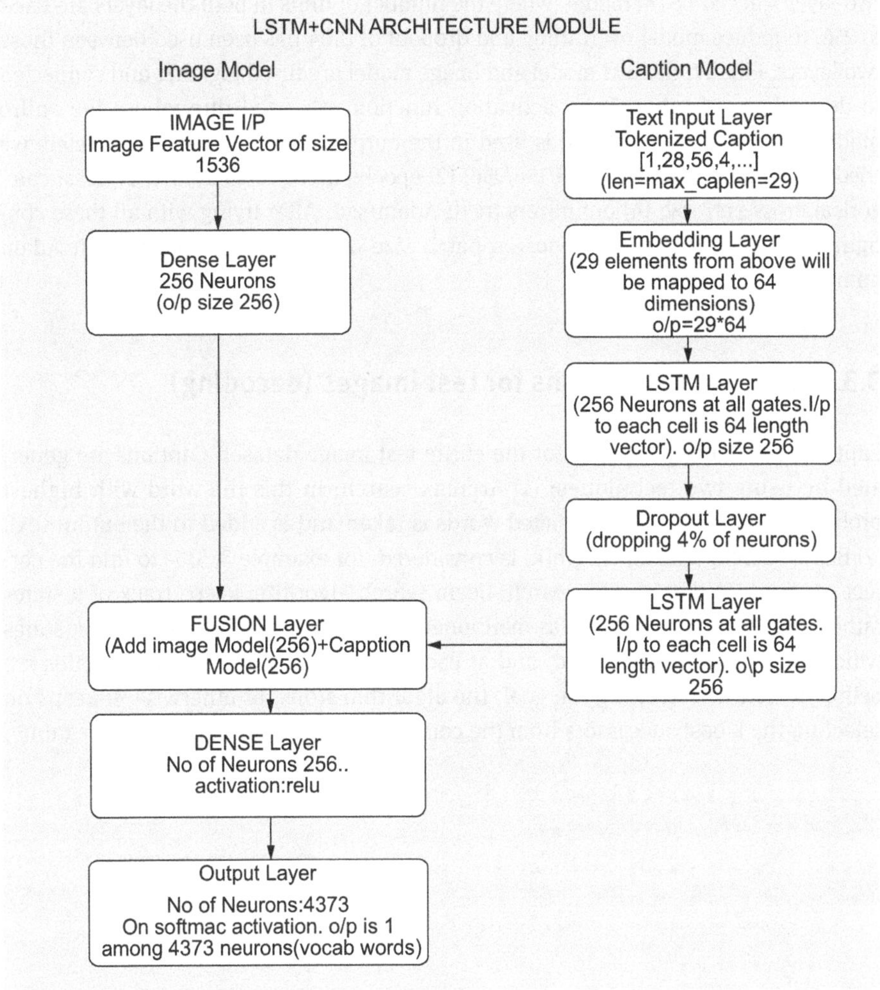


**Figure 3.4:** CNN-LSTM deep learning architecture.

We give both the image feature vector and captions represented as list to the model. The image feature vector is sent into the input layer of image model, which then as depicted in Figure 3.4. passed into a dense layer. Parallelly, the text list is sent into the input layer of text model, which is then passed into an embedding layer where all the words are mapped into a predefined vector space, and then these embeddings are passed into our LSTM layers. Here, we used stacked LSTMs.

It is one of the greatest broadly used algorithms in sequence modeling and text generation. Gates are used for controlling the movement of information in the repetitive computations. It successfully overcomes the drawbacks of conventional RNN models. In order to get good captions, we can stack LSTM layers. Here we used two-layer stacked LSTM model, where the number of units in both the layers are fixed to 256, to reduce model overfitting, and dropout of 0.04 has been used between those two layers. Finally, the text model and image model are fused together and connected to dense layers further. ReLu activation function was used throughout the entire model, and SoftMax activation is used in the output layer. The hyperparameters we tried are (1) batch sizes: 64/128/196/256; (2) epochs tried: 10/15/20/35; (3), loss: categorical cross-entropy; (4) optimizers tried: Adam/sgd. After trying with all these configurations, we decided to choose a batch size of 196 and 35 epochs with Adam optimizer.

### 3.3.9 Generating captions for test images (decoding)

Captions are being generated for the entire test image dataset. Captions are generated by using two techniques: (1) Argmax search; in this the word with highest probability among all the predicted words is taken and is added to the output text. (2) Beam search, a beam length(k) is considered, for example 3, 5, 7, to find the correct caption. Unlike Argmax search, beam search algorithm keeps track of k states rather than tracking one state as mentioned in Figure 3.5. It starts with the k states which are randomly generated, and at each step, the inheritors of all k states are being yielded. On reaching the goal, the algorithm stops, or otherwise, it keeps on selecting the k best successors from the complete list and starts the process again.

```
------------------ARGMAX & BEAM SEARCH-------------------------

ARGMAX SEARCH: BEAMSEARCH WITH LENGTH=1

          l=1-  l=2    - l=3------------------------------------l<=max_length/reaches endseq word
startseq  a  -man(80%) - eating(60%) (will give 1 more outcomes)

output caption from ARGMAX search:

          a  man  eating  next_word(max_probabilty_among 27 outputs)------

BEAM SEARCH with length 3:

l=1-  l=2    - l=3--------------------------------------------l<=max_length/reaches endseq word
a  -man(80%)  - eating(60%)   (will give 3 more outcomes)
              - laughing(35%) (will give 3 more outcomes)
              - staring(22%)  (w ill give 3 more outcomes)

   -boy(72%)  - biting(67%)  (will give 3 more outcomes)
              - eating (58%) (will give 3 more outcomes)
              - dances(4%)   (will give 3 more outcomes)

   -woman(70%)- eating(65%)  (will give 3 more outcomes)
              - staring(12%) (will give 3 more outcomes)
              - jumps(0.4%)  (will give 3 more outcomes)
--------------------------------------------------------------------------

output caption from beam search 3:

a  man  biting  word(max_probabilty_among 27 outputs)----------------------
```

Figure 3.5: Argmax and beam search.

### 3.3.10 Model evaluation

**(i) Evaluation using BLEU score:**

BLEU score, which is the acronym of Bilingual Evaluation Understudy, is an algorithm which is used to gauge the accuracy of sentences that are generated by the machines. BLEU score value is present between 0 and 1, where 0 means mismatch and 1 means perfect match. The score value indicates how similar the candidate text is to the output text. BLEU computes the score using n-grams, for BLEU-1, for 2-grams BLEU-2, for 3-gram score BLEU-3, for 4-gram BLEU-4 score. The longer n-gram (BLEU-3/4) scores justify the quality and articulacy of the translation and to what level it declaims is good English.

We evaluated the model using BLEU score in two ways: the first way is on the corpus-level evaluation in which the predicted caption's n-grams are compared with the n-grams of all five candidate captions and the score is generated, and the second way is sentence-level evaluation where the generated caption is compared with the corresponding caption (which we filtered in caption selection module) and the score is produced.

**(ii) Evaluation using METEOR score:**

METEOR score, which is acronym of Metric for Evaluation of Translation with Explicit Ordering, is one of the popularly used metrics for machine translation's evaluation method. This overcomes the problems in BLEU metric, and yields good association with the human decision at the sentence level, whereas the BLEU seeks association at the corpus level. This score also considers the amount of information that is matched in the output sentence and it is also more effective.

METEOR score is evaluated just like how BLEU score is evaluated before, that is, on both sentence level and corpus level. The scores are automated to store in Google Sheets, by connecting these two modules to the sheet's API in Colab. This helps in identifying the correct configurations for the model that is giving better BLEU and METEOR scores.

**(iii) Web application:**

A web application is designed in order to show the working of image captioning model, where the user can directly interact with the app and can able to view the generated captions which is shown in Figure 3.6. It has both well-designed front end and back end.

- **Front end:** We used HTML and CSS to design the front end, where the user will upload any image, and AJAX Http Call is made internally to the back end in order to fetch the predicted captions.
- **Back end:** We used Flask webserver and ngrok.io library for running flask on Colab. The image is extracted from the Ajax Http Call's body and sent to the loaded .h5 model which was already trained before and saved in Google Drive along with captions context. The model predicts the captions for image and

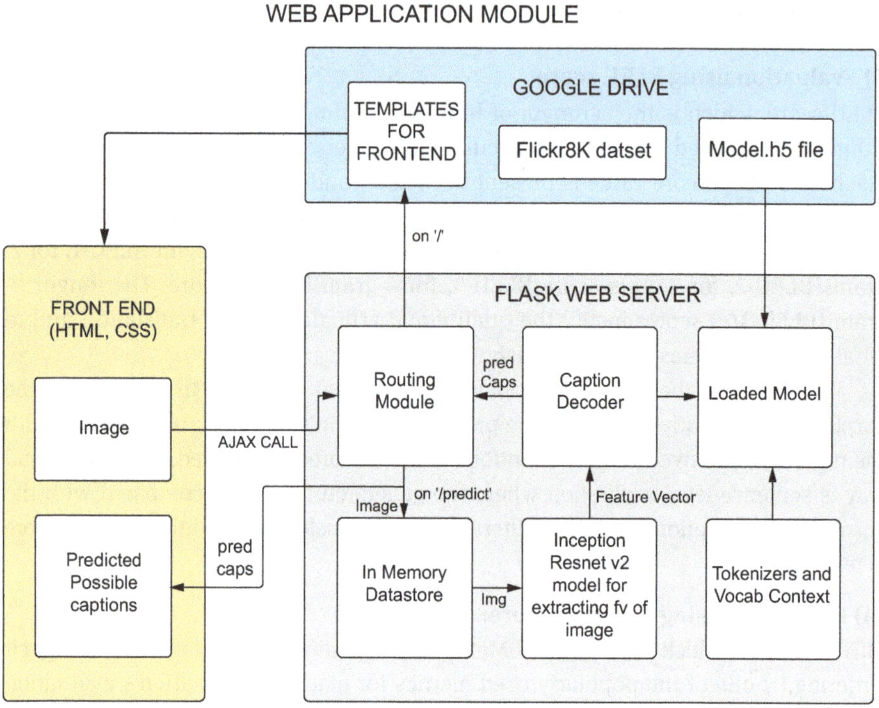


**Figure 3.6:** Web application architecture.

sends them to UI. An efficient routing mechanism is designed, which makes us to consider the server as a web API.

- **Google Drive (database):** The entire dataset along with its captions is stored in Google Drive for easy retrieval during development. It also stores the vocabulary context, tokenizers, saved h5 model files generated in transfer learning stage and the template files for user interface.

## 3.4 Results and discussions

We tried various configurations and tweaked some hyperparameters and recorded few observations, where we found that these are the highest scores obtained:

Obtained BLEU-4 score (sentence level): 38.66681524
Obtained BLEU-4 score (corpus level): 56.74269185
Obtained METEOR score (sentence level): 17.04744747
Obtained METEOR score (corpus level): 29.41817676

We can see that BLEU4 score, both on sentence and corpus level, is more than whatever they obtain in base paper. METEOR score is more when we evaluated the predicted captions on corpus level (five captions based). BLEU-4 is only considered here because the weights considered here are 25% of the 1-gram, 2-gram, 3-gram and 4-grams, which will produce an unbiased result.

### 3.4.1 EDA module output

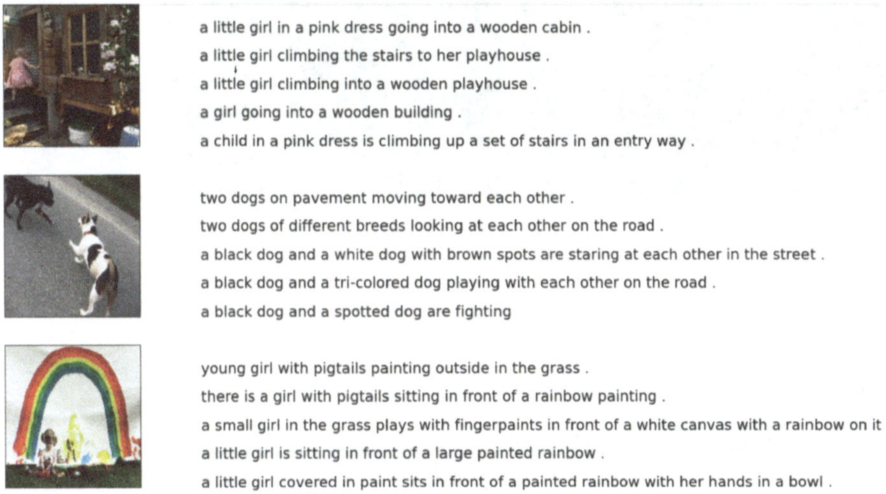


**Figure 3.7:** Images–captions.

Figure 3.7 shows the result of image loader and visualizer module. Here the contents of the dataset have been shown that the proposed approach is used. We used Flickr-8k dataset, where there are 8096 images and 5 captions for every image. So, we mapped the five captions related to every image and visualized the first five of them. The objective of this module is to understand the contents of the dataset.

This is shown under the EDA module of the architecture. The main purpose of showing this is to understand the context of the captions. Word cloud is a visual representation of words, used to highlight phrases based on relevance and frequency as given in Figure 3.8. They provide us with quick and simple visual insights that can lead to more in-depth analyses. Without considering stop words, we can see words like Man and Dog are highly occurring in the captions.

This is also shown under the EDA module of the architecture. The main purpose of showing this is to know which words are more frequently occurring in caption dataset and least frequently repeating in the captions. We plotted a histogram in Figure 3.9 to show the 50 most frequently occurring and 50 least frequently occurring phrases (including stop words like in, are, the).

**Figure 3.8:** Word cloud.

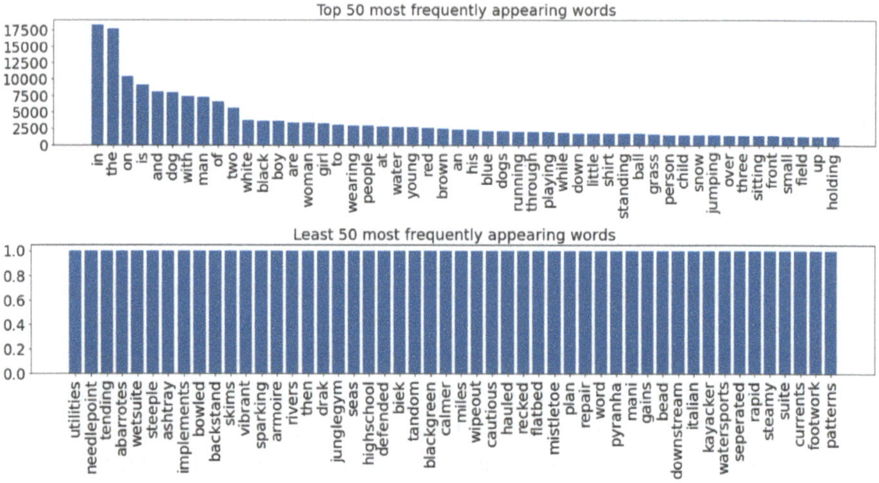


**Figure 3.9:** Frequency histogram.

### 3.4.2 Training output

The architecture shown in Figure 3.3 is constructed. Figure 3.10 is the output obtained at every epoch during training the model. We used 80% of the dataset for training the model with a batch size of 196 with 35 epochs. So, from the image we can see that the accuracy of the model is increasing on every epoch with a decrease

```
330/330 [==============================] - 42s 128ms/step - loss: 1.3406 - accuracy: 0.6127
epoch 18 running...
330/330 [==============================] - 42s 127ms/step - loss: 1.2624 - accuracy: 0.6306
epoch 19 running...
330/330 [==============================] - 42s 127ms/step - loss: 1.1801 - accuracy: 0.6541
epoch 20 running...
330/330 [==============================] - 42s 129ms/step - loss: 1.1117 - accuracy: 0.6706
epoch 21 running...
330/330 [==============================] - 43s 129ms/step - loss: 1.0405 - accuracy: 0.6914
epoch 22 running...
330/330 [==============================] - 42s 128ms/step - loss: 0.9779 - accuracy: 0.7083
epoch 23 running...
330/330 [==============================] - 42s 127ms/step - loss: 0.9184 - accuracy: 0.7239
epoch 24 running...
330/330 [==============================] - 43s 129ms/step - loss: 0.8582 - accuracy: 0.7426
epoch 25 running...
330/330 [==============================] - 42s 127ms/step - loss: 0.8030 - accuracy: 0.7569
epoch 26 running...
330/330 [==============================] - 42s 128ms/step - loss: 0.7500 - accuracy: 0.7735
epoch 27 running...
330/330 [==============================] - 43s 129ms/step - loss: 0.7020 - accuracy: 0.7873
epoch 28 running...
330/330 [==============================] - 42s 129ms/step - loss: 0.6574 - accuracy: 0.8011
epoch 29 running...
330/330 [==============================] - 42s 128ms/step - loss: 0.6104 - accuracy: 0.8146
epoch 30 running...
330/330 [==============================] - 43s 130ms/step - loss: 0.5784 - accuracy: 0.8260
epoch 31 running...
330/330 [==============================] - 43s 130ms/step - loss: 0.5393 - accuracy: 0.8374
epoch 32 running...
330/330 [==============================] - 42s 128ms/step - loss: 0.4992 - accuracy: 0.8499
epoch 33 running...
330/330 [==============================] - 42s 128ms/step - loss: 0.4688 - accuracy: 0.8583
epoch 34 running...
330/330 [==============================] - 42s 128ms/step - loss: 0.4462 - accuracy: 0.8643
epoch 35 running...
330/330 [==============================] - 44s 132ms/step - loss: 0.4131 - accuracy: 0.8751
```

**Figure 3.10:** Training statistics.

in the loss. But we did not limit the estimation metric just to accuracy because it is not sufficient enough for sequential models like this. Figure 3.11 depicts a plot between the accuracy versus number of epochs and loss versus number of epochs. It is evident that the model is getting accurate enough and loss is decreasing over training. Figure 3.12 and 3.13 discloses the BLEU score for the generated captions by the proposed method.

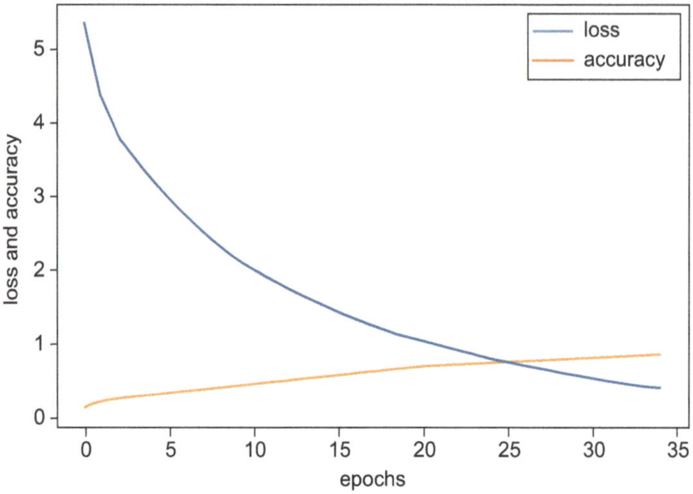


Figure 3.11: Epochs accuracy/loss graph.

beam_len=7 the black dog is running in the field

beam_len=5 black dog running in the field

beam_len=3 two dogs are running on grassy field

startseq two dogs run on grassy field endseq

beam_len=7 the man in the red shirt is walking down the street

beam_len=5 the lady in the red shirt is walking down the street

beam_len=3 the lady in the red shirt is walking down the street

startseq the lady in the red shirt is walking up the street endseq

beam_len=7 the man in the red sweater is eating snack

beam_len=5 the man in the red sweater is eating snack

beam_len=3 the man in the red sweater is eating snack

startseq the man is wearing bandanna on his head endseq

Figure 3.12: Generated captions.

| | BLEU_using_Single_Sentence | BLEU_using_all_Corpus |
|---|---|---|
| Individual BLEU-1 | 0.178052 | 0.432651 |
| Individual BLEU-2 | 0.480868 | 0.528749 |
| Individual BLEU-3 | 0.610848 | 0.787996 |
| Individual BLEU-4 | 0.175497 | 0.41634 |
| Cumulative BLEU-2 | 0.25238 | 0.42178 |
| Cumulative BLEU-3 | 0.331724 | 0.49776 |
| Cumulative BLEU-4(imp*) | 0.386668 | 0.567427 |

**Figure 3.13:** BLEU score model.

## 3.5 Conclusion and future plan

In this proposed approach, an image caption generating system using stacked LSTM network and CNN using transfer learning has been developed. It is embedded into a web application in order to make it available and test online.The entire work has been incorporated in a modular way, each person in the team contributing to specific model implementation. This approach is able to generate a unique URL for using the web application, where the user can upload an image and see the generated captions. Still in this work there are few limitations. The generated captions are based on the context and vocabulary that we provide to the model while training, so if there is no word to say "nature" in the context, then the predicted captions can never have that word. Therefore, it is indeed to increase the captions and images while training the model in order to produce captions for images having different contexts. In future, better transfer learning CNN models can be developed in order to get good object classification, which helps in producing good-quality captions.

This work can be further scaled up to use in specific scene recognition, task-specific caption generation, automatic subtitling system for videos and so on. We need to improve the output rendering time by reducing the response time.

## References

[1] Goh, H., Thome, N., Cord, M., Lim, J.H. (2014). Learning deep hierarchical visual feature coding. IEEE Transactions on Neural Networks and Learning Systems, 25(12), 2212–2225.

[2] Zhang, N., Ding, S., Zhang, J., Xue, Y. (2017). Research on point-wise gated deep networks. Applied Soft Computing, 52, 1210–1221.

[3] Papa, J.P., Scheirer, W., Cox, D.D. (2016). Fine-tuning deep belief networks using harmony search. Applied Soft Computing, 46, 875–885.
[4] Bai, S. (2017). Growing random forest on deep convolutional neural networks for scene categorization. Expert Systems with Applications, 71, 279–287.
[5] Cho, K., Van Merriënboer, B., Gulcehre, C., Bahdanau, D., Bougares, F., Schwenk, H., Bengio, Y. (2014). Learning phrase representations using RNN encoder-decoder for statistical machine translation. arXiv preprint. arXiv:1406.1078, 1724–1734.
[6] Rashtchian, C., Young, P., Hodosh, M., Hockenmaier, J. (2010). Collecting image annotations using Amazon's Mechanical Turk. Proceedings of the NAACL HLT 2010 workshop on creating speech and language data with Amazon's Mechanical Turk. Association for Computational Linguistics, 139–147
[7] Young, P., Lai, A., Hodosh, M., Hockenmaier, J. (2014). From image descriptions to visual denotations: New similarity metrics for semantic inference over event descriptions. Transactions of the Association for Computational Linguistics, 2, 67–78.
[8] Gupta, N., Jalal, A.S. (2017) A comparison of visual attention models for the salient text content detection in natural scene. 2017 Conference on Information and Communication Technology (CICT), 1–5. IEEE.
[9] Farhadi, A., Hejrati, M., Sadeghi, M.A., Young, P., Rashtchian, C., Hockenmaier, J., Forsyth, D. (2010). Every picture tells a story: Generating sentences from images. In: European Conference on Computer Vision. Berlin, Springer, 15–29.
[10] Ordonez, V., Kulkarni, G., Berg, T.L. (2011). Im2text: Describing images using 1 million captioned photographs. Advances in Neural Information Processing Systems, 1143–1151
[11] Gupta, A., Verma, Y., Jawahar, C.V. (2012). Choosing linguistics over vision to describe images. AAAI, 1.
[12] Kulkarni, G., Premraj, V., Ordonez, V., Dhar, S., Li, S., Choi, Y., et al (2013). Babytalk: Understanding and generating simple image descriptions. IEEE Transactions on Pattern Analysis and Machine Intelligence, 35(12), 2891–2903.
[13] Mitchell, M., Han, X., Dodge, J., Mensch, A., Goyal, A., Berg, A., et al (2012). Midge: Generating image descriptions from computer vision detections. Proceedings of the 13th Conference of the European chapter of the Association for Computational Linguistics, 747–756. Association for Computational Linguistics.
[14] Ushiku, Y., Yamaguchi, M., Mukuta, Y., Harada, T. (2015). Common subspace for model and similarity: Phrase learning for caption generation from images. Proceedings of the IEEE international conference on computer vision, 2668–2676.
[15] LeCun, Y., Bengio, Y., Hinton, G. (2015). Deep Learning. Nature, 521(7553), 436–444.
[16] Socher, R., Karpathy, A., Qv, L., Cd, M., AY, N. (2014). Grounded compositional semantics for finding and describing images with sentences. Transactions of the Association for Computational Linguistics, 2(1), 207–218.
[17] Karpathy, A., Joulin, A., Fei-Fei, L.F. (2014). Deep fragment embeddings for bidirectional image sentence mapping. Advances in Neural Information Processing Systems, 1889–1897.
[18] Girshick, R., Donahue, J., Darrell, T., Malik, J. (2014). Rich feature hierarchies for accurate object detection and semantic segmentation. Proceedings of the IEEE Conference on Computer Vision and Pattern Recognition, 580–587.
[19] Ma, L., Lu, Z., Shang, L., Li, H. (2015). Multimodal convolutional neural networks for matching image and sentence. Proceedings of the IEEE International Conference on Computer Vision, 2623–2631.
[20] Kiros, R., Salakhutdinov, R., Zemel, R. (2014). Multimodal neural language models. International conference on machine learning, 595–603.

[21] Mao, J., Xu, W., Yang, Y., Wang, J., Yuille, A.L. (2014). Explain images with multimodal recurrent neural networks. arXiv preprint. arXiv:1410.1090.
[22] Krizhevsky, A., Sutskever, I., Hinton, G.E. (2012) ImageNet classification with deep convolutional neural networks. Advances in Neural Information Processing Systems, 1097–1105.
[23] Vinyals, O., Toshev, A., Bengio, S., Erhan, D. (2015). Show and tell: A neural image caption generator. Proceedings of the IEEE Conference on Computer Vision and Pattern Recognition, 3156–3164.
[24] Simonyan, K., Zisserman, A. (2014). Very deep convolutional networks for large-scale image recognition. arXiv preprint. arXiv:1409.1556.
[25] Gupta, N., Jalal, A.S. (2018). A robust model for salient text detection in natural scene images using MSER feature detector and Grabcut. Multimedia Tools and Applications, 78(8), 10821–10835.
[26] Gerber, R., Nagel, H.H. (1998). (Mis?)-Using DRT for generation of natural language text from image sequences. In: Burkhardt, H., Neumann, B. (eds) Computer Vision — ECCV'98. ECCV 1998. Lecture Notes in Computer Science, vol 1407. Springer, Berlin, Heidelberg. pp. 255–270. https://doi.org/10.1007/BFb0054746.
[27] Arens, M., Gerber, R., & Nagel, H. H. (2008). Conceptual representations between video signals and natural language descriptions. Image and Vision Computing, 26(1), 53–66.

Indira Linginani, Akkalakshmi Muddana*

# 4 Computer vision for healthcare applications

Computer vision for healthcare applications

**Abstract:** Automatic medical diagnosis is a surfacing center of the rate of interest in computer vision as it gives unobtrusive unbiased details on a client's disorder. The detection of oddities or even irregular functions is actually of utmost usefulness when it pertains to medical diagnostics. Computer vision utilizes algorithms to learn styles as well as anomalies in images to help make much faster as well as more exact medical diagnoses than a doctor could make. With a repetitive learning procedure aided by neural networks, computer vision identifies, examines and analyzes images. Advanced computer vision strategies provide resources to assist usual human demands, like psychological performance, personal mobility, sensory functionalities, daily staying tasks, image processing and pattern recognition. Technology is affecting our lives on a massive scale and its expansion into the medical and healthcare industries is a welcome development. People may now readily connect to these healthcare applications and know what is happening to them, unlike a few years ago when pale-faced patients could be observed waiting in long lines for their doctors.

The chapter discusses the importance of computer vision in healthcare, different artificial intelligence techniques and computer vision methods with scenarios in healthcare. This chapter also presents the use cases and future research scope.

**Keywords:** computer vision, image processing, deep learning, object detection, image classification

## 4.1 Introduction

With the unceasing improvement in living standards and the mounting problem of the aging population, there is a burning necessity for technologies that can advance medical care. To meet this demand, medical technology has been lumped with the knowledge of artificial intelligence (AI). One such part of AI is computer vision which deals with the topic of how a computer or systems can achieve a high-level perception from images. It attempts to give machines the ability to perceive and understand the world

***Corresponding author: Akka Lakshmi Muddana**, Department of CSE, GITAM Deemed University, Hyderabad, India, e-mail: lakshmi.muddana@gitam.edu
**Indira Liginani,** Department of CSE, GITAM Deemed University, Hyderabad, India

https://doi.org/10.1515/9783110756722-004

around them in the same way that humans do. One of the fields where this technology is becoming increasingly relevant is healthcare. Due to its ability to diagnose diseases and conditions utilizing various scans such as X-ray, MR or CT (computed tomography), AI-powered solutions are becoming increasingly popular among doctors and opens up a world of possibilities for training computer vision algorithms to improve patient care and the overall efficiency of the healthcare business. Automating activities that rely on image recognition can improve care while reducing the need for human interaction. Machines must acquire, analyze, evaluate and comprehend images in order to mimic human vision. Computer vision accomplishes this by leveraging image processing AI algorithms that process images in healthcare with the goal to produce findings that are both faster and more accurate than those produced by doctors.

### 4.1.1 What is computer vision and its advantage in healthcare?

It is the subset of AI which can empower the computers for the extraction of meaningful information from digital content and the other graphical inputs and make results established from the information. Computer vision refers to a computer's ability to see, interpret and comprehend visual data from digital movies, photographs and a few other graphical inputs and make references based on that data. Its methods have shown great promise in a range of medical settings, including medical imaging and surgical planning [1]. Computer vision enables computers to perform these activities in a much shorter length of time using cameras, data and algorithms. The more data there is, the better computer vision functions get. Currently, the most common use cases for computer vision in healthcare are in radiology and imaging. Physicians who use computer vision technology will be able to examine a wide range of health and fitness parameters to make better medical judgments. Today, such gadgets are used by hospitals to measure blood loss during surgery, such as C-sections. Furthermore, the technique may be used to calculate people's body fat percentages using photographs captured with conventional cameras. Its goal is to learn about, automate and mimic the human visual system to produce findings that are faster and more accurate than those produced by physicians. One of the driving causes behind the growth of computer vision is the amount of data being collected today, which is subsequently utilized to train and improve computer vision. It focuses on teaching intelligent computer algorithms to comprehend and interpret visual data without having to be explicitly written in a specific program or language. As a subset of AI, it aims to teach computers how to see and interpret objects in front of them in the same way that humans do. And, with rapid advances in deep learning and increasing computing power, computers are becoming increasingly capable of handling the work at hand.

The responsibilities involved with computer vision are:
1. Obtaining pictures
2. Processing data in the pictures
3. Extraction of multidimensional data

This data is then processed to uncover patterns, identify distinct items and trends in the data, and spot anomalies in the images fed to it. Healthcare, which has a lot of clinical imaging data, is using computer vision to help doctors make better decisions and improve patient outcomes.

#### 4.1.1.1 Advantages toward computer vision in healthcare

Even minor details matter in the world of healthcare and medicine, and can mean the difference between life and death in many cases. Computer vision has a wide variety of utilization in health and wellness surveillance, bringing specialists closer to their individuals, saving them an opportunity on image analysis, as well as providing more dependable information to partner with. While manual examination of this clinical imaging is time-consuming and error-prone, clever computer vision-based systems evaluate medical images considerably more quickly and flag suspicious areas to warn a clinician. Computer vision models that have been trained with deep learning algorithms detect diseases automatically. The visual pattern recognition capabilities of AI-driven computer vision technologies can make a difference in the following ways:

(i) **An additional accurate meter:** The use of computer vision in healthcare applications leads to faster and also a lot more precise diagnostics. Even more, the data the system secures for protocol instruction, the greater is the precision costs. This suggests that neural network formulas can be trained to identify items along with higher preciseness than doctors and also to identify designs that the individual eye will typically look over. Other examples of strengthened precision feature X-rays, ultrasound examination, CT scans and MRI.

(ii) **Very early disease recognition:** A lot of ailments respond to medical therapy simply in their early stages. Computer vision technology permits medical professionals to step in early through detecting signs that are certainly not however apparent. This creates a substantial distinction in the treatment of people who would not or else acquire the necessary treatment. Identifying early-onset diseases enables doctors to offer medicines to help combat the health conditions or may execute procedures earlier and also sparing lives. AI's precisely calibrated pattern recognition capacity with high confidence quickens the medical diagnosis method and also helps make therapy a lot more successful.

(iii) **Enhanced medical procedure efficiency:** Computer vision is well-known not only for its analysis accuracy but also for its overall efficiency for both people and healthcare specialists. Computer-aided medical diagnosis, particularly, reduces doctor–patient interaction. This decline is especially favorable given the projected doctor scarcity. Subsequently, the treatment of computer vision on digital photographers can justify the crew's workflow and, as a result, decrease expenses. Most of the heavy hauling is done through state-of-the-art AI protocols, allowing the specialist to pay attention to relationship building, info extraction, therapy, as well as monitoring.

(iv) **Automatic age of medical reports:** The development of computer vision progression has made it feasible to create significant use of medical imaging information for even more exact health problem prognosis, treatment and prophecy. Making use of computer vision procedures, the doctor may collect boosted medical info that can be utilized not merely to establish a diagnosis and administer medication, yet likewise to expect disease and also create rational records. Healthcare experts may make use of computer vision to automate the production of medical reports.

(v) **Interactive medical imaging:** Computer vision in medical imaging permits active and also comprehensive 3D visual images. Deep learning protocols have significantly helped medical photo review over the last few years. These developments got rid of the path for computer vision to become extra productive in healthcare image processing. Deep Learning and Computer Vision might right now be taken advantage of to do graphic studies of dynamic 3D designs to offer even more exact medical diagnoses. Therefore, 3D breast imaging powered through modern-day Computer Vision units, for example, proves a lot more useful for cancer protection at the beginning of disease progress.

### 4.1.2 How computer vision improves healthcare?

Computer vision systems offer precise diagnoses minimizing false positives. In many instances, computer vision algorithms can be even more effective than human physicians. This is because these algorithms are trained on thousands of medical images presenting a given disease or anomaly. Computer vision in healthcare can significantly improve many fields of modern medicine, that is, X-ray radiography, magnetic resonance imaging (MRI), ultrasound examination, endoscopy and thermography.

(i) **Increased diagnostic precision:** In the field of medicine, accuracy of diagnosis is critical for some reasons. Most medical procedures can be costly for individuals and their families, especially in nations lacking universal healthcare. As a result, patients must be fully convinced of a condition's diagnosis before proceeding to pay for surgery or other forms of treatment. False positives for serious, potentially fatal illnesses can put people under excessive financial and mental strain. Computer vision systems' accurate diagnosis reduces false positives and may even eliminate the need for unneeded surgical procedures and costly medicines. False negatives can be just as bothersome as false positives, and they can occasionally be even more dangerous. Computer Vision algorithms that have been trained on a huge amount of training data may detect even the tiniest signals that a condition is there which human doctors might normally overlook due to their sensory limitations. As a result, the use of computer vision in healthcare diagnosis can provide extremely high levels of accuracy, which might eventually reach 100%. This might possibly save hundreds of lives each year, as well as thousands of dollars.

(ii) **Early detection of illnesses:** Most lethal conditions, such as cancer, can be treated with a higher chance of success if caught early. However, the limitations of most contemporary diagnostic methods may make it difficult to detect certain disorders until it is too late. Due to AI's finely honed pattern recognition capacity, the use of computer vision in healthcare can enable the detection of early symptoms with high accuracy. This can lead to prompt treatment and, in the long run, save many lives.

(iii) **Lab tests automation:** AI is additionally worked within blood counts, tissue cell analysis, adjustment monitoring and various other laboratory screening. Computer vision blood analyzers either take photographs of bloodstream samples or even receive logical input in the form of a picture of a recently readied slide consisting of a bloodstream film. Typically, such images are captured by skilled specialists using a specially developed camera mounted to a standard microscope. The system then examines the information and automatically finds particular abnormalities in blood samples using image processing and computer vision technology.

(iv) **Increased efficiency in medical processes:** Doctors can spend more time with patients and provide individualized and helpful advice if they use computer vision to reduce the time they spend examining reports and photographs. In addition to improving the quality of physician–patient interactions, computer vision in healthcare can assist physicians in providing consultation to an increasing number of patients.

(v) **Increased surgery:** AI is becoming increasingly significant in modern surgery, particularly in minimally invasive surgery. Robot assistants are now commonplace in operating rooms, assisting doctors in their work. Medical robots use 3D, high-definition imagery, which improves the vision of the operating field and allows for depth perception. As a result, surgical procedures are more precise and time-efficient.

(vi) **Improves clinical trials:** Clinical trials are the most costly stage of medication development. Finding eligible individuals is one of the most important aspects of conducting successful clinical tests. Candidates must reflect a diverse range of races, ethnicities, ages and genders (ideally all).

Furthermore, all patients participating in the studies must undergo a comprehensive examination before being allowed to participate. Many applicants are turned down before the clinical trial even begins. It is an expensive and time-consuming operation. AI applications can assist in the selection of appropriate candidates as well as a variety of other tasks.

(vii) **Automatic medicinal diagnosis:** A significant application of computer vision is automatic medical diagnosis. The automatic diagnosis system can analyze patients with significant success allocating a coefficient on each one of the risk factors which consist of any medical illness. People may now connect to the healthcare applications and know what exactly is happening to them, in contrast to patients waiting in lines for the doctors. The solicitations of the technology for therapeutic use have to be accomplished by extending existing ways that it is now being used and putting on a layer of creativity and resourcefulness. The automatic diagnosis uses a machine learning algorithm to discriminate features

## 4.1.3 Applications of computer vision in health monitoring

Computer vision has a plethora of real-world applications. Healthcare is one of the industries where this technology is becoming increasingly important. Its techniques have shown considerable promise in a variety of medical contexts, including medical imaging and surgical planning. It reduces the amount of invasive surgeries that were previously required to diagnose health conditions. Figure 4.1 shows some of the applications of computer vision.

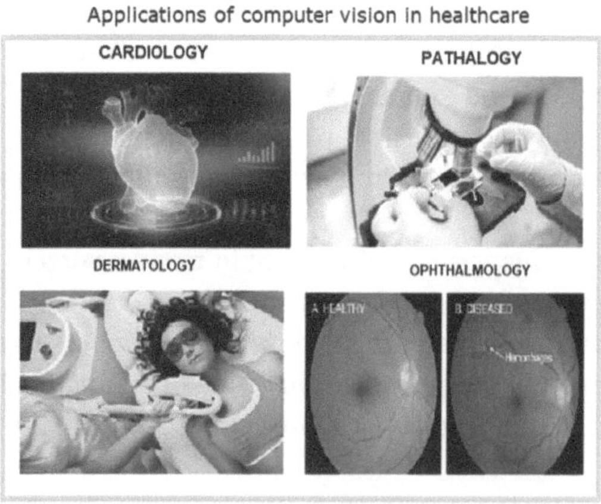


**Figure 4.1:** Applications of computer vision.

(i) **Cardiology:** Cardiac imaging is increasingly being employed in a wide range of medical diagnostics. Deep learning's key medical applications include image screening and diagnosis. The echocardiogram, often known as cardiac ultrasound, is the most widely used imaging technique in cardiovascular medicine [2]. It is a low-cost, nonradioactive approach. Because of the ease of data capture and interpretation, echocardiography is unquestionably ideal for deep learning. It is commonly utilized in many critical inpatient institutions as well as for emergency uses. Furthermore, 3D imaging systems such as CT and MRI are primarily employed to comprehend the anatomy of cardiac systems and CT segmentation procedures for coronary artery

(ii) **Pathology:** Pathology is a field of medication that analyzes the factors, roots and similar components of sickness. Thus, concerning investigation study and also to diagnose the disorder, it includes the assessment of tissues, physical body organs, physical liquids and also corpses. Computational pathology in addition to utilizing AI for cell analytics is, in fact, taking off, along with the feasibility to transform pathology together with requests that accelerate the method, strengthen diagnosis and improve patient outcomes. Pathological inquiry based on graphical examination of tissue samples using microscopes is often self-contained. Changes in visual perception and medical education can result in disparities in analytical and anticipatory judgments. In these circumstances, computer vision can assist with acute medical tasks such as diagnosis, prognosis of therapy results and reactions, illness monitoring and pathology dissection. Micron-level stiffness tissue scanners that take Gig pixel whole slide pictures have been implemented in recent years. This advancement, in conjunction with advancements in computer vision, has directed research and investment activities in AI-focused digital diagnosis [3].

(iii) **Dermatology:** Dermatology is primarily interested in an aesthetic assessment of the individual's skin layer. The most essential medical tasks for computer vision in dermatology include lesion-specific differential diagnostics, detecting lesions among several benign lesions, and tracking lesion evolution over time. A series of processes have demonstrated that CNNs can link the presentation of skilled dermatologists to categories of malicious skin lesions from benign ones [4, 5].

(iv) **Ophthalmology:** CNNs can identify a huge number of nonhuman explicable types in the eyes, which are suggestive of expressive clinical data. It has been demonstrated that CNN can organize a significant number of cardiac and vascular illnesses, as well as diabetic risk factors, using fundus images incorporating age, gender, hemoglobin A1, body mass index and blood pressure. CNNs can also detect anemia and chronic renal disease in fundus pictures. This provides an exciting opportunity for future AI investigations calculating nonoptical data from eyesight. It may result in a typical shift in care in which eye checkups monitor you for the occurrence of both optical and nonoptical disorders that are presently limited to human surgeons [6, 7].

## 4.2 Role of AI in healthcare

Even though the accumulation of many technologies with a distinctive approach towards processing a program and supporting the specified task, they tend to be related closely to the medical field. Studies on AI have shown that for such tasks, AI has been on par with the performance of a human or even better in certain situations.

AI uses complex algorithms that are designed to review the task in an automated fashion. The algorithms constructed for the tasks take data from the medical analysts, doctors or scientists, and review it or interpret the data or come up with a solution for a problem. Some of the applications in the current day with AI associated with healthcare are medical diagnostics, drug discovery, clinical trials, pain management and improving patient outcomes. Applying AI in the medical field has improved the process of gathering and processing the information of the patient data. This has enabled the program of surgeon robots from the data accumulated [8].

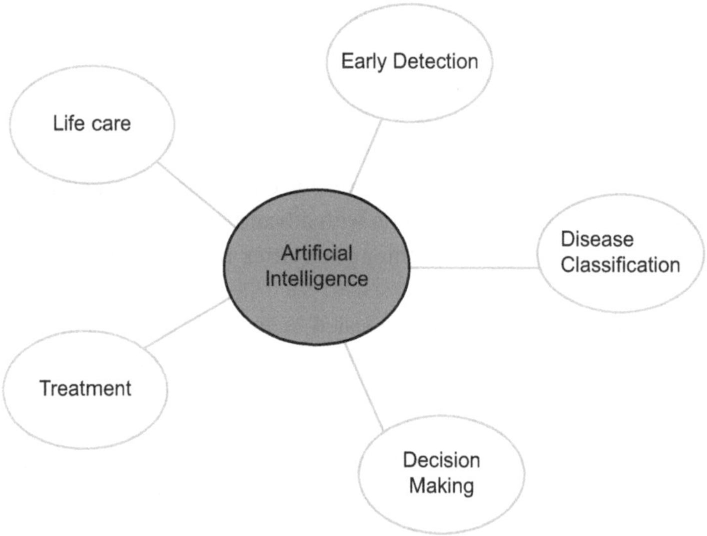


**Figure 4.2:** Role of AI in healthcare.

Figure 4.2 shows the five ways in which AI improves the treatment process:

(i) **Detecting diseases:** The application of AI to medical care prognosis possesses many advantages in the healthcare area that is providing. The software application powered through AI can find a certain health problem of a person even before the indicators have appeared. In Google.com's most recent examination, it has been shown that a semantic network can be informed to identify bronchi cancer cells even before the radiologists. This software must undergo testing, validation on a bigger audience, and a number of tests. The mere thought of doing this has maintained interest in using AI to find cancer in its earliest stages. Simply, the idea of the species has always kept the interest in operation of AI in acknowledging cancer tissues at the initial stage.

(ii) **Sickness classification:** The range of sharp finding out present-day technologies in examining graphics and recognizing patterns opens a possibility for producing formulas that help the physician in the clinical diagnosis of particular health and wellness disorders much faster and likewise with much higher preciseness. These solutions develop consistently with understanding brand-new things. AI-powered plans can be configured to locate the clues of specific sickness adequately in medical graphics like MRI, CT checks, in addition to X-rays. AI is presently being taken advantage of in cancer by handling pictures of sores of the skin layer. Promoting the very best essential treatment in detecting individuals has become simpler for the doctor.

(iii) **Improving decision-making approach:** Procedure along with diagnostics has consistently not been simple. There are numerous aspects that doctors need to have to think about just before resolving a person like signs, existing technique methods, achievable adverse effects and diseases that possess similar signs. Professional decision-making is being affected with Expert system. AI is aiding medical professionals in hammering the study obstacles and in handling comprehensive records rapidly. To pinpoint conditions in addition to reviewing healthcare choices, AI units might eat and examine, in addition to document large quantities of reports all over numerous techniques. AI apps can effortlessly handle the massive amounts of documents generated in the medical sector and also reveal fresh particulars that will certainly or else be decreased in the avalanche of healthcare files.

(iv) **Solutions for AI-based treatment:** While the disorder is recognized, the treatment procedure can conveniently rationalize extra issues. A service preparation additionally includes recommending medication operations along with exercises. Nonetheless, in addition to parallel care plans, assistance preys to regulate their solution planning, and additionally bear in mind the risk of damaging festivity. Modern AI procedures currently assist doctors to establish a full procedure for condition surveillance. In addition, they are often used to improve important robots that accomplish enormously complex techniques.

(v) **Making people live longer:** AI is usually anticipated to be a necessary advancement to assist individuals to stay longer as well as additionally reduce the necessity for hospitalization. First, methods might refine all the particulars regarding our health and wellness, way of living and additionally ambience our professionals to reside in for our business. As a result, they may anticipate our natural ageing and plan the steps our organisation needs to take to continue to be well-planned. AI is already included in addition to growing old study to generate an aging time clock, which is seeing help keep track of immunosenescence amounts and also identify brand-new treatments made to enhance the body's immune system in the aged ones.

### 4.2.1 Medical images toward machine learning algorithms

Machine learning is a pattern recognition technology that can be used on medical images. It is a strong instrument that can aid in healthcare prognosis. The machine learning algorithms uncover the ideal image component combination for sorting the photo or even figuring out some statistics for the given area of an image. The image features that are regarded to be essential in generating the forecast or even analysis of the rate of interest are often calculated by the equipment finding out the protocol system. Afterward, the equipment finding out the algorithm system determines the greatest combo of these photo attributes for identifying the photo or even generating a metric for the specified picture region [9]. In AI, the initial measure is to draw out the features that keep the information that is utilized to make judgments. There are numerous formulas for determining the proper component body weights. These formulas are based upon a variety of strategies for affecting attribute weights as well as records presumptions. Neural networks, k-nearest neighbors, support vector makers, decision trees and the naive Bayes algorithm are some of the absolute most utilized methods.
In Figure 4.3(a), the machine knowing protocol system utilizes a collection of input photographs for instruction to calculate the photo associates that, when administered, would certainly cause the right classification of the graphic, that is, whether it illustrates a curable or malignant tumor as opposed to the supplied tags for these input pictures.

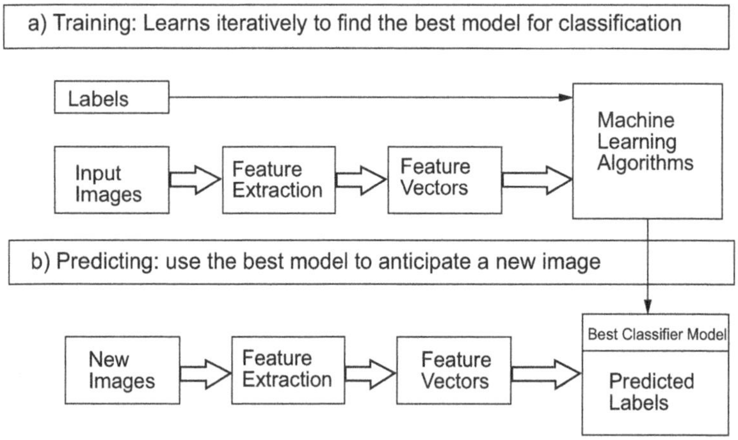


**Figure 4.3:** Development of a Machine learning model and an application model for medical image categorization.

In Figure 4.3(b), once the system has discovered how to classify photos, it administers to find out a model to brand- new graphics to help radiologists figure out the tumor style.

Before the outcome is recognized, different phases are carried out on medical photographs. The medical graphic is first fed right into the equipment finding out protocols. The graphic is then taken into parts to zoom in on the place of enthusiasm. The features are actually removed after using information retrieval methods from these portions. Next, the intended attributes are selected, and also the history sound is removed. Finally, the classifier is used to categorize the drawn-out information and create prophecies based on it. Every machine learning experiment follows these measures. The primary classifications of AI protocols are managed, semisupervised, not being watched, supported and also active learning algorithms [10].

#### 4.2.1.1 Methods of machine learning for medical images

In medical imaging, machine learning algorithms are highly useful for studying specific disorders. Different types of entities in medical image processing, such as lesions and organs, can be too sophisticated to be shown correctly by a simple mathematical solution. Suzuki [11] employed pixel-based analysis to look for disorders in medical photos. In medical image processing, pixel analysis in machine learning was introduced, which uses specific values in images as input data rather than extracting features from chunks. For some problems, the implementation of this method may be superior to that of basic feature-based classifiers. The investigation of the qualities of a low-contrast image is a difficult task. Unlike traditional classifiers, which prevent errors caused by faulty segmentation and feature calculation, pixel-based machine learning does not require feature calculation or segmentation. Because of the high dimensionality of data (a big number of pixels in an image), pixel analysis takes a long time to learn. Agarwal et al. [12] chose low-contrast medical images for analysis. Histogram equalization is the most effective technique for improving contrast. The authors offered "Modified Histogram-Based Contrast Enhancement with Homomorphic Filtering" (MH-FIL) as a technique. It used a two-phase handling procedure, and the first of which boosted global contrast by modifying the histogram. For picture sharpening, second-phase homomorphic filtering is also planned. The investigation looks at 10 medical photos of low-contrast chest X-rays. In comparison to other approaches, the MH-FIL has minimal values in all 10 photographs. Radiologists have the highest responsibility for medical image clarification, with assignments comprising both high-quality images and their analysis. For numerous years, computer-aided design has progressed. Machine learning approaches such as linear discriminant analysis, support vector machines, decision trees and others are used to evaluate medical images. Nanni et al. [13] employed machine learning algorithms to evaluate medical images. They employed local binary patterns widely among texture descriptors in particular. In addition, a new trial using many low binary pattern descriptors of biomedical images was investigated. Beginning with face descriptions, the dataset of neonatal facial photos was used to categorize pain syndromes. Specifically, the results on the meticulously planned 2D-HeLa dataset, and the suggested descriptor, which includes all of the various texture descriptors, achieve maximal implementation. On the 2D-HeLa dataset and the PAP dataset, a linear support vector machine classifier is used. The accuracy of 92.4% was obtained, which is the highest result among the other descriptors on the dataset. In medical pictures, the neural network approach is utilized to explore disease features [14]. For cancer research, the neural network groups are kept. It is used to criticize when a cell is normal with an excessive amount of certainty, where each different network has just two outcomes: normal or malignant cell. A popular method, plurality voting, is used to combine the forecasts of these cells' network. The results revealed that the neural network as a whole had a high rate of accuracy and a low false-negative analytical value. A statistical analysis is a solid criterion for analyzing image features in medical image analysis. The channelized Hoteling observer (CHO) is a nuclear medicine imaging tool that is commonly utilized. The thought of amenable subjects in the human visual structure excites the channels. This approach is utilized to detect picture quality evaluation, and the CHO has a logical and beneficial impact on medical imaging.

### 4.2.2 Deep learning-enabled computer vision in healthcare

Deep learning is a department of machine learning and is also practically a three- or even more-layered semantic network. These neural networks make an effort to develop specific mind actions. Although a single-layer semantic network may easily still produce imprecise forecasts, the included concealed levels can effortlessly aid in boosting along with improving reliability. Several AI applications in addition to business depend upon deep learning to improve hands-free operation and also perform sensibly as well as additional physical exertions without individual interaction. Deep learning procedures are used to deal with a variety of complications in the field of medical imaging. In radiology, one instance is real to feel health problems or even quirks in X-ray images and also determine each one of all of them to rely on to a disease kind or perhaps severeness. It suggests only how to find functionalities that may quickly recognize relevant information for a specific condition. This idea goes to the spirit of many deep learning approaches [15]. The styles included several degrees that translated input photographs into results involving information health problems using incrementally incorporating top-level attributes. **Convolutional neural networks (CNN)** are the best perfect type of these types for image evaluation. CNNs are made up of numerous layers that use convolution filters to convert the input.

CNNs have three main layers:

- Fully connected layer
- Pooling layer
- Convolutional layer

The first layer of a convolutional network is the convolutional layer. Although further convolutional or grouping layers can be added after the convolutional layer, the fully connected layer is the final layer. CNN will become more complicated with each layer, recognizing a larger portion of the image. The previous levels concentrate on simple elements such as colors and borders. As the visual input travels through the CNN layer, it learns to detect larger parts or features of the item, eventually recognizing the expected object. The CNN architecture is built by repeatedly alternating these convolutional and pooling layers. Fully connected layers are affixed at the end of the CNN design and also provide a final decision for the category or even regression duties [16]. Convolutional coatings as well as up-sampling levels, on the other hand, are affixed at the end of merging layers in the division process to rebuild the size of the input photo. Thus, training loss is calculated by matching up the classified cover-up picture to the rebuilt result image utilizing CNN.

### 4.2.3 Reinforcement learning in healthcare

The application of reinforcement learning (RL) in healthcare goes to the crossroads of computer science and medicine. RL delivers a mathematically as well as a practically sound service to ideal decision creating a selection of healthcare jobs that are complicated through raucous, multidimensional and incomplete records, complicated as well as nonlinear aspects, especially sequential selection procedures along with assessment reviews that are delayed. It deals with consecutive decision-making difficulties through providing tried out, evaluative, as well as delayed input at the same time. As a result of these setting apart attributes, the RL strategy is an exceptional alternative for creating strong options in a stable of healthcare sectors where detecting selections or treatment programs are normally defined through an extensive and sequential procedure.

In RL troubles, a broker selects an activity based upon its existing state at each opportunity action as well as receives evaluative feedback and also the brand-new condition from the environment. The agent's objective is to cultivate a superior policy (i.e., an application from states to activities) that maximizes the built-up perks eventually. Therefore, agents certainly do not receive clear guidelines on which actions to do. Instead, they should discover which acts are most effective using trial-and-error interactions with the environment. Unlike every other control-based technique [17], RL carries out, certainly does not need, a well-represented algebraic model of the setting as well as rather generates a command policy directly from adventure to forecast conditions and also benefits during a learning procedure. The previously mentioned characteristics typically produce RL as an attractive technique for building effective policies in numerous healthcare places where the decision-making method is generally specified by an extensive timeframe or even sequential strategy. A medical or professional therapy regimen is normally composed of a collection of selections to

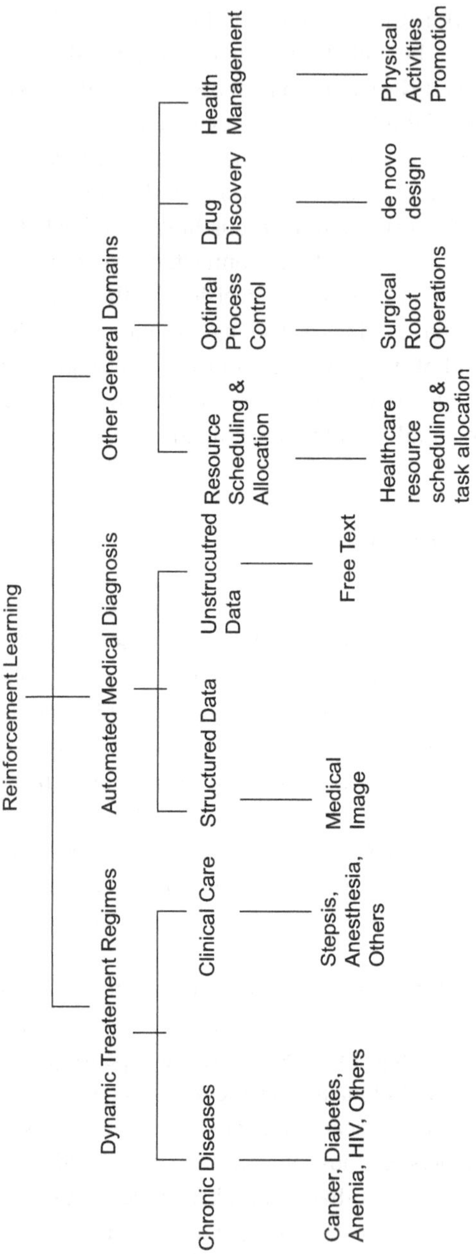


**Figure 4.4:** RL in healthcare: outline of application domains.

identify the course of decisions like procedure kind, medicine dosage or even re- examination timing of each time factor based upon a private patient's present wellness status and also prior procedure history if you want to promote the individual's lasting advantages.

The RL uses domain names that can be sorted out into three main kinds: Dynamic treatment regimes in severe disease or even crucial care, automated medical diagnostics and other extensive industries like health and wellness source allotment as well as booking, ideal procedure control, drug finding and growth, and also health administration. One objective of healthcare decision-making is actually to develop helpful therapy programs that can dynamically get used to transforming professional conditions as well as boost people's lasting perks. Dynamic therapy regimes are the most widely used approaches for many diseases such as cancer and sepsis. Figure 4.4 shows the outline of application domains.

## 4.3 Techniques of computer vision

The study and comprehension of usable data from an image or a series of correlated images is important to computer vision. Video sequences, various camera perspectives or multidimensional information from a medical digital scanning are all examples of image data. Methods for obtaining, processing and interpreting images, as well as data extraction in order to provide numerical or visual information are all parts of computer vision tasks.

During the epidemic, for example, we come across thermal screening image processors in colleges, schools, malls and other public places, which are used to detect a person's body temperature as well as whether or not they are wearing a mask. It must first load a mask classifier from disk, then recognize faces in an image or video stream, extract each face region of interest (ROI), apply a face mask classifier to each face ROI to identify whether a mask should be applied and finally display the results. Before we can do that, we must first load the face mask dataset into the system, then train the face mask detector and deserialize it to the disk. It requires a large amount of tire pictures and tire-related objects to study the distinctions and differentiate a tire, particularly one without flaws [18].

Different computer vision techniques are

(i) Image classification
(ii) Object detection
(iii) Object tracking
(iv) Semantic segmentation
(v) Instance segmentation.

## 4.3.1 Image classification

Image classification is a computer vision process that may classify a graphic based on its graphic info. The objective of image classification is actually to determine and also depict the attributes in an image about the things they show on the ground. Among the purposes are automatic graphic organising, inventory photography and online video websites, graphic search for greater product discoverability, sizable aesthetic data banks, picture and image recognition on social media networks, and a variety of additional uses. Medical image classification's primary goal is to determine which body areas are affected by an illness, not just to achieve high accuracy. Viewpoint variance, scale variance, intraclass variance, picture contortion, picture occlusion, illumination disorders and also backdrop clutter are actually only a few of the challenges connected with this activity. That is why classifiers are required in order to get the highest level of accuracy.

The steps involved in image processing are as follows:
(i) Image preprocessing
(ii) Object detection
(iii) Training and feature extraction
(iv) Classification of the objects

### 4.3.1.1 Image preprocessing

The goal of the step is actually to improve graphic functions through restraining undesirable misinterpretations and boosting some crucial image residential or commercial properties to make sure that computer vision models can easily perform along with much better information [19]. Graphic preprocessing entails a variety of procedures, including resizing the graphic's dimension, reviewing the graphic as well as data augmentation (gray scaling of image, representation, Gaussian blurring, Pie chart, equalization, rotation as well as interpretation and trimming).

### 4.3.1.2 Object detection

Detection pertains to the localization of an item, which involves segmenting the graphic as well as recognizing the site of the object of enthusiasm.

### 4.3.1.3 Training and feature extraction

Feature extraction is an intervention in the dimensionality reduction procedure, which splits and also decreases a sizable set of uncooked information into smaller collections. Consequently, handling will be less complex. The truth that these enormous datasets have a large number of variables is the most important feature. To process these variables, a huge volume of computational power is needed. As a result, function extraction helps in the extraction of the most ideal attribute coming from huge datasets through selecting and also blending variables right into functions, properly decreasing the amount of data. These components are straightforward to make use of while still correctly as well as exclusively illustrating the real data specified.

#### 4.3.1.4 Object classification

This stage compares the digital image patterns to the target pattern and classifies the identified into predetermined classes using an appropriate classification technique.

### 4.3.2 Detection of object

Object detection is a computer vision technique that allows our crew to recognize as well as likewise track attributes in images and videos. Object detection may be utilized to count products in a setting, determine and likewise track their exact areas, and properly categorize all of them utilizing this type of recognition and localization. Object detection, in most cases, generates bounding boxes around the items that are to be discovered or are already being detected. The bounding box may or may not be accurate in locating the object's location. Because object detection and picture recognition are frequently confused, it is crucial to understand the differences between the two before moving on. If there is an image of a tree with a man underneath it, the object detector will create a box around the objects and label it with words like "man" and "Tree." The model forecasts the location of the object and the label that should be applied [20]. Object detection, in this sense, gives more information about an image than recognition. Following are the some of the object detection algorithms:

Fast R-CNN
Region-based **convolutional neural networks (R-CNN)**
Histogram of oriented gradients (HOG)
Region-based fully convolutional network (R-FCN)
Single-shot detector (SSD)
YOLO (you only look once)
Spatial pyramid pooling (SPP-net)

### 4.3.3 Tracking of object

Object tracking refers to the technique of following a single or several objects of interest. It has traditionally been used in video and real-world interactions when observations are performed after the first detection of an object. According to the observation model, it can be split into two groups. The generative method, for example, employs a generative model to describe the visible properties. The discriminative approach can also be used to distinguish between the object and its surroundings. Its performance is more reliable, and it is gradually replacing other tracking methods.

### 4.3.4 Semantic segmentation

The method of categorizing each pixel as belonging to a specific label is known as semantic segmentation. It is consistent across multiple occurrences of the same thing. For example, if a picture contains two birds, semantic segmentation assigns the same label to both birds' pixels. The most informative of these three is semantic segmentation, in which we want to classify each and every pixel in the image.

### 4.3.5 Instance segmentation

Object detection, also known as localization, is a step on the way from coarse to fine digital picture inference. It not only gives the picture object class but also tells where the image object is. Bounding boxes or centroids are used to indicate the location. By predicting labels for each pixel in the input image, semantic segmentation provides fine inference. Each pixel is labeled with the object type that it is contained within. Instance segmentation continues this progression by assigning distinct labels to various instances of objects belonging to the same class [21]. As a result, instance segmentation may be defined as a technique for handling both the object detection and semantic segmentation problems at the same time.

## 4.4 Computer vision in healthcare: use cases

Computer vision possesses a lot of real-world applications. Healthcare is one of the areas where this technology is coming to be especially relevant. The usefulness of computer vision technology in healthcare may not be overemphasized. Its approaches have shown great requests in a lot of medical cases, like medical imaging, medical organizing and remain to gain more common usage. Smart computer vision algorithms are capable of learning to determine complex patterns with instruction on recently diagnosed situations. Today, computer vision is released in a developing number of medical fields and also proceeds to create good changes to healthcare.

### 4.4.1 Medical image analysis

Computer vision applications are helping in minimizing the variety of intrusive surgical operations that are required to become done previously for detecting wellness problems. Arteries have qualified their computer vision software program to be able to sense oddities in the scans of critical organs like the cardiovascular system, liver, as well as lungs. Their software program can be installed on a normal MRI machine. It presents the scans of the patient's organs in a 3D format that can easily then be checked on the computer monitor [22]. This provides the radiologists with an effective understanding of the patient's problems without them needing to go through intrusive surgical procedures. The system can additionally manage to identify irregularities that will otherwise certainly not have been diagnosed due to the individual eye, therefore boosting the reliability.

### 4.4.2 Diagnostics of disease

Experts are administering deep learning as well as natural language processing (NLP) systems to collect patient-relevant details, determine a person's comments and also minimize the medical prognosis in a preappointment appointment. The system sends this searching to the physician before the private receives the go-to.

Ellie, a Computer Vision unit established through pros at the Concept of Creative Technologies at the College of Southern The golden state, was built to carry out this. Ellie inquires the person a collection of concerns and also utilizes the integrated webcam and sensing unit, and scans the client's face to determine their skin and body movements to develop a possible diagnosis. By comparing spoken clues together with undervalued facial along with body language with a dataset of bunches of managements, Ellie might discover folks alongside psychological health problems including clinical depression with anxiousness. While this system does certainly not change a specific medical professional, it gives refined facts that medical professionals may undoubtedly not promptly create, as a result strengthening the prognosis of a specific sickness.

Babylon Wellness, a British technician start-up, is furthermore building NLP as well as deep learning devices that make use of speech and foreign language handling to extract indicators and also bodily looking for that are actually important to make a prognosis. This information is forwarded to a physician to assess before checking an individual.

These algorithms certainly not just utilize these client records to help make diagnoses but they likewise give tailored wellness info as well as learning in greater particular than a medical professional ever could.

### 4.4.3 Cancer classification

Computer vision and also machine learning algorithms have provided a pledge in picking up precancerous lesions using the minimum particulars in cell images, enhancing the level of sensitivity as well as likewise dependability of cancer assessment examinations [23]. In filtering for skin coating cancer, for instance, AI analysts have created computer vision designs that might examine images as well as likewise biopsy tastings of a person's skin level for cancerous adjustments a lot quicker in addition to much more efficiently than a physician would certainly. Diagnosing skin cancer could be a tough job as a result of the minute irregularity in the look of cancerous skin alterations [24]. Making use of deep-rooted **CNNs,** pros at the Stanford College Artificial Intelligence Research laboratory generated a layout that reviews skin images against a dataset of a lot more than 120,000 skin cancer images. The results uncovered that this CNN version uncovers and likewise identifies skin layer cancer as properly as recognized skin professionals.

Likewise, computer vision might be made use of in the seizure cancer testing process, in addition to algorithms that are proficient in acknowledging and also grouping cancerous modifications stemming from millions of mammogram images exposing both well-balanced and identified examples. Carrying out these algorithms in reviewing the location of a visual is the absolute most enhanced malignant or perhaps precancerous pattern within seconds, offering remarkable second details for doctors. MIT's Infotech as well as Artificial Intelligence Research Laboratory and additionally Massachusetts General Healthcare Center (MGH) analysts have cultivated a brand-new deep learning design that can effortlessly determine a person's 5-year danger of breast cancer coming from a mammogram graphic. The design is taught on recognizing precancerous cell patterns in the bosom utilizing far more than 90,000 mammogram images of MGH clients. Utilizing this data, the style finds precancerous tit tissue patterns that are elusive to the human eye. Detection of bone tissue cancer has also been simplified taking advantage of machine learning models. Remote Sensing And Image Processing (RSIP) concept constructed an image processing innovation that might center regions of major and also subsequent cancerous sores in bone images to help doctors detect bone cancers a lot previously as well as faster. This accurate information likewise aids medical professionals supply customized treatment for people.

### 4.4.4 Covid detection

The two essential diagnostic strategies for coronavrus disease 2019 (COVID-19) detection are CT as well as trunk radiography. Computer system eyesight protocols have currently validated performance for both. Personal CT is dealt with some of the greatest efficient analysis approaches for COVID-19 detection. Time-consuming hands-on CT scan analysis could cause stress to the lives of those living in the present. Making an effort to minimize the situation, Chinese scientists occupied the challenge and cultivated a serious knowing variation that may easily portion contaminated locations when analyzing CT checks coming from healthy along with COVID-19-positive individuals. A collection of hands-on CT scan reviews executed through pro-radiologists is performed as a model standard. When the style satisfied higher accuracy, it was discharged at the Wuhan University Renmin Medical resource as well as also made open source for swifter diagnostics of brand-new situations at damaged Internet sites worldwide.

Computer dreams may aid certainly not merely in the coronavirus detection but also with detecting COVID-19-related pneumonia. In this regard, it could likewise outmaneuver professional radiologists. Mandarin professionals made a machine learning committed device that examines individuals' medical records (grow older, sexual activity, and more) and also CT scans. When the researchers matched up the system efficiency keeping that of expert radiologists, a shocking particular visited. The blended model took care of to notice COVID-19 ailment in roughly 70% of people whose CTs were flagged as regular through medical specialists. This truth enables our staff to recommend that deep-seated discovering styles possess the probability to recognize complicated concepts that go hidden because of the individual's eye, having said that trained. Though CT is an important analysis method, its personal utilization has some disadvantages. Firstly, it is the radiation that individuals go through. What is extra, it is costly, which may impair its routine in remote along with backwoods. Right here, the torso radiography could be much more secure along with a lot more cost-efficient alternative.

In the beginning, automated services that examined X-ray images cannot determine the coronavirus primarily. Still, they revealed efficiency in examining the three forms of pneumonia by similarly establishing the one generally triggered by the coronavirus. Having mentioned that, the situation has changed now. Darwinia has partnered along with Reddish Hat to adjust an entire room of solid semantic networks to acknowledge the coronavirus in boot X-rays. Their devices may simply stratify the results through illness development and severeness and also have finally shown legitimate. Moreover, the suite is open source and also accessible to specialists and professionals worldwide. When used in diagnostics, computer vision supplies top quality results in the shortest opportunity feasible. This makes it possible for a doctor to the extra initiative and also time on hands-on examination evaluation and also concentrate on giving suitable treatments.

With the virus spreading around the world, the Earth Health Company established suggestions targeted at decreasing its escalate, including utilizing private preventative devices and also protecting social assortment [25]. Personal computer sight has shown efficiency for watching on and also tracking the recommendation transgressions, as well as a lot of Asian countries followed the trigger to boost surveillance and also to monitor along with the power of AI. At the starting point, personal computer eyesight thinking for such surveillance was reasonably straight. The suggestions claim people need to put on a safety face mask whenever they are in a public location as well as machine learning versions were actually to discover simply that. Along with options, nevertheless, computer system sight alternatives for COVID-19 command have established extra stylish. For instance, an option via RayReach Technologies takes advantage of closed-circuit TV (CCTV) digital video camera structures to take note of and also track a whole collection of tip violations, coming from crowding along with the lack of whitewashes to inadequate sanitizer use. If the deep learning concept locates a violation, it sends information to committed CCTV surveillance groups.

Ensuring adherence to suggestions is certainly not the only technique to quit the coronavirus array; a group of Malaysian researchers offered an alternative. Personal computer sight software program was positioned in cameras at the doorways of shopping centers, banks as well as numerous other social spots to make sure continuous monitoring. The semantic networks inspect guests for adhering to suggestions in addition to for any kind of COVID-19 symptom (high temperature, hacking, and so on), and after that breaks website visitors right into Higher Alarm, Keep Distance, and also Safe teams. If the system chooses a guest who shows a threat, it notifies each one of all of them and also a physician. Researchers presume that this solution will aid the authorities to make certain ongoing surveillance in social spots as well as strengthen control over the spreading of infection.

## 4.5 Conclusion

This chapter is a collection of introductions to computer vision, AI and how it works for computer vision applications. It also explained about the various applications and techniques of computer vision. Machine learning algorithms including deep learning in healthcare and RL in healthcare are also discussed. Techniques of computer vision like image segmentation, object segmentation, instance segmentation and object tracking are explained. This chapter comprises a study on computer vision, its techniques, use cases and applications in healthcare using machine learning and deep learning algorithms.

## 4.6 Future research scope

The rise of AI-based solutions has given many benefits toward the improvisation of the healthcare industry. Essentially, computer vision-based techniques withhold the potential support to deliver life-saving expediencies for patients in many different applications. Today, computer vision assists many doctors and other medical staff in diagnosing better, monitoring the disorders and prescribing the right treatments. Computer vision technology helps medical practitioners and professionals to automate routine tasks and saves time. This can help them to spend more time with patients. Also, computer vision-enabled machines can host many benefits to the health industry by implementing an unmanned way of reception by following object detection, human pose recognition, action and activity recognition, face recognition, image retrieval, semantic segmentation and so on. However, each category has both advantages and disadvantages, as the healthcare industry holds more innovations and state-of-the-art technologies that will direct its gaze to computer vision. In the future, we are going to envisage an increasing number of healthcare organizations experimenting with computer vision to deliver better services.

## References


[1] Gao, J., Yang, Y., Lin, P., Park, D.S. (2018 04 Mar). Computer vision in healthcare applications, 2018.

[2] Bizopoulos, P., Koutsouris, D.D.G. (2018 December). Deep learning in cardiology. IEEE Reviews in Biomedical Engineering, PP(99):1-1

[3] Serag, A., Ion-Margineanu, A., Qureshi, H., McMillan, R., Saint Martin, M.-J., Diamond, J., O'Reilly, P., Hamilton, P. (2019 01 October). Translational AI and deep learning in diagnostic pathology. Frontiers of Medicine.

[4] Chan, S., Reddy, V., Myers, B., Thibodeaux, Q., Brownstone, N., Liao, W. (2020 Jun). Machine learning in dermatology: Current applications, opportunities, and limitations, 10(3).

[5] De, A., Sarda, A., Gupta, S., Das, S. (2020). Use of artificial intelligence in dermatology. Indian Journal of Dermatology, 65(5), 352–357.

[6] Alyoubi, W.L., Shalash, W.M., Abulkhair, M.F. (2020). Diabetic retinopathy detection through Deep Learning techniques: A review. Informatics in Medicine Unlocked, 20.

[7] Dai, L., Wu, L., Li, H., et al. (2021). A deep learning system for detecting diabetic retinopathy across the disease spectrum. Nature Communications, 12(3242). https://doi.org/10.1038/s41467-021-23458-5.

[8] Mishra, S.G., Takke, A.K., Auti, S.T., et al. (2017). Role of artificial intelligence in health care. Biochemistry: An Indian Journal, 11(5), 120.

[9] Erickson, B.J., Korfiatis, P., Akkus, Z., Kline, T.L. (2017 Feb 17). Machine learning for Medical Imaging. RadioGraphics, 37(2). https://doi.org/10.1148/rg.2017160130.

[10] Wernick, M.N., Yang, Y., Brankov, J.G., Yourganov, G., Strother, S.C. (2010 Jul). Machine learning in medical imaging. IEEE Signal Processing Magazine, 27(4), 25–38. 10.1109/MSP.2010.936730.

[11] Suzuki, K. (2012). Pixel-based machine learning in medical imaging. Journal of Biomedical Imaging, 2012, 1.
[12] Agarwal, T.K., Tiwari, M., Lamba, S.S. (2014). Modified histogram based contrast enhancement using homomorphic filtering for medical images. In Advance Computing Conference (IACC), IEEE International.
[13] Nanni, L., Lumini, A., Brahnam, S. (2010). Local binary patterns variants as texture descriptors for medical image analysis. Artificial Intelligence in Medicine, 49(2), 117–125.
[14] Shi, Z., He, L. (2010). Application of neural networks in medical Image Processing. In Proceedings of the second international symposium on networking and network security. Citeseer.
[15] Yazhini, K., Loganathan, D. (2019). A state of art approaches on deep learning models in healthcare: An application perspective, 2019 3rd International Conference on Trends in Electronics and Informatics (ICOEI),IEEE Access.
[16] Subiksha, K.P. (2018). Improvement in analyzing healthcare systems using deep learning architecture; 4th International Conference on Computing Communication and Automation (ICCCA), IEEE Access.
[17] Chennam, K.K., Uma Maheshwari, V., Aluvalu, R. (2022). Maintaining IoT healthcare records using cloud storage. In: IoT and IoE Driven Smart Cities. Cham, Springer, 215–233.
[18] Uma Maheswari, V., Aluvalu, R., Chennam, K.K. (2021). Application of machine learning algorithms for facial expression analysis. Machine Learning for Sustainable Development, 9, 77.
[19] Nanni, L., Lumini, A., Brahnam, S. (2010). Local binary patterns variants as texture descriptors for medical image analysis. Artificial Intelligence in Medicine, 49(2), 117–125.
[20] Lin, T.Y., Goyal, P., Girshick, R., et al. (2017). Focal loss for dense object detection. Proceedings of the 2017 IEEE conference on Computer Vision and pattern recognition; 2017 Jul 21–23; Honolulu (HI), USA, 2980–2988.
[21] Moeskops, P., Wolterink, J.M., van der Velden, B.H., et al. (2016). Deep learning for multi-task medical image segmentation in multiple modalities. Proceedings of the 19th International Conference on Medical Image Computing and Computer-Assisted Intervention; 2016 Oct 17–21; Athens, Greece. Springer, 478–486.
[22] Ker, J., Wang, L., Rao, J., et al. (2017). Deep Learning applications in medical image analysis. IEEE Access, 6, 9375–9389.
[23] Nguyen, P.T., Nguyen, T.T., Nguyen, N.C., Le, T.T. (2019). Multiclass breast cancer classification using convolutional neural network. 2019 International Symposium on Electrical and Electronics Engineering (ISEE), IEEE Access.
[24] Hartanto, C.A., Wibowo, A. (2020). Development of mobile skin cancer detection using faster R-CNN and MOBILENET v2 model. 2020 7th International Conference on Information Technology, Computer, and Electrical Engineering (ICITACEE), IEEE Access.
[25] Kuflinski, Y. (2021 February 5). Computer vision in the fight against the COVID -19 pandemic, The conference for Machine learning Innovation.

Gagan Dubey

# 5 Mental disorder prediction using facial expression based on machine learning techniques

**Abstract:** Mental disorders are the most common illness in today's community. An individual's emotional, psychological and social conditions are characterized by their mental health. It affects how a person is going to think, feel or treat a situation. Positive mental health allows a person to function and reach their full potential productively. Mental well-being is important at any stage of life, from infancy to adulthood. Mental disorders, such as depression, anxiety disorders and eating disorders are extremely predominant and have been shown to influence the physical disposition of an individual. Methods of machine learning have recently been applied to support mental health domains. The aim of this chapter is to propose a novel architecture to identify the presence of mental disorder based on facial expressions that are extracted from different images captured during the human–computer interactive evaluation process. The proposed architecture can be used in the near future as a basis for the various mental illness prediction systems based on machine learning techniques.

**Keywords:** mental disorder, ML, facial expression recognition

## 5.1 Introduction

Machine learning (ML), a technique of knowledge analysis within which computers "learn" to independently modify or adapt their actions (e.g., build predictions) to provide additional correct choices and results, has emerged as a strong analytic tool for giant and sophisticated datasets. As such, ML lends itself to the process of unwellness biomarkers and has been enforced in medical diagnostic tools starting from the detection and classification of varied unwellness. ML strategies are providing higher prediction accuracy than older strategies like applied mathematics techniques or physician-based professional systems. In parallel, ML techniques are applied to look at emotional show variations exhibited throughout feeling states, like facial features and vocal prosody, through audio- and video-based analysis. These advances have generated a replacement field of analysis that has successfully used ML techniques, like support vector machines (SVMs), regression and neural networks, for automatic

**Gagan Dubey,** IT, Manipal University, Jaipur, Rajasthan, India, e-mail: gagan.dubey23@gmail.com

https://doi.org/10.1515/9783110756722-005

recognition of feeling victimization of visual information from typical databases. ML has conjointly been extended to research verbal and nonverbal emotional abnormalities related to mental disorders and has gone on with success to classify those presenting with, and while not, a given designation .We can improve the potency of the system with the employment of varied ML techniques like deep learning. ML model explores multiple regions of the face with a theme to fuse the response from completely different facial regions, and nonverbal factors may be evaluated like voice for generating additional correct results. Early mental health diagnosis is crucial for fully knowing the conditions of mental illness and ensuring better care for patients. Problems are a key step in the process. The human face image includes details that can be used to establish the facial features and identity. There are a number of uses for facial features and identification determination. The information obtained during the facial expression identification process may be used to identify the existence of mental disorder.

### 5.1.1 Overview

Facial expression is one of nonverbal communication's most essential types. The feelings of an individual are emitted by facial expressions, and it allows others to judge the person. Some can to some degree understand facial expressions of underlying emotions and based on that a mental state of an individual can be predicted. Recognition of facial expression (FER) is a method for identifying expressions from the face of a person. It plays an important role in the fields of mental illness diagnosis and human social/physiological interaction identification in today's world. Various FER methods exist nowadays. These facial expression techniques can be combined to ML approach to detect specific mental disorder for an individual. ML-based techniques show incredible promise for psychiatric diagnostics through harnessing observable affect-related behaviors through highly objective methods. In fact, observable affect-related facial expressions are commonly used by mental health professionals to assist in psychiatric diagnosis.

The assessment of mental health is extremely significant, and it is important to identify and recommend treatments for patients with an abnormal mental behavior. Most people are vulnerable to stress, while some are affected by anxiety for different reasons [1]. There is a paradigm shift in healthcare providers to include the mental fitness outline of an affected person as it would be in the coming years. It has been made mandatory to offer improved medicines, and to encourage quick healing as well. Since similar aspects and indications may point to multiple indications, the identification of mental health problems is a difficult process and often a misdiagnosis can occur. The patient has to collaborate to detect a problem efficiently. Mental detection problem is a daunting job since a misdiagnosis can cause severe problems. The chapter is focused toward accurately predicting mental disorder categories based on facial expression inputs that are extracted from various images. The proposed framework provides a simple architecture for extracting facial expressions and from input videos to compare them with available labels and predict mental disorder. The presented approach is representing a revised approach of facial expression recognition. The presented approach can be used as a base for the proposed system to predict a mental disorder based on image data and FER techniques.

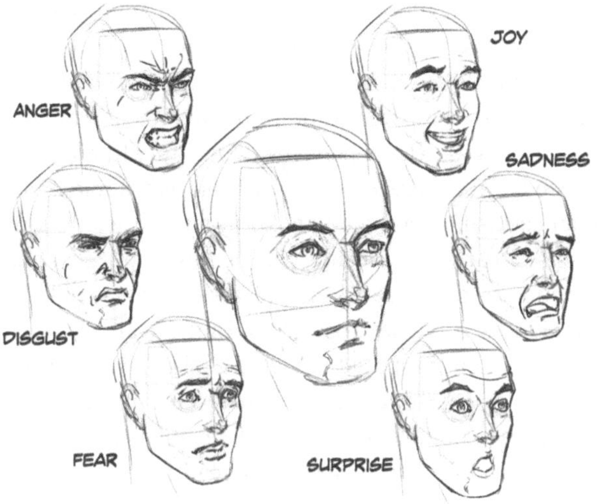


**Figure 5.1:** Various facial expressions.

### 5.1.2 Organization of the chapter

The remainder of the chapter is structured as follows. The employment of image knowledge, similarly as varied techniques for extracting facial expressions, is explored in Section 5.2. Section 5.3 provides a high-level summary of ML techniques' role in mental state prediction. Section 5.4 provides a planned approach for the proposed model, which is introduced to identify the presence of mental disorder based on image and feature data with an ML technique. The chapter is concluded in Section 5.5.

## 5.2 Facial expression recognition algorithms

Expressions are an important way of communicating humanity. Emotions play an effective role in this, and emotions are an effective method of nonverbal communication [2]. Psychologist Mehrabian indicated that the verbal part (i.e., spoken words) of a message contributes only for 7% to the effect of the message as a whole, the vocal part (e.g., voice intonation) contributes 38%, while facial expression of the speaker contributes 55% to the effect of the spoken message [3]. Dampened facial expressions, avoiding direct eye contact are few symptoms of depression [1]. Ekman, a pioneer in studying facial expression, has classified them as various states like happiness, anger, sadness, surprise, disgust and fear [4], which are given in Figure 5.1. In order to provide a better classification, extracting features from one face to another is a complex and responsive task. In 1978, Ekman and Freisen [5] are among the first scientists interested in facial expression who developed Facial Action Coding System, in which facial movements are described by action units (AUs), which are broken down into 46 AUs. Each AU is coded with one or more facial muscles. The automatic FER is the most studied by researchers compared to other modalities to statistics made by Philipp et al. [6]. Research suggests that a person with mental disorders has a bias for the treatment of adverse facial expressions, and this can play a role in decreased social functioning, often coexisting with a diagnosis of mental health. This research investigated correlations in early adolescence between psychological and somatic disorders and facial emotion recognition, as any processing biases could be an early predictor of later mental illnesses. We have used some of the available facial expression recognition methodology in our proposed model to explain and measure facial expression and included steps to identify various types of facial expression and selection of features. We have proposed a model based on ML techniques like deep learning for the prediction of mental disorder based on selected features.

### 5.2.1 Steps involved in feature extraction

1. Different inputs are facial images captured from different video and datasets. Inputs that are static (frame-based) and dynamic (video-based) are used often to derive different characteristics.
2. Analyze the facial expressions captured by images using facial expressions. Landmarks, namely nose alae, eyebrow corner and mouth corners, are the face's emotional highlights and differ confounding factors of manifestations of emotions such as illumination variance during signal acquisition, subject dependency, and a shift in head-pose affects the FER system's reliability.
3. Extracting useful information from the image/input signals is called feature extraction. Gabor feature extraction, local binary pattern, optical flow method, feature point tracking and many such techniques are employed for the same. These are classifiable based on geometry and appearance. FER geometric structures not only extract form and places but also angles between the different elements of the face, such as the eyes, ears, mouth and nose.
4. To extract features, common classifiers used are k-nearest neighbors, SVM and probabilistic neural networks. But ML techniques such as deep learning can also be applied to serve the purpose of feature extraction. Refer Figure 5.2 for steps taken to extract features.

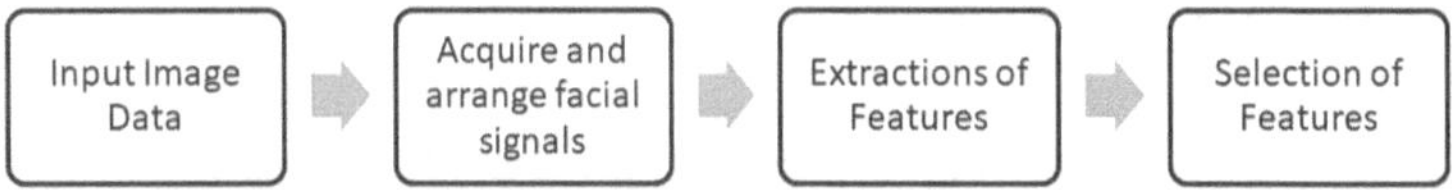


**Figure 5.2:** Steps taken to extract features.

## 5.3 Machine learning in mental illness prediction

The medical sector has been introduced to ML as a way to provide diagnostic methods that are able to boost precision and accuracy while minimizing laborious tasks that need intervention by humans. There is a mounting evidence that the ML-fueled technology can recognize and dramatically enhance the treatment of complex psychiatric illnesses, such as depression. Diseases of mental well-being have also been one of the most critical and prevalent issues affecting public health. Depression is a leading cause of disability, for instance, and may contribute to an increased risk of suicidal ideation and attempts to commit suicide [7]. To truly understand the circumstances of mental well-being and provide better treatment for patients, early mental health detection problems are a critical move. Separate from the diagnosis of other chronic conditions that are based on laboratory tests, mental disorders are usually diagnosed with measurements, based on a self-report from an individual to unique questionnaires built for the identification of patterns of feelings or social relationships [8]. ML-based methods demonstrate enormous potential for psychiatric diagnostics by using highly analytical approaches to harness measurable affect-related behaviors. In fact, observable habits related to impact are widely used by mental health practitioners to support in psychiatric diagnostics, mostly by non-structured approaches that result in qualitative, general methods data. The bulk of current algorithms, however, still require some human involvement standard, such as labor-intensive manual labeling or classification of hands data to derive useful characteristics prior to analysis [9]. Data of mental health collected with questionnaires and posts pooled will be used to extract relevant features. Then, statistical analyses will be executed to detect significant corre lations between levels and features, which will be selected to build predictive models. ML algorithms will be used to train predictive models to detect users with poor mental health [10]. Based on findings, we suggested a way to use the power of ML for the recognition of mental illnesses for early detection in the proposed architecture.

## 5.3.1 Machine learning techniques

The aim of ML is to develop process algorithms or applied math models which will infer hidden patterns mechanically from knowledge [10]. Recent years have seen a growing number of ML models. However, conventional ML approaches need a substantial quantity of feature engineering – a move that is acceptable for optimum performance to attain smart performance for many application scenarios that are often time overwhelming and resource overwhelming. Because the newest wave of technologies for ML-associated computing approaches aim to determine an end-to-end approach, a mechanism that directly maps the raw options of the input into the outputs through a multilayer network structure that is able to capture the hidden patterns among the information. This section explores a number of ML-based techniques like recurrent neural network (RNN), convolutional neural network (CNN) and deep learning.

## 5.3.2 Recurrent neural network (RNN)

RNNs were developed to examine sequential data including natural language, video and voice. When given a sequence as input, one RNN element processes one element of the sequence at a time by feeding it to a recurrent neuron. Every recurrent neuron receives the input element at the corresponding time point and the neuron's output at the previous time stamp and will also be provided with the output to encrypt historical data along the series. The neuron is stamped again the next time (this is also where the word "recurrent" comes from). Figure 5.3 [11] demonstrates an example of RNN architecture.

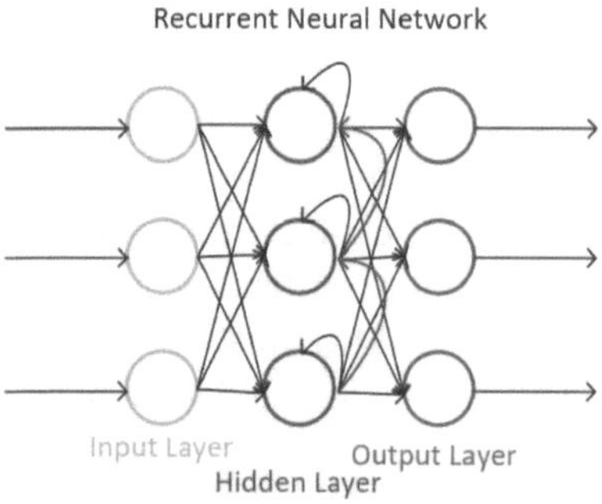


**Figure 5.3:** Recurrent neural network.

## 5.3.3 Convolutional neural network (CNN)

The method of extracting CNN features involves alternating convolutional layers and then pooling them together. During CNN model propagation, the feature map after sequential operations (convolution–ReLU (rectified linear unit)–Maxpool) decreases sequence by sequence. Following that, the system isolates the coarse features from the data patches.

The convolution layer has a large number of channels that can be configured. Each channel is converted across the entire region to deliver distinct feature maps. In the convolution layers, the feature maps from the previous layer are combined with several convolutional kernels (to be learned in the training process). The performance of the convolution operators, along with a bias (to be learned), are transferred through an activation function [12] to form the feature map for the next layer. Figure 5.4 will represent functioning of CNN.

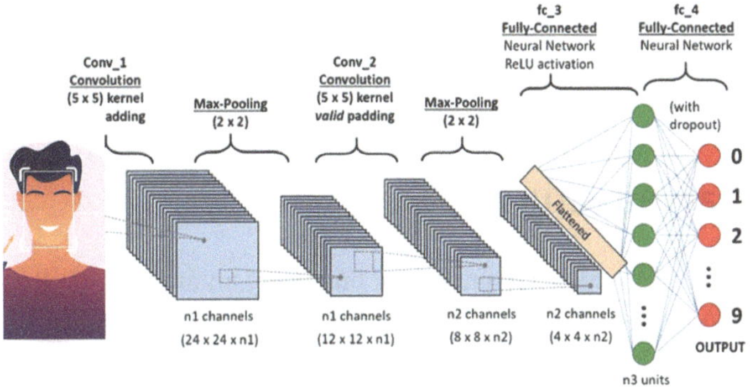


**Figure 5.4:** Convolutional neural network.

## 5.3.4 Deep learning

The basic feature of an artificial neural network (ANN) is represented in Figure 5.6, and it is proposed with the aim of mimicking how the human brain functions. An artificial neuron is a nonlinear transformation device that takes the weighted sum of all inputs and feeds it to an activation function like sigmoid, rectifier (i.e., ReLU) or hyperbolic tangent. An ANN is made up of multiple artificial neurons with different link architectures. The feedforward neural network (FNN) is the most basic ANN architecture. It stacks neurons layer by layer in a feedforward fashion, with neurons across adjacent layers completely linked to each other. The input layer of the FNN is the first layer, which receives one dimension of the data vector for each unit. The output layer is the final layer, which shows the probabilities that a subject belongs to various classes (in classification). The secret layers are those that exist between the input and output layers. A deep FNN (DFNN) is a type of neural network that has multiple hidden layers. Figure .5.5 Represents deep Feedforward Neural network.

In this section, we have shown over all of the common methods and techniques for image processing and classification of image features. The proposed method suggests that by integrating ML and facial expression recognition techniques, we can diagnose mental illness at an early level.

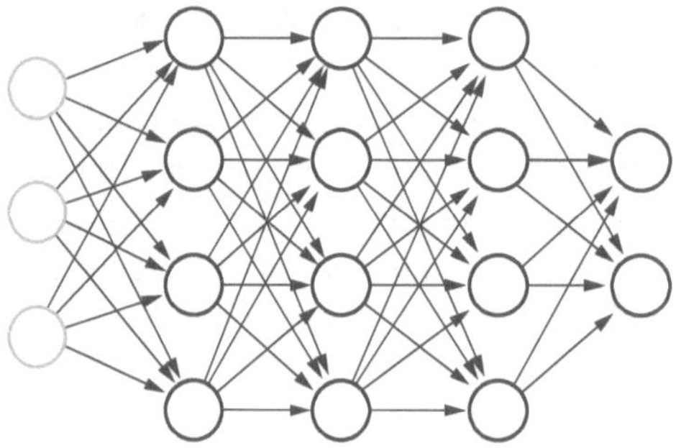

**Figure 5.5:** Deep feedforward neural network.

# 5.4 Proposed architecture

Based on earlier discussion in this chapter, we are proposing an architecture that can predict mental disorder based on different facial expression with the use of ML methods.

The architecture uses image data obtained from different sources as an input. Features are extracted for the further process by applying different extraction techniques, which are passed on to the ML model. In the next step, an ML model is deployed to test the collected features using that model. Model training is carried out on the basis of samples and then the result is used for the proposed model test process.

The proposed model has a capability to utilize the power of any of the ML techniques listed in Section 5.3. The architecture suggested can be used as the primary model to detect the presence of mental disorder in the early stages and has the capability to suggest the diagnostic method that can be used on the early stages of any mental illness.

Applying appropriate classification or regression method can achieve desired accuracy for the proposed models. A complete list of classifiers working in the test domain is available in Figure 5.8 and has also been ranked. SVM is the most common machine by far for mental disorder categorical evaluation and is well suited for binary problems of high dimensionality such as the distinction of low symptom of mental illness to the presence of high symptom for mental illness.

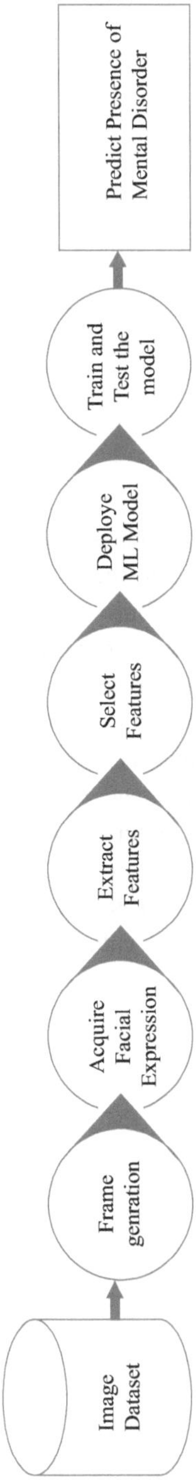


**Figure 5.6:** Proposed model.

**Table 5.1:** Classification Algorithms.

| |
|---|
| SVM |
| Nearest neighbor |
| Gaussian mixture models |
| Neural networks |
| Naïve Bayes |
| Maximum entropy model (MaxEnt) |
| Random forest (RF) |
| Relevance vector machines (RVM) |
| Logistic regression (LogR) |
| Hidden conditional random fields (HCRF) |
| Hidden Markov model (HMM) |
| Coupled hidden Markov model (CHMM) |
| Stacked denoising autoencoders (SDA) |

Table 5.1 represents various Classification Algorithms. Selection of classification algorithm is based on the data and complexity of the methods applied. Furthermore, a variety of neural networks have been employed, including FNNs, probabilistic neural networks, restricted Boltzmann machines, radial-basis functions, multilayer perceptron and extreme learning machines. We can utilize any of the available neural network architecture to strengthen the proposed methodology. The CNN (Figure 5.9) is a category of neural network that has been proven very efficient in image classification. The CNN learns the filters that traditional algorithms were hand-engineered. Using convolutional network, we can improve the efficiency of the proposed architecture.

Accuracy is a measure of how close the measurements are to each other, not necessarily to a specific value in ML. To put it another way, if a set of data points from repeated measurements of the same quantity are close to the true value of the

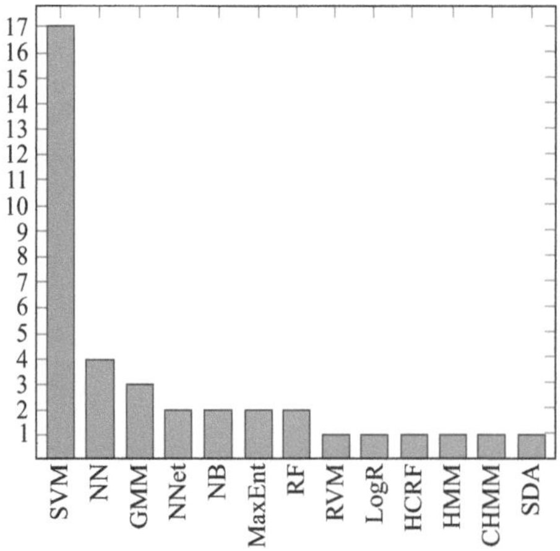


**Figure 5.7:** Listing and ranking of various classification algorithms.

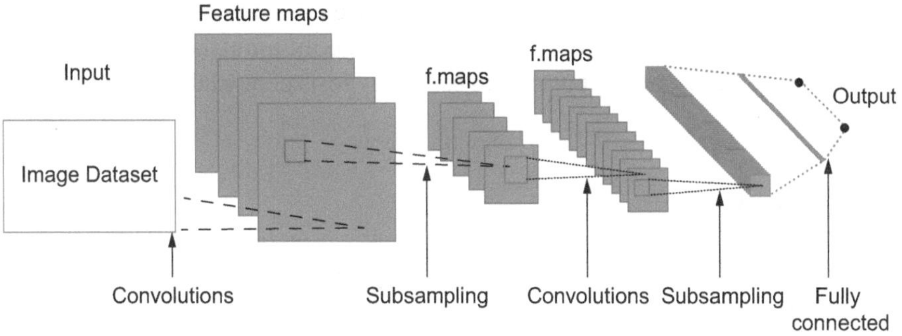


**Figure 5.8:** CNN for image classification.

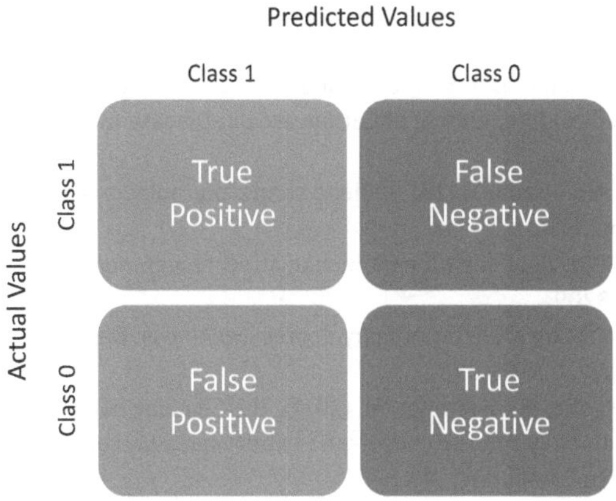


**Figure 5.9:** Confusion matrix.

quantity being measured, the set is said to be correct. Figure 5.7 represents ranking of various classification algorithms. On the other hand, if the values are identical to each other, we call the set precise. Since the two definitions are independent of one another, the dataset may be reliable, precise, both or neither. The accuracy of the proposed model can be evaluated based on available accuracy measures such as confusion matrix (Figure 5.9).

## 5.5 Conclusion

The use of available image recognition methods to derive different facial expressions from image inputs is demonstrated in this chapter. Furthermore, we have proposed a model for predicting the existence of mental illness that is based on a capable ML system that can be trained using the data gathered during the image recognition process. The proposed model can be used as a basis for smart systems that can predict mental illness at the primary stage. In a nutshell, the proposed model reflects an innovative approach to early diagnosis of mental illness.

## References


[1] Kim, D.H., Baddar, W., Jang, J., Ro, Y. (2017). Multi-objective based spatio-temporal feature representation learning robust to expression intensity variations for facial expression recognition. IEEE Transactions on Affective Computing.

[2] Ekman, P., Friesen, W.V. (1978). Facial Action Coding System: A Technique for the Measurement of Facial Movement. Palo Alto, Consulting Psychologists Press.

[3] C. 2019. « Survey on AI-based multimodal methods for emotion detection », in High-Performance Modelling and Simulation for Big Data Applications: Selected Results of the COST Action IC1406 cHiPSet. *J. Kołodziej et H. González-Vélez*, 324, 307–8209.

[4] Kalyani, R., Thej, M., Prabhakar, K., Kiran, J. (2011). Single dominant left coronary artery: An autopsy case report with review of literature. Journal of Cardiovascular Disease Research, 2(2), 130–132.

[5] Alkawaz, M.H., Mohamad, D., Basori, A.H., Saba, T. (2015). Blend shape interpolation and FACS for realistic avatar. 3D Resolution, 6(1).

[6] Rouast, P.V., Adam, M., Chiong, R. (2018). Deep learning for human affect recognition: Insights and new developments, 1, 1–8209.

[7] Hamilton, M. (1967). Development of a rating scale for primary depressive illness. British Journal of Social and Clinical Psychology, 6, 278–296.

[8] Mckeown, G., Valstar, M., Cowie, R., Pantic, M., Schroder, M. (2012). The Semaine database: Annotated multimodal records of emotionally colored conversations between a person and a limited agent. IEEE Transaction Affective Computing, 3(1), 5–17.

[9] Murphy, K.P. (2012). Machine Learning: A Probabilistic Perspective. MIT Press.

[10] Biship, C.M. (2007). Pattern Recognition and Machine Learning (Information Science and Statistics). Springer-Verlag.

[11] Lecun, Y., Bottou, L., Bengio, Y., Haffner, P. (1998). Gradient-based learning applied to document recognition. Proceedings of IEEE, 86, 2278–2324.

G. Malini Devi

# 6 Neonatal monitoring system using IoT and object detection

**Abstract:** It is very important for a parent or guardian to monitor a newborn/infant every moment, which includes checking if the infant is safe in the placed environment, actions that infants do and so on. Sudden infant death syndrome (SIDS) is the main reason of sudden and unexplained death of an infant younger than one-year-old. Rural as well as urban healthcare centers lack in capacity to curtail or minimize the time taken to react or respond to SIDS instances. A recent study states that a quarter of reported neonatal death cause is early preterm birth. The project proposes a learning/monitoring/detection facilitated through a video in amalgamation with the concept of connecting the device(s) to the Internet of things. If any abnormal event arises, this is noticed and a caveat in the form of an alarm is passed on to caretakers/guardians/parents. The newborn baby's health strictures like body heat, oxygen levels and moisture or body wetness are noted in real time, which are gathered and stored on cloud platform. The system detects an increase or decrease in predefined normal levels; it sends a message alerting the caretakers/guardians/parents. This alerting system helps in timely remedial action, if need be, for child's safety. Advantages of this system are increased response time of the guardian in emergencies, monitoring the environment in which the infant is placed, keeping track of the actions infants do and notifying when an abnormal action is detected to respond. This inexpensive baby's physical condition monitoring arrangement overcomes the inhibitive preexisting incubator cost.

**Keywords:** IoT (Internet of things), object detection, sensors

## 6.1 Introduction

Rural/urban areas face the deficit of latest technology, which led to highly probable lack of ability to minimize the reaction time in sudden infant death syndrome cases. There are around 4 million infant mortality occurrences in the first month of birth and around a million on the day of birth. A study noted that a quarter of infant deaths are due to early preterm births. In India, it is noted that parents/guardians' work and home time commitment impacts newborn/infants caretaking abilities leading to more pressure and stress, which is dominantly reflected on female

**G. Malini Devi,** Dept. of CSE, G. Narayanamma Institute of Technology and Science (For Women), e-mail: gmalini12@gnits.ac.in

https://doi.org/10.1515/9783110756722-006

counterparts. It will be an inspirational miracle if any peripheral help comes to minimize the pressure of newborn heath care which in turn can lead to nonstressful parent life. This new model is proposed to screen child's activity and notify any abnormal event on time, relieving parent's/guardian's anxiety [1].

Infant incubators in hospitals are used to provide a warm environment for babies born prematurely or for other infants who are unable to maintain a normal body temperature or at risk of other medical vulnerabilities. The infant incubator is a relatively small, glass-walled box that may have portholes fitted with long rubber gloves through which nurses can handle and care for the infant. Most infant incubators are fitted with microcontrollers and sensors that can control the concentration of oxygen inside the incubator; this is necessary because some infants need either greater or lesser amounts of oxygen owing to particular diseases they may have. Infant incubators also regulate the humidity inside the enclosure [2].

### 6.1.1 IOT module

There are four sensors which help detect infant health and use Internet of things (IoT):

- Temperature sensor
- Humidity sensor
- Moisture sensor
- Ultrasonic sensor (for detection of distance between the baby and cradle)

The system that facilitates observation of infant's health by just visiting an application is designed. The present trend of user-friendly apps which use the IoT is much beneficial, and this system also uses the same technology to observe infants' health through apps in android devices [3]. An android mobile device with Internet and Wi-Fi is essential for the system to run perfectly [4].

The connection to Wi-Fi network is done with Wi-Fi module through a microcontroller. Wi-Fi network is essential for the project to run. The created Wi-Fi zone with either a Wi-Fi module or a mobile device enables hotspot. Sensors sent signals to a microcontroller which is then routed through cloud and later defined the uniform resource locator/Internet protocol address. Prior systems' precincts are overcome by this new planned methodology. The infant's body heat and heart rate are tracked along with the sweat absorbed in bedding. An open-source microcontroller board, in this case Arduino Uno (low-cost, flexible and easy-to-use programmable), helps in conversion of analogue to digital [5]. Enabling Internet connectivity to embedded applications, the microcontroller communicates with Wi-Fi module using UART interface [6].

### 6.1.2 Object detection module

Region-based convolutional neural networks (CNNs), a family of machine learning models, are used for specifically object detection; here, it targets regions of interest having higher probability of object enclosure [7]. YOLO is a clever CNN for object detection in real time, which is incredibly fast and accurately captures total image and predicts multiple bounding boxes and class probabilities [8]. Using YOLO algorithm, image classification goes through levels of incremental complexity and it is also a clever CNN for doing object detection in real time [9].

### 6.1.3 Existing system

Prior process is comprised of sensors and hardwares along with cloud-based servers and source application, tracking the heart rate and blood gas analysis with sensors and Arduino microcontroller. The stored cloud data lacks security which is a main concern.

### 6.1.4 Proposed system

Here, the project tracks the heart rate and blood gas analysis and moisture with real-time storage in cloud (secured with data encryption). Infant's actions are observed by recognizing action, any variations from predefined levels, and send an automated alert message to the caretaker/parent/guardian. Essentially infant observation system works with sensors and microcontroller.

## 6.2 Literature survey

YOLO is a clever CNN for object detection in real time, which is incredibly fast and accurate and captures total image and predicts multiple bounding boxes and class probabilities [10] [11]. In this chapter, instead of predicting one box per pixel, we divide an image into $S \times S$ grids and predict centered object per grid. Officially defined confidence scores are Pr(Object) × IOUtruth pred, which indicate the exist objects' likeliness (Pr(Object) ≥ 0) and prediction confidence (IOUtruthpred). Simultaneously, irrespective of volume of boxes, C conditional class probabilities (P r(Classi |Object)) predicts object in the grid. Grid cell contribution of object is calculated:

$$\Pr(\text{Object})*\text{IOUtruthpred}*\Pr(\text{Classi}|\text{Object}) = \Pr(\text{Classi})*\text{IOUtruthpred} \tag{6.1}$$

The scores enciphered both the probability of that class objects shown in the box and predicted box fitness in comparison with the object.

Given a cell i, (xi, yi) reflects a box's center in relation to the grid cell, (wi, hi) regularized measurements in relation to the size of an image, confidence scores are represented by Ci, object's existence is indicated by 1 obj i, which in turn signifies prediction performed by jth bounding box predictor. It is noted that loss function penalizes errors of classification only if the object is available in a grid cell. Likewise, bounding box coordinate errors penalize when predictor is accountable for ground truth box (predictor's highest IoU is achieved in a grid cell).

FC layers and 24 conv layers are there in YOLO; here, few conv layers create a troupe of inception modules having 1 × 1 lessening layers pursued by 3 × 3 conv layers. Image processing happens at 45 FPS in a given network, and with fast YOLO, 155 FPS can be achieved, which are much better than real-time detectors.The main idea of YOLO is given in Figure 6.1.

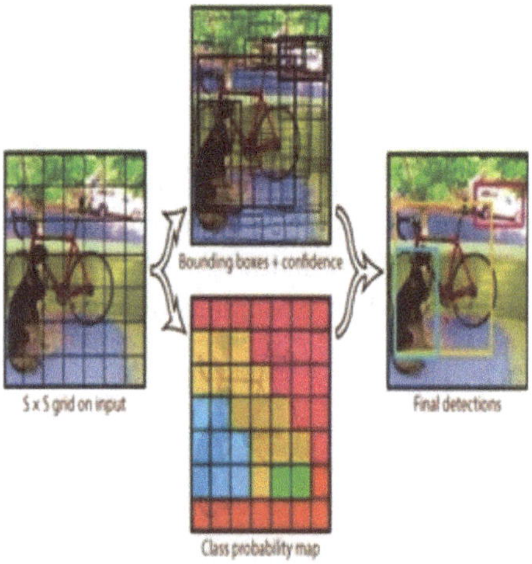


**Figure 6.1:** The main idea of YOLO.

## 6.3 Methodology

A contextual framework for research, a system and theory of analysis applied field study methods, including theoretical analysis toward process and philosophies pertaining to a branch of knowledge, and also technical concepts, theoretical models and quantitative or qualitative techniques.

### 6.3.1 Architecture of system

In the system, IoT module consists of three sensors, that is, temperature sensor, ultrasonic sensor and moisture sensor [12]. The values from each sensor is converted into the form of node js which has its own URL for connecting to the MIT app [13]. The MIT app is an application used by the user like parents/caretakers which shows all the values of the sensor reading, and if any threshold value is surpassed, an alert to the user is sent which is also a feature of the application [14].

The object detection module consists of a camera which captures the object and detects if it is baby or not. The camera is connected to camera software that is YAW Cam (Yet another webcam) which has its own URL/IP address which is used for connecting to the MIT app. The yaw cam is connected to the REST API which is used for resource transferring between the user and server when the user requests resources [15]. The rest API is connected to MIT app where the baby's actions are observed and monitored on the screen [16]. With the help of ultrasonic sensor and object detection as baby, if any abnormal actions of baby are detected, an alert to the user is sent by the MIT app [17]. The architecture of the system is shown in Figure 6.2.

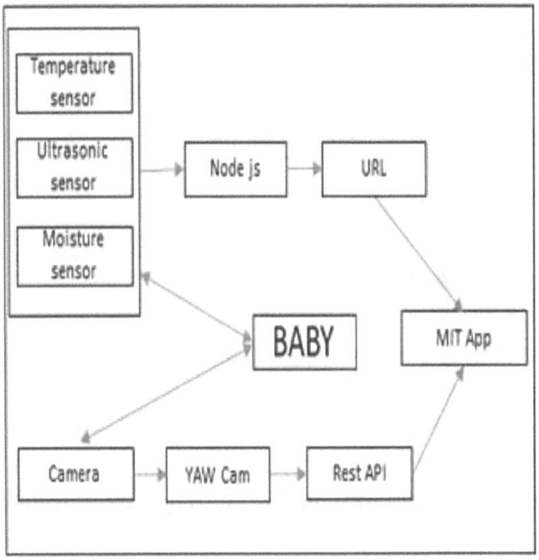


**Figure 6.2:** Architecture of the system.

### 6.3.2 Modules

Modules assist to organize the course content by time or subject, and can also cater for diverse organizational structure. Modules effectively craft a one-directional linear flow for students to do in a course. This project is a fusion of three modules:

A. Sensor module
   - This module is comprised of components that monitor the environmental parameters of an infant.
   - This module monitors the infant's condition by fetching the data from the hardware units called sensors in every fraction of a second. These collected data are sent to the cloud for verification of parameters against the threshold.
B. Object detection module
   - This module is comprised of software which is triggered to detect the captured object by the surveillance camera.
   - The project helps to send updates about the infant's position and alert when the position is trivial.
   - These collected data from object detection module is sent to the cloud.
C. App module
   - This module comprises the actual interface to the end user.
   - App module helps to notify the user about infant's whereabouts.
   - The integrated modules and components can be accessed through this module.

## 6.4 Results and discussion

This section depicts the outcomes of the system design, architecture, modules and system implementation.

### 6.4.1 Microcontroller Arduino collected data

The Arduino Uno Genuino acts as a central unit for controlling operations of hardware components involved in collecting the environmental parameters. To trigger the action, some trivial conditions are set on encountering the value above or below the threshold values.

Table 4.1 shows the sample dataset required to refer for the triggering of trivial conditions under surveillance of environmental parameters continuously. The dependency of each parameter on the final result can be visually depicted by a graph. The result shows the status of observed parameters like not bad, worst, very good, mediocre, not good, good, partly better, better, very bad and partly bad. The parameter dataset is shown in Table 6.1.

Figure 6.3 shows a visual representation, that is, of moisture, against the final result. The graph shows the dependency of each parameter over the final result and sends the final alert to the microcontroller which in return triggers the corresponding event.

**Table 6.1:** Parameter dataset.

| Humidity | Temperature | Moisture | Result |
|---|---|---|---|
| 51 | 17.33 | 0.51 | Partly bad |
| 62 | 10.42 | 0.62 | Very bad |
| 66 | 9.91 | 0.66 | Very bad |
| 71 | 17.36 | 0.71 | Better |
| 74 | 17.20 | 0.74 | Better |
| 96 | 7.31 | 0.96 | Partly better |
| 93 | 7.24 | 0.93 | Partly better |
| 85 | 10.44 | 0.85 | Very bad |
| 93 | 9.52 | 0.93 | Partly bad |
| 90 | 8.40 | 0.9 | Partly bad |
| 88 | 14.37 | 0.88 | Good |

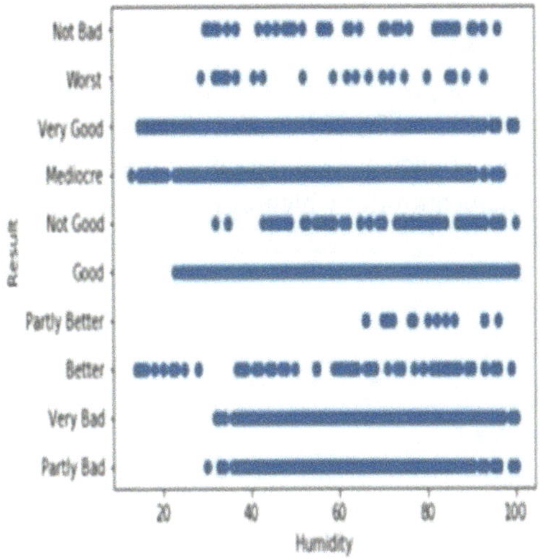


**Figure 6.3:** Moisture versus result.

Figure 6.4 shows a visual representation, that is, of temperature against the final result. The respective graphs show the dependency of each parameter over the final result and sends the final alert to the microcontroller which in return triggers the corresponding event.

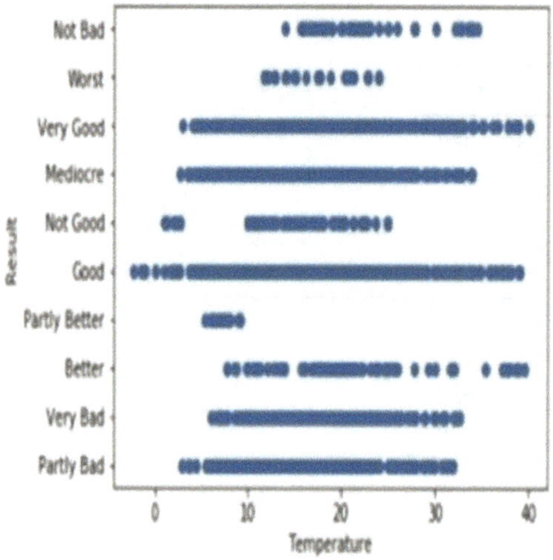


**Figure 6.4:** Temperature versus result.

Figure 6.5 shows a visual representation, that is, of humidity against the final result. The respective graphs show the dependency of each parameter over the final result and sends the final alert to the microcontroller which in return triggers the corresponding event.

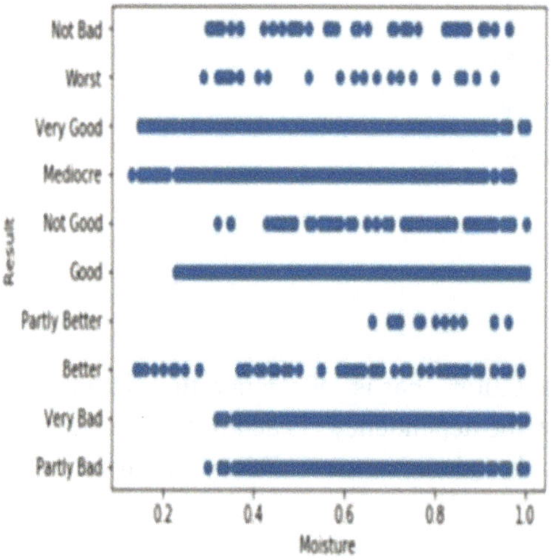


**Figure 6.5:** Humidity versus result.

## 6.4.2 Camera capturing infant (sample)

Figure 6.6 is an image of a baby captured by a camera. The Yawcam software connected to the camera has an output window console in which the image of the baby can be seen, and using this input image the detection is done and whether the object shown in the window is baby or not is compared with datasets; finally, the output result is shown in the monitoring system. The image can also be observed in the MIT application by parents/caretakers in their personal devices.

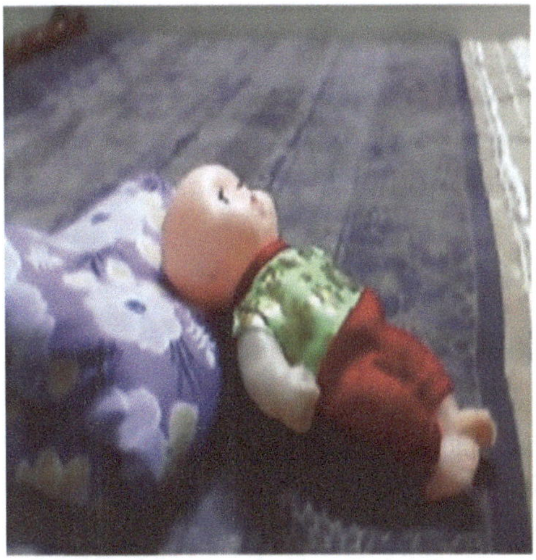

**Figure 6.6:** Camera capturing baby image.

## 6.4.3 Arrangement of prototype

Figure 6.7 is the arrangement of the prototype of the proposed system. The baby is assumed to be on a soft bed with IoT modules and object detection modules surrounded by it.

These modules capture and collect the data based on the behavior, movement and surrounding physical environment conditions, and send them to the monitoring system running at the off screen to detect any risks or abnormalities, alerting the parents/caretakers based on the input it gets.

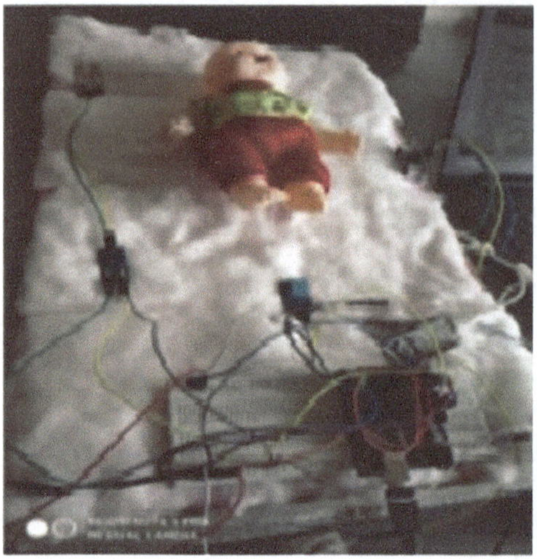

Figure 6.7: Final prototype.

### 6.4.4 The final output

Figure 6.8 is the MIT application of the baby monitoring system. It is used by parents and caretakers to monitor the surrounding conditions of the baby and observing the actions and movements of the baby to detect any risk for alerts.

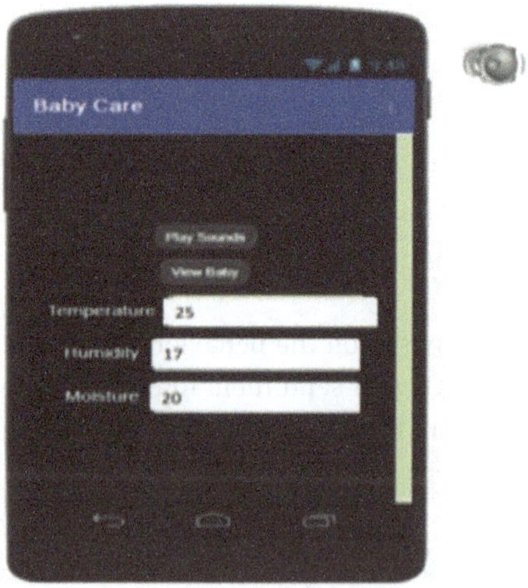


Figure 6.8: Final prototype.

There is a feature for playing sounds for the baby for calming baby in times of anxiety and while sleeping.

## 6.5 Conclusion

We have proposed an infant health monitoring system using sensors and is based on IoT. This proposed system offers numerous advantages in comparison to the traditional method. Apart from proposed sensors, more strictures can be integrated, and every sensor can be connected to a processing board via a wireless module. This system can also be applied to infant's bedding for warmth/moisture measurement. It is also more helpful to all those parents who are working and become convenient to observe each and every movement of baby along with their health.

## References

[1] Shrivastava, A.K., Verma, A., Singh, S.P. (2010 February). Distance measurement of an object or obstacle by ultrasound sensors using P89C51RD2. International Journal of Innovative Technology and Exploring Engineering (IJITEE), 2(1), 64–68.

[2] Symon, A.F., Hassan, N., Rashid, H., Ahmed, I.U., Taslim Reza, S.M. (2017 September). Design and development of a smart baby monitoring system based on raspberry Pi and Pi camera. International Journal of Innovative Technology and Exploring Engineering (IJITEE), 117–122. https://doi.org/10.1109/ICAEE.2017.8255338.

[3] Nirmala, G., Jeyashree, S., Lakshmi, B.M., Nandhini, K. (2018 March). A secure IoT based baby healthcare monitoring and maintenance system in cloud. IEEE, 5(3), 1–5.

[4] Holla, S., Katti, M.M. (2012 March). Android based mobile application development and its security. International Journal of Computer Trends and Technology (IJCTT), 3(3), 486–490.

[5] Srivastava, D., Kesarwani, A., Dubey, S. (2018 December). Measurement of temperature and humidity by using Arduino tool and DHT11. IEEE, 5(12), 876–878.

[6] Louis, L. (2016 April). Working principle of Arduino and using it as a tool for study and research. International Journal of Control, Automation, Communication and Systems (IJCACS), 1(2), 21–29. 10.5121/ijcacs.2016.1203.

[7] Uma Maheswari, V., Aluvalu, R., Chennam, K.K. (2021). Application of machine learning algorithms for facial expression analysis. Machine Learning for Sustainable Development, 9, 77.

[8] Ziganshin, E.G., Numerov, M.A., Vygolov, S.A. (2010 September). UWB baby monitor. IEEE, 6(10), 48–52. https://doi.org/10.1109/UWBUSIS.2010.5609156.

[9] Tavagad, S., Bhosale, S., Singh, A.P., Kumar, D. (2016 February). Survey paper on smart surveillance system. International Research Journal of Engineering and Technology (IRJET), 3(2), 315–318.

[10] Uma Maheswari, V., Prasad, G.V., Viswanadha Raju, S. (2021). Facial expression analysis using local directional stigma mean patterns and convolutional neural networks. International Journal of Knowledge-based and Intelligent Engineering Systems, 25(1), 119–128.

[11] Patel, A., Devaki, P. (2019 April). Survey on node MCU and raspberry pi: IoT. IEEE, 6(4), 5101–5105.
[12] Vacher, M., Jean, M., Serignat, F., Castell, E. (2014). Smart sound sensor for perceptive spaces dan instate. IEEE, 15.
[13] Amasha, M.A., Al-Omary, S. (2017 June). Quizrevision: A mobile application using the Google MIT app inventor language compared with LMS. (IJACSA) International Journal of Advanced Computer Science and Applications, 8(6), 313–320. https://dx.doi.org/10.14569/IJACSA.2017.080641.
[14] Barapatre, P., Patel, J.N. (2019 May). Determination of soil moisture using various sensors for irrigation water management. International Journal of Innovative Technology and Exploring Engineering (IJITEE), 8(7), 576–582.
[15] Fielding, R.T., Taylor, R.N., Erenkrantz, J.R., Gorlick, M.M., Whitehead, J., Khare, R., Oreizy, P. (2017 September). Reflections on the REST architectural style and principled design of the modern web architecture. ACM,ESEC/FSE', 17, 4–14. https://doi.org/10.1145/3106237.3121282.
[16] Patil, S.P., Mhetre, M.R. (2014 January). Intelligent baby monitoring system. International ITSI Transactions on Electrical and Electronics Engineering (ITSI-TEEE), 2(1), 11–16.
[17] Zhao, Z.-Q., Zheng, P., Xu, S.-T., Wu, X. (2019 January). Object detection with deep learning: A review. IEEE Transactions On Neural Networks And Learning Systems, 1–21. http://dx.doi.org/10.1109/TNNLS.2018.2876865.

Kiran P. Kamble, Vijay R Ghorpade, Sonali Patil, Pratibha Garje, Pooja D. Mundada

# 7 Video surveillance fire detection and quick alarm systems

**Abstract:** Fire disaster events are often uncertain and lead to a huge economic, ecological and human loss if not detected in the early stages. It is crucial to find precise and possibly portable solutions to detect fire. Several efforts are already done to design a fire-detection system that will be effective and has an appropriate cost. However, most of them demonstrate a trade-off between model size and performance. In the proposed system, we designed, and have tested, to solve both these issues. They designed a model that will be both lightweight and has high performance than the previously existing system. Smart IoT devices are capable of detecting fire in the frames captured through the video surveillance system. By inferencing, the deep learning-based SpNet model designed using CNN with the durable Jetson Tx2 kit high reliability is achieved. The network-designed SpNet neural network on the firedb data set is better in performance. The model is designed based on a foundation of state-of-the-art VGG16 architecture.

**Keywords:** Fire detection, deep learning, CNN, CCTV video analysis

## 7.1 Introduction

IoT-based devices have good abilities to capture and process frames in the surveillance system. Their capabilities are increasing and smart enough to detect objects in video surveillance. Fire disasters are abnormal and can occur anytime, so the challenge is that they must be notified earlier. In India, about 7,000 fire events are detected per year in residential buildings, and the minor fire events count further increases. Fire accidents are occurring very often in residential and office buildings. These buildings have a preexisting CCTV system. Using this system reduces the number of fire accidents, thereby saving from the losses can be done using CNN-based deep learning model. The model we designed for this purpose is based on

**Acknowledgment:** We are grateful to all teams of the department of computer science and engineering, WCE who densely evaluated us and provide guidance. The work is also endorsed by students and alumni of WCE Sangli. We express our gratitude to all.

**Kiran P. Kamble,** Shivraj University. Kolhapur, e-mail: kirankamble5065@gmail.com
**Vijay R Ghorpade, Sonali Patil, Pratibha Garje, Pooja D. Mundada,** Walchand College of Engineering, Sangli

https://doi.org/10.1515/9783110756722-007

image classification of frames received from the camera. This system helps to detect fire at an early stage in an efficient manner. The available fire-detection system is based on physically mounted sensors such as heat and smoke detection sensors [1, 2]. However, previous existing systems for fire detection were inefficient, as they often lead to false alarms. Those previously proposed IoT sensor-based detection prototype is not very reliable and with low accuracy for fire detection. For instance, quite often, there are cases of false triggering with smoke detectors, as it cannot detect and identify whether the smoke is resulting due to fire. Heat sensors can falsely measure the temperature of the surroundings as fire and smoke sensors can falsely measure the dense smoke. As sensors are installed at high heights, it requires a large time to detect fire. The number of sensors that need to be installed is also more; for large buildings, they are not cost-friendly [3]. The proposed system is aimed at using surveillance cameras to accurately detect and alert the occurrence of fire continuously. Cameras were already installed in places; our system is aimed at reducing the disadvantages of false alarm, by developing the system reliable. It also enhances the accuracy of fire detection. Even after the fire is detected, proper action needs to be taken. Therefore, when the alarm rings, messages are sent to the respective people in charge of the place to handle the case if anyone is not there in that area when the fire is detected.

## 7.2 Literature survey

Detection of fire is a major issue, the majority of the research till now is focused on traditional features extraction methods.

a. **IOT-based fire detection:** Early systems such as fire and smoke detection systems mostly use temperature sensors, smoke sensors or both. The smoke sensors measure false density of air to detect it as the smoke which often gives false results. The heat sensors can falsely detect fire by measuring the surrounding high temperature. These sensors installed at heights take too much time to detect fire. They need to be inclined with each other. This system has very little accuracy. A large number of sensors need to be installed on each floor of the building, which is not cost-effective [4].
b. **Image processing technique:** Most of the approaches proposed are based on image processing techniques. Visual-based fire-detection systems have many advantages over hardware-based alarm systems in terms of cost, accuracy, robustness and reliability [5]. In order to manage false alarms of the conventional fire-detection system, a computer vision-based fire detection algorithm is proposed. Fire detection algorithms consist of two main parts: fire color modeling and motion detection. The algorithm can be used in parallel with the conventional fire detection model to reduce false alarms in real-time deployment. Also can be deployed

as a live system in order to predict fire using video images acquired through camera-based smart edge equipment. But the accuracy achieved is not yet satisfying.

c. **VGG16 based model:** Decade time the manually visual fire detection approaches and also other deep learning approaches, which depicts not up to the satisfactory level in terms of accuracy and false alarming [4, 6]. They can be replaced by Logistic Regression based on VGG architecture model, which is higher in performance in terms of varying metrics. The VGG16 model built had successfully overcome the drawbacks of previous approaches. The model has been trained and tested on an artificial data set. Deep CNN works as a feature extractor.VGG-16 model has been reported to take up to 550 MB of machine memory [7]. The modified model works as per the following steps.
d. **Feature extraction and fine-tuning:** We have used a pretrained network model VGG16 and used the method of fine-tuning in which the network model is used as a feature extractor for our data set. The fine-tuning of weights and bias parameters is a process of transfer learning.
e. **Image classification:** The main task is to classify an image as fire or non-fire. We have trained our own classifier over the top of the features extracted. The classifier is based on logistic regression is trained and tested for classifying fire and non-fire images. The accuracy of this classifier is 98%. The researcher also experimented CNN image-based classification on fake currency detection [8] where the data set was divided into two categories, fake and real currency. CNN-based image classification is also carried out for lie detection in the criminal investigation of crime branch, where the data set is created using a thermal camera focusing on periorbital tissue and cutaneous muscles [7] for image classification. Classification using machine learning is also executed for face reorganization and expression detection in video scenes [9, 10] disadvantage of this approach is less accurate due to handcrafted features. Moreover, the facial expression analysis part is executed using local directional stigma mean patterns and convolutional neural networks [11, 12].

## 7.3 Approach

The particularly developed software is embedded into a conventional CCTV system which then constantly checks the video images to detect fire and instantly alarm an emerging fire. The system after detecting the fire through the camera the appropriate actions will be taken. The approach we followed is as follows:

a. **Step 1:** Data set collection and segregation: We have collected our data set containing a total of 4,172 images
b. **Step 2:** Feature extraction using transfer learning and modification in layers structure as per need. Treating networks as arbitrary feature extractors. We have used a pretrained CNN to extract.

c. **Step 3:** Training is performed on extracted features. We have built logistic regression classifier and trained it by using feature extracted.
d. **Step 5:** Testing a classifier: We have tested the performance of the classifier on testing images. Images are classified into the fire and non-fire classes.
e. **Step 6:** Action to be taken when the fire is detected: The alarm will be sound, and messages are sent to the respective in charge of that place.

## 7.3.1 Data set description

We have collected our data set containing a total of 4,172 images [13]. These images are divided into two classes: fire and non-fire. There are 2,245 fire images and 1,927 non-fire images. Some of the images in our data set are shown below. Sample images are shown in Figure 7.1.

**Figure 7.1:** Sample images from firedb data set.

## 7.3.2 Method

The primary objective of the system is to detect the fire in its early stage. The frames in the videos recorded through camera systems of WCE are checked for fire. The deep learning-based approach for fire detection is implemented as follows. Initially, we extracted features from the firedb data set. For extracting features from the data set, we converted our data set to HDF5 format. For extracting, we used transfer learning and fine-tuning of pretrained CNN VGG16. Transfer learning is a process where we can use an existing pretrained classifier as a starting point for the new classification task. We have done fine-tuning of VGG16. We have taken VGG16 model that has already been trained and used it as a feature extractor for our data set. The convolutional neural network is an end-to-end image classifier. We input an image to the network. The image forward propagates through the network and passed through various layers of CNN. The different layers through which it is passed are shown in Figure 7.2.

### 7.3.3 Use of HDF5 format

HDF5 is a hierarchical data format used to store large numerical data sets on disk which facilitates easy access and computation on the rows of the data set. The features extracted from the data set are stored in the form of an HDF5 file. The HDF5 file contains the label name, label and features of each image. While extracting features, it is passed through the following layers.

### 7.3.4 Convolution layers

Convolution is the starting of the convolution neural network to get extracted features from a given input scene in image format. CNN layers perform convolution operation image matrix and the filter as an input and the output as a feature set. This operation is to learn the features of the image without changing the connection between the pixels present. Following the scoring function, $f$ maps input to the correct class labels as follows. Here, $x$ is filter matrix, $b$ is the bias vector and $W$ is the weight matrix.

$$\boldsymbol{f}(\boldsymbol{x}_i, \boldsymbol{W}, \boldsymbol{b}) = \boldsymbol{W} * \boldsymbol{x}_i + \boldsymbol{b}$$

CNN layers learn the following feature from the given image sample, (1) edge part object present in image. (2) In subsequent layer, CNN learns the shape of the object present an image with reference to edge. (3) In next layer, it learns more high-level features such as facial structure, nose shape and eye shape.

Figure 7.2 depicts the working of convolution operation, which indicates its mathematical operation as product and summation between tiny matrix and larger matrix (image).

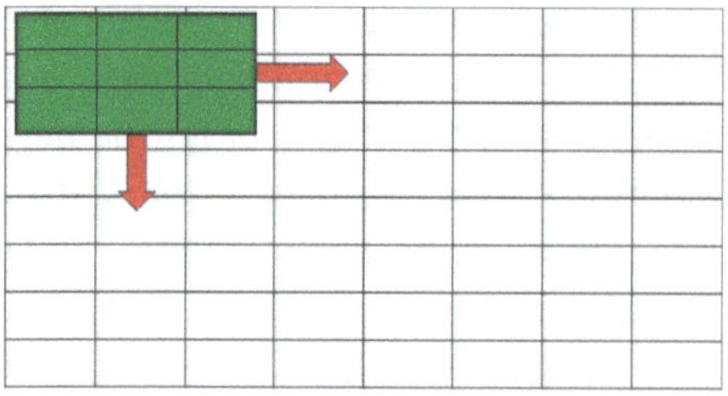

**Figure 7.2:** A small, tiny matrix highlighted with green color slides across, from left-to-right and top-to-bottom, of a larger matrix (image). At each pixel in the input image, the neighborhood of the image is convolution operation is performed using tiny matrix and the output as feature set.

## 7.3.5 Pooling layers

The pooling layer minimizes the available pixel value. Spatial pooling called as down-sampling which reduces the dimensions of each feature map but retains important information. Spatial pooling can be of different types – max pooling, average pooling and sum pooling. Max pooling is one of the most common approaches used. The output of the final pool layer is (512*7*7) which, when flattened yields a feature vector of length of 2,5088. Therefore, when using VGG16 for feature extraction we will set dimensions = (N*2,5088) where *N* be the total number of images in our data sets. We can perform any of the following pooling operations as per application or problem requirement, such as: (1) *Max pooling*: Extract max value using given tiny kernel dimension. (2) *Min pooling*: Extract max value using given tiny kernel dimension. (3) *Mean pooling*: Extract average value using given tiny kernel dimension and elements of a respective large matrix. Figure 7.3 indicates the working of pooling operation with max pooling mathematical function, here we have used Rectified linear unit as activation function formulation of ReLU activation function is max pooling operation, such as max (−249, 0) = 0, max(−91, 0) = 0, max(250, 0) = 250.

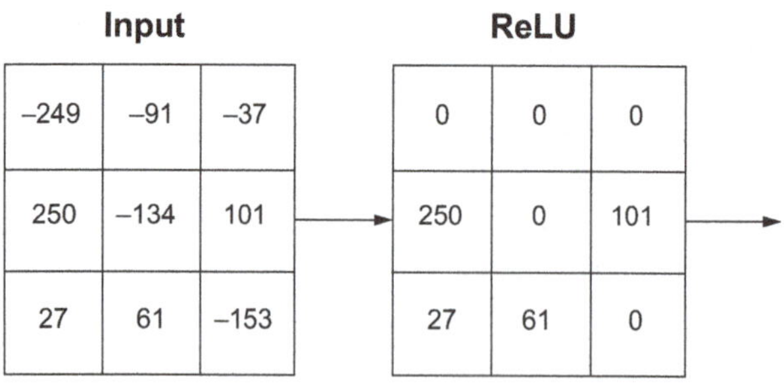


**Figure 7.3:** Here is example of max pooling operation performed.

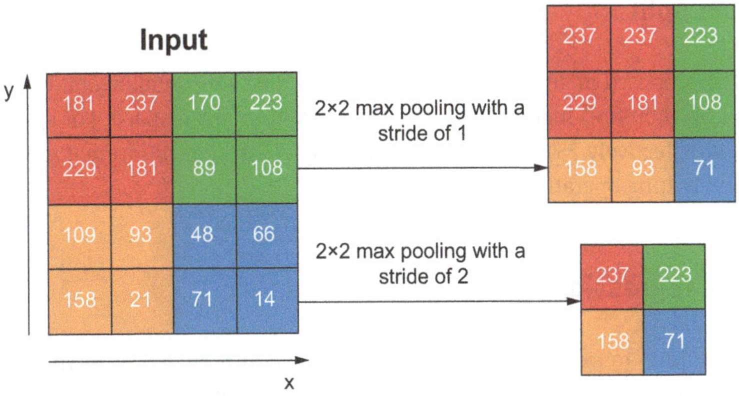


**Figure 7.4:** Max pooling operation with stride 1 and 2.

## 7.3.6 Flatten layer

Layer wise the feature is extracted, all extracted features added into flatten layer as single vector. In flatten layer, matrix is transformed into a single vector whose output is then added to the full dense connection layer for prediction. Figure 7.5 depicts the 3 X 3 matrix/image flatten into 1D vectors.

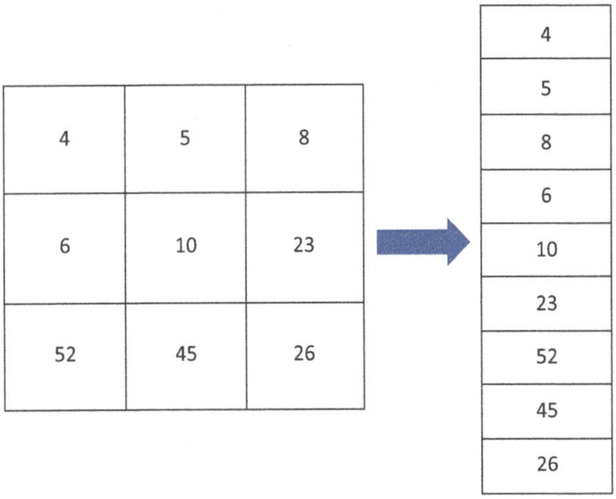


**Figure 7.5:** Working of flatten layer.

### 7.3.7 Data set split

We evaluated the proposed methods on our data set. Our data set contains total 4,172 (2,245 fire and 1,927 non-fire) images that are enough to fit in given space. For splitting of entire data effectively 75% and 25% we have used HDF5 file [14] format to reduce reading and writing operation overhead.

### 7.3.8 Logistic regression

Classification algorithm to separate out two classes based on provided observation. The method we adopted is binary classification and multilinear functions. The probability of different classes is computed using the sigmoid function. We select a threshold value values greater than that value we will consider in class 1 and below which in class 2.

$$\boldsymbol{pr} \geq \mathbf{0.5}, \boldsymbol{class} = \mathbf{1}$$

$$\boldsymbol{pr} < \mathbf{0.5}, \boldsymbol{class} = \mathbf{2}$$

We deployed our classification model to the JETSON TX2 kit [15] and interfaced the kit with fire alarms. Our system overcomes the problem associated with the sensor-based fire-detection system. Overview of the complete system: (a) camera, (b) NVIDIA Jetson Tx2 Developer Kit, (c) fire alarm and (d) end-users for receiving fire alerts in both visual and textual form. For end-to-end detection, we used Twilio [16] for sending SMS/MMS to the end-users which is one of the messaging services. Figure 7.3 shows an overview of the complete edge device enabled fire detection unit capable of distinguishing between fire and smoke. Microcontroller is installed on JETSON TX-2 Kit will trigger the sound alarms if the fire is detected from the camera on JETSON Kit.

## 7.4 Result

We evaluated the proposed methods on our data set. We have a total of 1,043 (561 fire and 482 non-fire) images for testing. This data set includes the pictures often found in real-world fire strokes that are commonly detected with cameras attached to edge devices such as JETSON KIT. To evaluate and show the performance of our data set, we trained our model in the most realistic situations on a deep learning server with configuration 128GB ram, V100 Nvidia [17] architecture. The proposed model we executed using logistic regression classifier on the features extracted via VGG16 [7] network on our data set and after testing it on 25% split of our data set and the performance obtained in the form of a confusion matrix is presented in Figures 7.4 and 7.5, which depicts all performance measurement parameters such as accuracy in percentage,

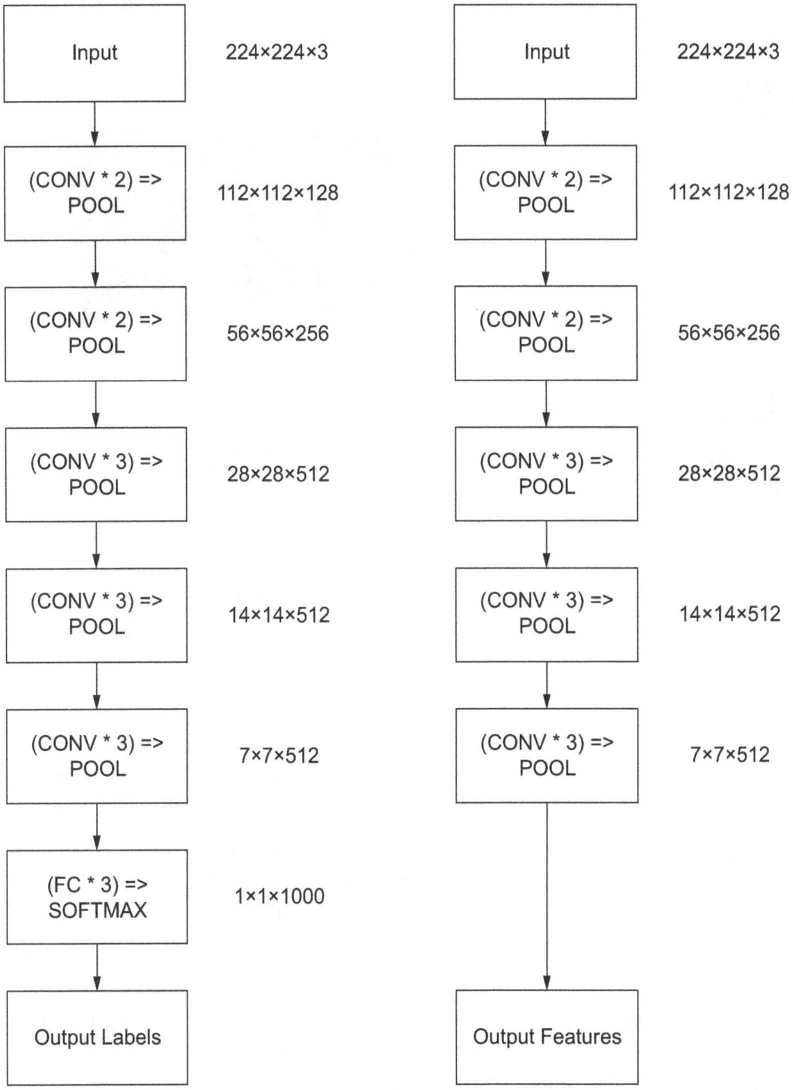


**Figure 7.6: Left**: The original VGG16 CNN architecture that outputs between 0 and 1. **Right:** Eliminating the fully connected layer from base VGG16 and returning the output of final pool layer instead of the same. Given output served as features.

precision vale, recall vale and finally F1 score is computed as the output. We have depicted a confusion matrix showing the false and true positives and false and true negatives for fire detection. Table indicated in Figures 7.4 and 7.5 have used metrics of confusion matrix such as accuracy, precision, recall and F1-score in order to make sure the complete reliability and analysis. As can be seen from Figure 7.5, the output values obtained for each metric are pretty much up to the mark on both data sets. In

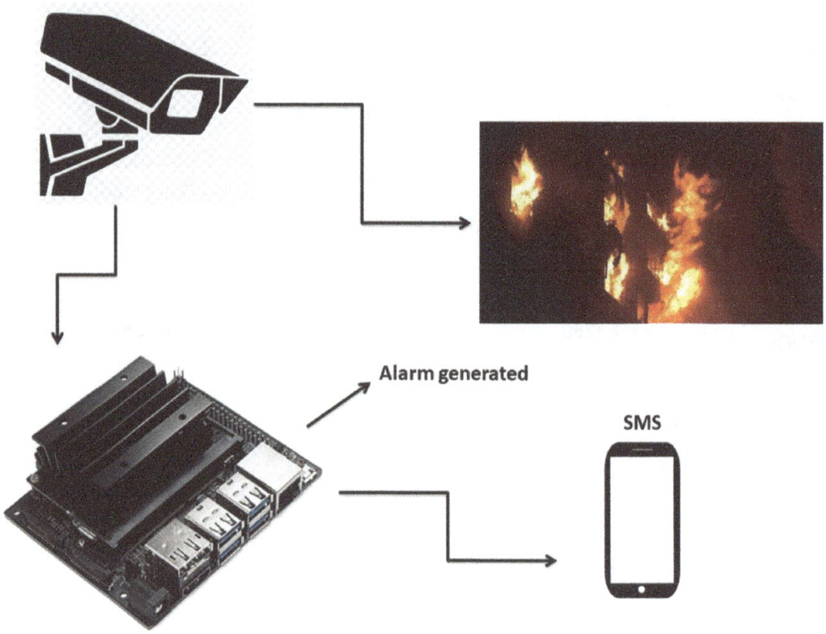


**Figure 7.7:** Overview of complete unit.

order to further make sure the reliability of the given model, we have shown the training process using the training graphical curve. Figure 7.6 depicts the training and validation loss view computed while in the training process, whereas the training and validation accuracy flow computed during the given process. As it can be seen from Figure 7.6, the trained model is able to generalize significantly on the validation set. Also, note that we have used a 70% and 30% split between the training and validation set. Figure 7.7 represents the complete unit, Figure 7.8 represents the confusion matrix used for the experiment. Figure 7.9 represents the text performance. Figure 7.10 depicts the graph for accuracy and loss model. Moreover, there is no overlap between the two respective sets. The advantage of the network is that it is lightweight so that it can run- on low-cost hardware such as Raspberry Pi 3B [18].

| | | Actual | |
|---|---|---|---|
| | | Fire | Non fire |
| Predicted | Fire | 550 | 10 |
| | Non fire | 11 | 472 |

**Figure 7.8:** Fire detection – confusion matrix for experiment.

| Metrices: | Precision | recall | Fl-score |
|---|---|---|---|
| Fire | 0.98 | 0.98 | 0.98 |
| Non-fire | 0.97 | 0.98 | 0.98 |
| Accuracy | 98% | 98% | 98% |

**Figure 7.9:** Test performance.

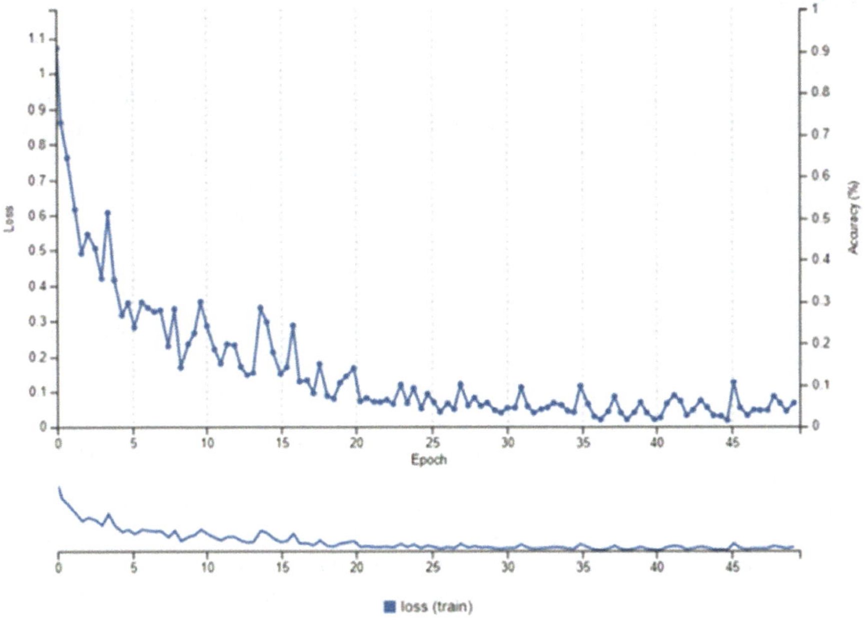


**Figure 7.10:** Training curves for model accuracy and model loss.

## 7.5 Conclusion

Fire is the most hazardous abnormal action, as failing to handle it at an initial level can stop in huge disasters, leading to human, ecological and economic losses. The proposed work is motivated by the high accuracy over convolution neural network. In the proposed work we worked on cost-effective and lightweight CNN based on the AlexNet [19] architecture for fire detection. Our approach can identify the fire in the object under surveillance. Our aim is to develop an Internet of things (IoT) based fire-detection system using CNN based deep learning model. It is also a big replacement for the currently sensor-based system and can reduce the problems associated with false and delayed triggering with such systems. Furthermore, the proposed CNN-based system balances both accuracy of the model and the space required on the edge device. In view of the CNN model's reasonable accuracy, precision, recall and F-measure for fire detection and the system can avoid false alarms and be helpful to disaster management teams in controlling fire disasters in a timely manner, thus avoiding huge losses. Furthermore, fire-detection systems can intelligently observe and authenticate the video stream and initiate appropriate action, in an autonomous way.

## References


[1] Çelik, T., Özkaramanlı, H., Demirel, H. (2007). Fire and smoke detection without sensors: Image processing based approach. 2007 15th European Signal Processing Conference. IEEE.

[2] Ruminski, M., et al. (2008). Use of multiple satellite sensors in NOAA's operational near real-time fire and smoke detection and characterization program. Remote Sensing of Fire: Science and Application, Vol. 7089. International Society for Optics and Photonics.

[3] Maksimović, M., Milošević, V. (2016). Evaluating the optimal sensor placement for smoke detection. Yugoslav Journal of Operations Research, 26(1), 33–50.

[4] Jadon, A., et al. (2019). FireNet: A specialized lightweight fire and smoke detection model for real-time IoT applications. arXiv Preprint arXiv:1905.11922.

[5] Seebamrungsat, J., Praising, S., Riyamongkol, P. (2014). Fire detection in the buildings using image processing. 2014 Third ICT International Student Project Conference (ICT-ISPC). IEEE.

[6] Muhammad, K., et al. (2018). Convolutional neural networks based fire detection in surveillance videos. IEEE Access, 6, 18174–18183.

[7] Simonyan, K., Zisserman, A. (2014). Very deep convolutional networks for large-scale image recognition. arXiv Preprint arXiv:1409.1556.

[8] Kamble, K., Bhansali, A., Satalgaonkar, P., Alagundgi, S. (2019). Counterfeit currency detection using deep convolutional neural network, 2019 IEEE Pune Section International Conference (PuneCon), 1–4, 10.1109/PuneCon46936.2019.9105683.

[9] Uma Maheswari, V., Aluvalu, R., Chennam, K.K. (2021). Application of machine learning algorithms for facial expression analysis. Machine Learning for Sustainable Development, 9, 77.

[10] Maheswari, V.U., Varaprasad, G., Viswanadha Raju, S. (2021). Local directional maximum edge patterns for facial expression recognition. Journal of Ambient Intelligence and Humanized Computing, 12(5), 4775–4783.

[11] Maheswari, V.U., Prasad, G.V., Viswanadha Raju, S. (2021). Facial expression analysis using local directional stigma mean patterns and convolutional neural networks. International Journal of Knowledge-based and Intelligent Engineering Systems, 25(1), 119–128.

[12] Maheswari, V.U., Viswanadha Raju, S., Sridhar Reddy, K. (2019). Local directional weighted threshold patterns (LDWTP) for facial expression recognition. 2019 Fifth International Conference on Image Information Processing (ICIIP). IEEE.

[13] Zhang, Q., et al. (2016). Deep convolutional neural networks for forest fire detection. Proceedings of the 2016 international forum on management, education and information technology application. Atlantis Press.

[14] Folk, M., et al. (2011). An overview of the HDF5 technology suite and its applications. Proceedings of the EDBT/ICDT 2011 Workshop on Array Databases.

[15] Kodavade, P., Bhandigare, S., Kadam, A., Redekar, N., Kamble, K.P. (2021). Lie detection using thermal imaging feature extraction from periorbital tissue and cutaneous muscle. In: Saini, H.S., Sayal, R., Govardhan, A., Buyya, R., eds. Innovations in Computer Science and

Engineering. Lecture Notes in Networks and Systems, vol. 171. Singapore, Springer. https://doi.org/10.1007/978-981-33-4543-0/68.
[16] Venkatesan, S., et al. (2017). Design and implementation of an automated security system using Twilio messaging service. 2017 International Conference on Smart Cities, Automation & Intelligent Computing Systems (ICON-SONICS). IEEE.
[17] Yang, C. (2020). 8 Steps to 3.7 TFLOP/s on NVIDIA V100 GPU: Roofline Analysis and Other Tricks. arXiv Preprint arXiv:2008.11326.
[18] Wazir, S., et al. (2019). Performance comparison of MPICH and MPI4py on Raspberry Pi-3B Beowulf Cluster. arXiv Preprint arXiv:1911.03709.
[19] Iandola, F.N., et al. (2016). SqueezeNet: AlexNet-level accuracy with 50x fewer parameters and< 0.5 MB model size. arXiv Preprint arXiv:1602.07360.

Vaibhav Tasgaonkar, Ujwala Kshirsagar, Soham Hudnurkar, Pravin Sridhar

# 8 Face mask detection and crowd size estimation using IoT platform

**Abstract:** Due to the rise in coronavirus disease (CoVID-19) cases, it has been a significant concern for governments all over the world to control the spread of the virus. We aim to aid in controlling the spread of the CoVID-19 virus. The face mask detection and crowd counting model with the Internet of things (IoT) platform (Thingspeak) can be deployed in situations where there is a lack of workforce, and additional assistance is required. Using a camera, an area can be remotely monitored, restricting contact between people and the frontline workers, promoting their safety. The research conducted to date about this topic is solely on detecting facemask or doing crowd counting, respectively. We included both face mask detection and crowd counting, connecting both to the IoT platform, enabling us to access the results online. In previous research work, traditional/wired cameras were used. But in our research work, we have used the concept of a Wi-Fi camera, i.e., wireless camera, by which the camera can be positioned in places where the placement of wires is difficult. For face mask detection, we have created two data sets which include images of faces with masks and without a mask, and this data set is used to train the model. For training and testing the model, we have included TensorFlow, Keras and SKLearn libraries.

We have included different deep learning layers in training the model like Dense, Dropout and AveragePooling2D. In the test model, python code is connected to a wireless camera, which can capture live video feed and detect whether the person is wearing a mask or not. We have used OpenCV Histogram of oriented gradients (HOG) descriptor in crowd counting, often used to extract features from image data. Using an OpenCV HOG Descriptor, we can count the number of people present in a crowd. Face mask and crowd count is then sent to the IoT platform. Whenever the person is found wearing a mask, a green box appears around the face, and whenever the person is not wearing a mask, a red box appears around the face. When lots of people are detected by a crowd counting model, a green color box enclosing the entire body appears. This study helps in getting face mask detection and crowd size estimation on a single IoT platform, which helps monitor the data with great ease.

**Vaibhav Tasgaonkar,** Symbiosis Institute of Technology, Pune, Maharashtra, India, e-mail: tasgaonkarvaibhav@gmail.com
**Ujwala Kshirsagar,** Symbiosis Institute of Technology, Pune, Maharashtra, India, e-mail: kshirsagarujwala9@gmail.com
**Soham Hudnurkar, Pravin Sridhar,** Symbiosis Institute of Technology, Pune, Maharashtra, India

https://doi.org/10.1515/9783110756722-008

**Keywords:** face mask detection, crowd size estimation, CoVID-19 pandemic, IoT platform, wireless camera

## 8.1 Literature review

Deep learning is widely used to find any feature, given extensive data, using the method of nonlinear transforms. Convolutional neural network or CNN is a widely used technique in the field of image classification. Keras is popular for this application and is based on Python. Several batch normalizations and convolution layers aid in improving performance. The image classification performance was compared with other methods in machine learning, such as K-nearest neighbors and XGBoost, using the same data set. Keras helped in getting used to the methods of deep learning. Deep learning concepts were applied for parameter estimation and advanced techniques to improve the model [1]. Image processing using histogram of oriented gradients or HOG descriptor is used with OpenCV library. HOG was used with Raspberry Pi along with RaspiCam in various applications. Top view image samples were used for the predefined models with massive motion variations to identify the movement of humans through any entrance [2]. MobileNet is a lightweight DNN or deep neural network. Dense blocks from dense nets were introduced to MobileNet. In these models, the dense block maps form MobileNet.

A low growth rate reduces the number of parameters and computation costs. The proposed model can reduce the parameters using the growth rate of the hyperparameter. This also increases accuracy for classification [3]. To detect face masks, Keras, OpenCV and TensorFlow were installed. As a face detector, the SSDMNV2 method employs the single-shot multibox detector. The DNN module from OpenCV, which incorporates a "single-shot multibox detector" (SSD) object detection model, was utilized to improve mask detection accuracy without being overly resource intensive. The suggested model SSDMNV2 has completed both the training and development of the picture data set, which was split into two categories: mask-wearing individuals and without mask individuals [4]. Crowd counting has found importance in practical use. Deep learning has now become the most effective method for crowd counting. It helps in tackling various irregularities like perspective distortion and scale transformation. This makes it a field of interest for several researchers, trying to make it efficient and faster and constantly becoming innovative [5]. The OpenCV deep neural networks method employed in this model yielded positive results. For video streaming, the IP Webcam app is used. The video is transmitted through Wi-Fi via the app. For the investigation purpose, a local area network was created, using Wi-Fi to optimize its performance. After setting up the router with correct specifications, specific protocols were executed to know the disadvantages of video streaming. After the completion of the test, the wireless local area network had a delay in the video. Several client devices can obtain the data simultaneously, which is similar to broadcasting to multiple devices. A limited number of client support should limit the pool of IP addresses to the Wi-Fi router. By limiting the users, the quality of the service can be kept high. The bandwidth can be primarily diverted toward controlling video streams by keeping the network explicitly for streaming purposes. This makes the transmission less delayed and allows more clients to access multiple devices [6] simultaneously. IoT has immense capabilities and can be used in various disciplines, for example, smart homes, smart cities and agriculture. IoT creates a platform where everything can be controlled from the same application. In the last decade, ordinary homes had various electronic appliances that could be controlled manually or remotely. In today's era, most of these appliances can be controlled using IoT. Systems are automated with smart homes while keeping low power consumption in mind. ThingSpeak platform provides security, data management and data visualization. By the use of wireless networks, wiring and switching costs are reduced. All the data is sent to the cloud and real-time statistics are provided on the web page or application [7].

## 8.2 Introduction

Spanish flu was the biggest pandemic in the twentieth century, which proved to be highly deadly as there were not enough technological advances or communication media to prevent the spread and come up with a cure or a vaccine. In the case of CoVID-19, the situation is different as compared to the previous century. Technical advancements can be used to prevent the increase in the number of cases. Severe acute respiratory syndrome coronavirus, also known as SARS-CoV-2 (CoVID-19), is a respiratory disease first discovered in December 2019 in Wuhan, China. Some of the early cases were first linked to a seafood market. The cases then rapidly spread in China and then to the rest of the world. The World Health Organization (WHO) had announced it as an international concern on January 30, 2020. On March 10, 2020, it was declared a global pandemic [8]. The primary mode of spread of the virus is through tiny droplets/particles suspended in the air. The virus is released through mucus and saliva droplets every time an infected person sneezes. These particles could be inhaled by any unaffected person, which will lead to an increase in the number of CoVID cases. The governments have imposed specific rules to prevent the spread of the virus and made it mandatory to wear a mask. Due to the widespread use of mass media and highly developed technologies, we are now able to create awareness and develop vaccines with the combined efforts of scientists globally. According to our observation, some of the rules were not being followed by the people. There were still large gatherings (like marriages and parties) held, and no one to supervise them.

Moreover, some people still commute without a mask. This could cause another wave of pandemic. As most of the population of most countries is not fully vaccinated yet, there is still a need to implement preventive measures. One such preventive measure is automated detection of face masks through computer vision to ensure mask wearing. Face mask detection helps to detect whether the person is wearing a mask or not. Two different data sets are created for the face mask detection process, one for with mask and without a mask, respectively. For training the model and testing the model, two different data sets are used. The whole algorithm is divided into two sections. One section of the algorithm is dedicated to training the model, and the other to testing the model. After running the whole algorithm, a video frame appears, where the person is detected either with a mask or without a mask. This will help make it easier for the law enforcers to ensure that people are wearing a mask all the time. In our research work, we have also managed to use crowd size estimation. Crowd size estimation is used to detect the number of people present in the vision of the camera. For the crowd size estimation, the OpenCV histogram of oriented gradients (HOG) descriptor is used. There are two approaches for crowd size estimation, texture-based and model-based. The algorithms used are support vector machine (SVM), and region-specific HOG are used for histogram of oriented gradients. The F-measure, recall and precision for the two algorithms are measured. The experiment took place in an indoor environment with surveillance cameras for low and high crowd levels. The HOG with SVM has a high precision value but a lower TPR [9]. Research has been conducted to date in the domain of face mask detection and crowd size estimation, but our research manages both these processes simultaneously with the help of the IoT (Internet of Things). IoT platforms are used to link multiple devices and applications to a single server, which then can be accessed by a single device, which is linked to the same server. They are also used to collect and analyze the data and store the information on the respective servers. In this project, the IoT platform used is ThingSpeak by MATLAB. Each IoT platform has its application programming interface (API) key. These codes are used to validate the server and to make our server secure. Using the IoT platform, we can obtain the results for face mask detection and crowd counting on the same platform. In the past, researchers have used their device cameras or cameras, which usually need to be connected using a cable. But we have used a wireless camera, which is linked to the IoT platform. The video obtained from this wireless camera is directly shown on the IoT platform, along with the detection of masks and crowd counting output. This enables us to monitor the area without the need for extensive human supervision.

## 8.3 Methodology

This section provides a detailed overview of face mask detection and crowd size estimation using the IoT platform. For both face mask detection and crowd size estimation, we have used a wireless camera. The device used for our research work has the following specifications: CPU – Intel Core i5 8300 H CPU @ 2.30 GHz, GPU – NVIDIA GeForce GTX 1650, Memory – 8.0 GB, Wi-Fi – 2.5 G/ 5 G, Camera – 32/ 64 MP (mobile camera).

### 8.3.1 Face mask detection

Mask wearing is among the most effective solutions for preventing CoVID-19 spread, and most people are urged to wear a mask in public places. This creates the need to monitor people in public places through techniques of face mask detection. Deep learning is used to design most of the latest and advanced techniques of face mask detection. Two object detection models used to achieve this task are R-CNN and YOLOv3. A data set containing two categories of images, people with masks and people without a mask, was used to train the above models. If a person is wearing a mask, a green box will be drawn bounding their face. Otherwise, a red box will be drawn. A daily ratio of people wearing and not wearing a mask is recorded [10]. Figure 8.1 shows a flowchart for face mask detection. For the wireless camera, we have used a mobile phone which uses the IP Webcam app. The IP Webcam app is then connected to python code. For an AI model, a data set is required. We have created a data set using Google images and data sets created by previous researchers for our face mask detection model. We have integrated both face mask detection and crowd size estimation in a single program. For face mask detection, we have used Keras, TensorFlow, MobileNetV2 and OpenCV python libraries.

Tensorflow, being a machine learning tool, operates in various heterogeneous environments at a large scale. Dataflow graphs are used to represent the computation, operations and shared state. Dataflow graph nodes are mapped across several machines and computational devices. Several applications have TensorFlow support. Tensorflow is also used by various Google services [11]. Deep learning is a machine learning method that uses a nonlinear transformation to find features from large data. Keras is one of the tools for deep learning. It is an API for several neural networks, which are generally written in Python [12]. In face mask detection, we have first trained the model by importing a data set. The data set consists of images with masks and without a mask. By keeping Keras running on top of the machine learning platform TensorFlow, we have imported ImageDataGenerator, MobileNetV2, AveragePooling2D and Dropout. Later, imported images from data sets are converted to arrays, and after that, preprocessing is done. For data augmentation, a training image generator is constructed. The MobileNetV2 network is then loaded.

During the testing phase of the model, the dimensions of the frame are initialized. Later, the list of faces, their corresponding locations (within the photo), and a list of predictions from the face mask network is initialized. Some specific conditions are mentioned to the face mask network to start the detection process if faces are detected. PrototxtPath, weightsPath are loaded from the local disc. Labels are given as "Mask" for mask detected and "No Mask" for mask not detected to the output. To keep the video frame open for a specific interval of time, a while loop is used. Before the loop, "counter = 0" is initialized. For the loop, "counter < 250" is given as a condition to the while loop (by assigning a specific value to the counter, we can keep the video frame open for a specific interval of time). We have kept "counter = counter + 1" in the while loop to increment the counter value. When the condition becomes false, while loop ends.

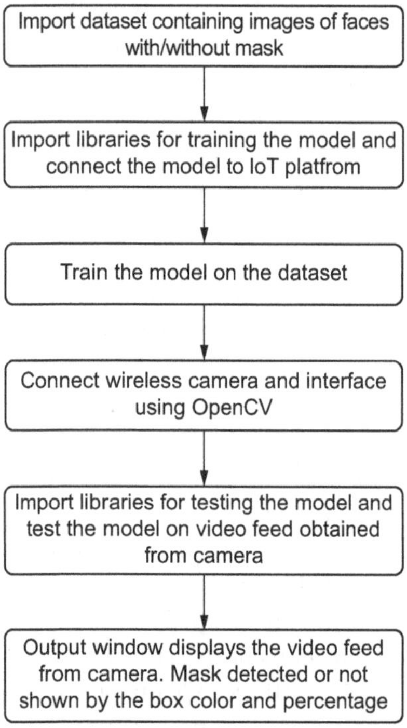


**Figure 8.1:** Face mask detection.

## 8.3.2 Crowd size estimation

Figure 8.2 shows a flowchart for crowd size detection/counting. We have used OpenCV HOG (histogram of oriented gradients) Descriptor to count the number of individuals in any given area for crowd size estimation. During the process of crowd size estimation, rectangular boxes appear around the human body. These boxes are defined with parameters x, y, w and h. For face mask detection and crowd size estimation, two different windows appear after a specific time interval, showing live video feeds of both detections. To keep the video frame open for a specific interval of time, a while loop is used. Before the loop, "counter = 0" is initialized. For the loop, "counter < 150" is given as a condition to the while loop (by assigning a specific value to the counter, we can keep the video frame open for a specific interval of time). We have kept "counter = counter + 1" in the while loop to increment the counter value. When the condition becomes false, while loop ends. Both these detections are connected to the IoT platform "ThingSpeak." The channel_id, WRITE_API and BASE_URL are mentioned to connect with the IoT platform. The number of faces detected in the camera video feed is counted and later updated on the IoT platform during face mask detection. The same process is followed for crowd size estimation too, and whenever the camera video feed detects a person, the count is updated on the IoT platform.

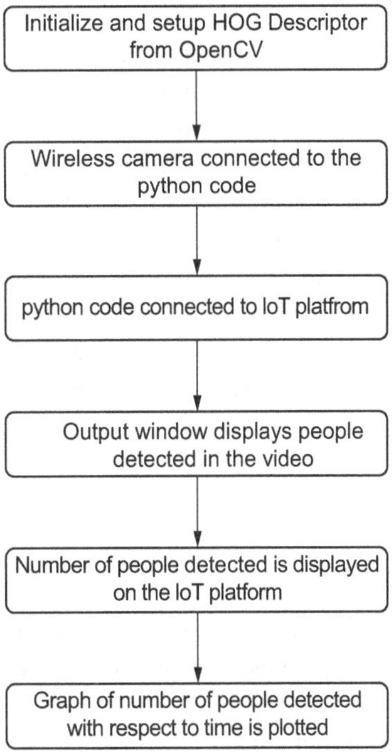


**Figure 8.2:** Crowd counting.

## 8.3.3 IoT

IoT platform called ThingSpeak is used to store and analyze data and visualize it on the cloud. The code is written in Python but can be written in any high-level language. Different formulae are used to get the numerical values from different sensors, which can then be visualized on ThingSpeak. One such example is an air quality monitoring system, which can be used in an indoor environment. Various sensors are used for this, which collect data from different environments. This data is then stored in the cloud to analyze any environmental condition [13]. ThingSpeak is the IoT platform that is being used in this project. Using this IoT platform, we get a detailed analysis in graphical representation and a counter. For a specific channel, a specific channel ID is given by the IoT platform. To use this IoT platform, a new channel is created, in which an application programming interface key (API key) is generated. An API key is used to prevent misuse of the channel. This key is unique to each channel. We can also generate a new API key for an already existing channel. API key and Base URL are both used to connect to the IoT platform.

### 8.3.4 IP Webcam

IP Webcam can be used as a wireless camera or a portable camera for some specific applications. One such application can be a real-time pattern recognition algorithm. An IP webcam app can use a real-time pattern recognition algorithm for additional operations through two six-sided dice [14]. IP Webcam is a mobile application that can be used as a wireless camera. It is used to replace traditional wired cameras with a wireless camera, which helps in mobility. IP Webcam can be connected to python code by using a server link. The server link can be accessed on any web browser, through which we can look at the video feed that is captured live by the mobile phone while using the application. When the IP Webcam app is accessed on a web browser, various features can be adjusted like zoom, stream quality, sensitivity and so on.

## 8.4 Results

For face mask detection, if a person is detected with a mask, a green color box appears around the human face (accuracy in percentage above the green box), as shown in Figure 8.3.

A red color box appears if the person is found without a mask (accuracy in percentage above the red box), as shown in Figure 8.4.

Faces counted during face mask detection are shown in Figure 8.5, with a detailed graph of face mask versus date on the IoT platform. The graph gets updated automatically as the count changes.

During crowd counting, when the person is detected, a box appears around his/her body, as shown in Figure 8.6.

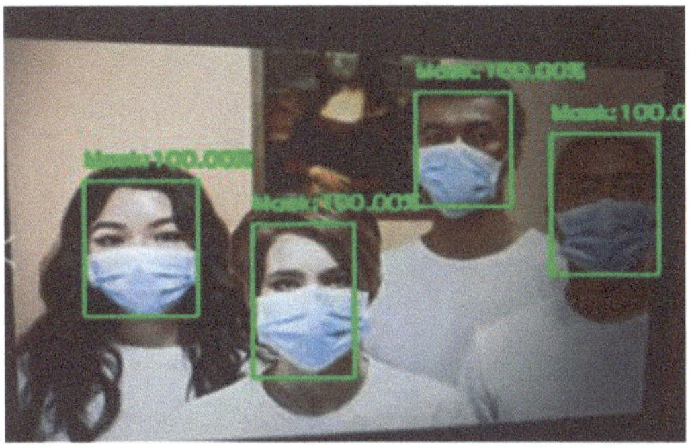

**Figure 8.3:** Face mask detection (people wearing mask).
Source: people-wearing-face-mask-for-protection-3957986-1600x1071.jpg (1600 × 1071) (thewire.in)

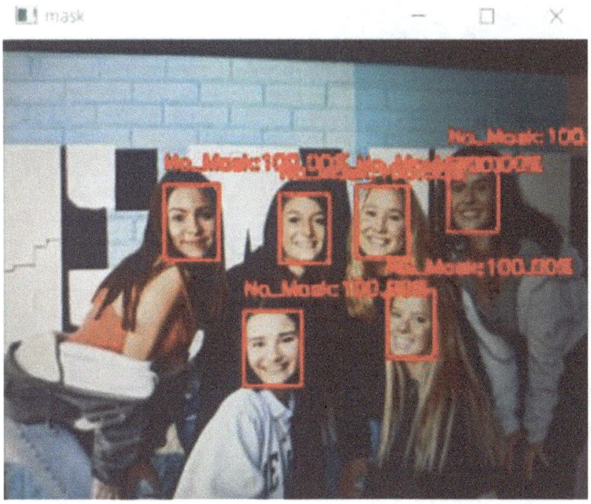

**Figure 8.4:** Face mask detection (people not wearing mask).
Source: https://images.unsplash.com/photo-1615332859295-aa9dc9f01fbc?ixid=MnwxMjA3fDB8MHxzZWFyY2h8OXx8Z3JvdXAlMjBvZiUyMGdpcmxzfGVufDB8fDB8fA%3D%3D&ixlib=rb-1.2.1&auto=format&fit=crop&w=500&q=60

People counted during crowd counting are shown in Figure 8.7, with a detailed graph of people versus date on the IoT platform. The crowd counting graph gets updated automatically as and when the count changes.

Figure 8.5: Output on IoT platform (face mask detection).

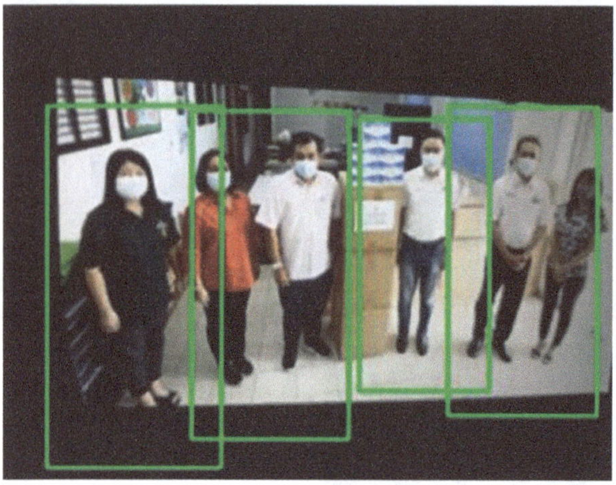

Figure 8.6: Crowd counting.
Source: C6F6CF7D-BC66-4F89-93E0-E6AE1F468D5C.jpeg.webp (1068 × 584) (borneotoday.net)

## 8.5 Conclusion

In this paper, computer vision applications are used to reduce human effort in the ongoing CoVID-19 pandemic. Face mask detection and crowd size estimation were used for this purpose and are achieved simultaneously in a single program. An overview of the results of the program can be visualized on the IoT platform (ThingSpeak). This can be used to monitor any area to see if the citizens are following the protocols/preventive measures issued by the government, such as wearing a mask and maintaining social distancing. For the purpose of monitoring an area, the

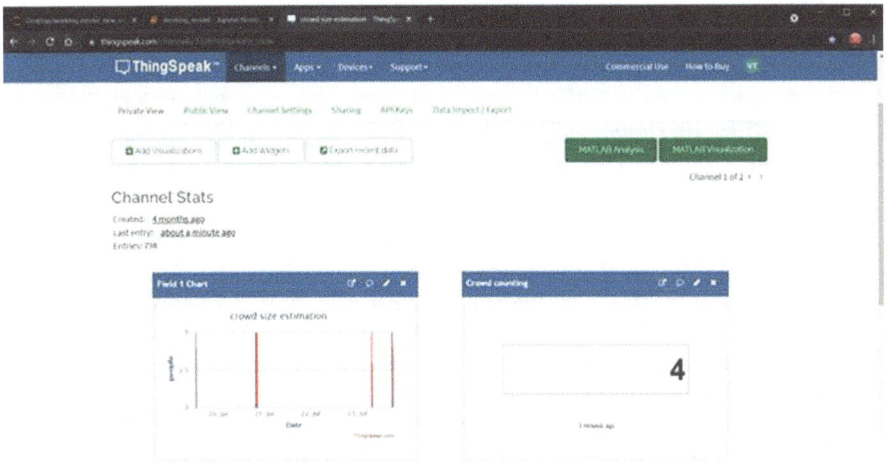


**Figure 8.7:** IoT platform output (crowd size estimation).

concept of a wireless camera has been used. This allows the placement of the camera with minimum effort, hence aiding in easy surveillance.

# References


[1] Lee, H.J. (2019 Nov). Introduction to convolutional neural network using Keras; an understanding from a statistician. Web of Science, 26. 10.29220.

[2] Shubham, M., Balaji, S., Sanyam, J., Kajal, C., Prabha, R.D. (2017). Human detector and counter using raspberry Pi microcontroller. Conference on Innovations in Power and Advanced Computing Technologies (i-PACT). Vellore, India: IEEE.

[3] Wang, W., Yutao, L., Ting, Z., Xin, W., Jieyu, Y. (2020). A novel image classification approach via dense-mobilenet models. Mobile Information Systems (Hindawi, Ltd Adam House, 3rd Flr, 1 Fitzroy Sq, London W1T 5HF, England), 2020. 10.1155/2020/7602384.

[4] Preeti, N., Rachna, J., Agam, M., Rohan, A., Piyush, K., Jude, H. (2021). SSDMNV2: A real time DNN-based face mask detection system using single shot multibox detector and MobileNetV2. Web of Science (Elsevierradarweg 29, 1043 NX Amsterdam, Netherlands), 66. 10.1016/j.scs.2020.102692.

[5] Zhi, L., Yong, L., Xipeng, W. (2020). Current researches and trends of crowd counting in the field of deep learning. In: Xu, B., Mou, K., eds. Proceedings of 2020 Ieee 4TH Information Technology, Networking, Electronic and Automation Control Conference (ITNEC 2020). IEEE, 1326–1329. 10.1109/itnec48623.2020.90849.

[6] Molina, M.G., Salgado, W.G., Hidalgo, J.I. (2017). Design of a local area network optimized for video streaming. Latin American and Caribbean Consortium of Engineering Institutions. doi:10.18687/LACCEI2017.1.1.39.

[7] Ali, M., Zarsha, N., Waqar, A., Javed Khadij, A., Maria, T., Muhammad, H., Aamir, H. (2020). An IoT based approach for efficient home automation with thingspeak. International Journal

of Advanced Computer Science and Applications (Science & Information Sai Organization Ltd), 11(6), 118–124.
[8] Jennifer, L., Vipin, V., Tariq, C. (2020). COVID-19 epidemiology, pathophysiology, transmission, symptoms. Critical Care Nursing Quarterly (Lippincott Williams & Wilkinstwo Commerce Sq, 2001 Market St, Philadelphia, PA 19103 USA), 43(4), 338–342. 10.1097/CNQ.0000000000000319.
[9] Joel, I., Macario, C. (2018). International joint conference on computer science and software engineering. In: Sawangphol, W., Mitrpanont, J., eds. 2018 15th International Joint Conference on Computer Science and Software Engineering (JCSSE). New York, USA, IEEE345 E 47TH ST, New York, NY 10017 USA, 163–167. 10.1109/JCSSE.2018.8457384.
[10] Sunil, S., Umang, A., Munish, K., Krishan, K., Sachdeva, M. (2021). Face mask detection using YOLOv3 and faster R-CNN models: COVID-19 environment. Multimedia Tools and Applications (Springer, Van Godewijckstraat 30, 3311 GZ Dordrecht, Netherlands), 80(13), 19753–19768. 10.1007/s11042-021-10711-8.
[11] USENIX Assoc. (2016). TensorFlow: A system for large-scale machine learning. Proceedings of OSDI'16: 12th Usenix Symposium On Operating Systems Design And Implementation. USENIX ASSOC. 265–283.
[12] Hagyeong, L., Jongwoo, S. (2019). Introduction to convolutional neural network using Keras; an understanding from a statistician. Communications For Statistical Applications and Methods (Korean Statistical Soc), 26(6), 591–610. 10.29220/CSAM.2019.26.6.591.
[13] Amita, B., Subhashini, J., Pasayat, A.K. (2019). Air quality monitoring system for indoor environments using IoT. Govindarajan, A., Sambath, P., Sangeetha, S., eds. AIP Conference Proceedings. Chennai: AMER INST PHYSICS. doi:10.1063/1.5112365.
[14] Suarez, O.J., Dfaz, N.H., Garcia, A.P. (2020). A real-time pattern recognition module via Matlab-Arduino interface. Proceedings of the LACCEI International Multi-conference for Engineering, Education and Technology. Guadalajara, Mexico: Latin American and Caribbean Consortium of Engineering Institutions. doi:10.18687/LACCEI2020.1.1.646.

J. Srinivas, Rajesham Gajula

# 9 Generative adversary networks novel trends in image processing

**Abstract:** Majority of researches are taking place in the image-processing dimension as the demand for image processing applications is widely increasing. Generative adversarial network is a state-of-the-art image-processing technology in recent improvement and development in the field of image processing. Generative adversarial network is a neural network architecture which is used as a generative model. Generative adversarial network is popularly known as GANs. GANs can be designed for generating scores of images from the data sets, text to image conversion, face aging, animations, storytelling and so on. GANs can also be used in the field of signal communication to attain optimization in combinational design. Based upon their design and applications, a number of variants of GANs have come into existence. The various classes of GANs include semantic enhancement GANs, resolution enhancement GANs, diversity enhancement GANs and motion enhancement GANs. Popular examples of GANS are deep convolution generative adversarial network (DCGAN), conditional generative adversarial network (cGAN), stack GAN, (discover cross-domain relations with generative adversarial networks (Disco GAN), InfoGAN Age-cGAN and so on. Basically, any GAN framework works on the adversarial principle of evaluating generative models. Models named generator and discriminator will be trained simultaneously. The role of generator is to synthesize a series of outputs and the role of discriminator is to differentiate the currently generated output from the previous outputs. This generation and discrimination process iterates until the discriminator declares the current output to be 100% different from the previous outputs. This chapter thoroughly introduces GANs and their latest trends for performing image processing and computer vision.

**Keywords:** generative adversary networks, text to image synthesis, deep learning, computer vision, image processing, neural network (NN)

## 9.1 Introduction

The past decade has witnessed a noteworthy improvement in the field of deep learning due to which computer vision systems have propelled exponentially. For instance, these systems can now analyze and identify diseases from medical images, aid in

**J. Srinivas,** Department of Information Technology, Matrusri Engineering College, Hyderabad.
**Rajesham Gajula,** CSED, Muffakam Jah College of Engineering and Technology, Hyderabad

https://doi.org/10.1515/9783110756722-009

autonomous driving vehicles, simulate images and videos from text. Many researchers [1, 2] consider discriminatory models as one of the most powerful advancement in deep learning where a class label is tagged to an input that has high dimensions and rich senses. The emergence of backpropagation and dropout algorithms can be considered as one of the main reasons for the success that deep learning models are enjoying currently [3]. Due to the unsolved challenges and its unparalleled potential, the area of computer vision still attracts many researchers. Over the years, GANs have shown a lot of potential in solving some of the computer vision challenges like text to image synthesis, image to image synthesis, image captioning [4], speech synthesis [5] and image in-painting [6]. GANs are robust models that can be used to synthesize data elegantly. They produce novel images that look almost similar to the training images.

Traditionally, GANs comprises of two NNs, namely the generator (G) and the discriminator (D). The generator can synthesize photos, whereas the discriminator is trained to discriminate the actual photos from the generated photos. The job of the generator is to fool the discriminator by making it believe the fake image as real, whereas the discriminator tries to prevent it from being fooled by finding out the differences between the real and synthesized images [7]. Both generator and discriminator comprise multiple convolution and/or fully connected layers. The two of them are pitted off for a two-player min-max game. If a collection of U data instances and V labels the generative models extract either the combined probability of P (U, V) or just P (U) when there are no labels on the other hand discriminative models extract the conditional probability P (V/U). Initially, the generator simulates an image of the training data which has a lot of differences when compared with real image; the discriminator easily classifies the image as fake. After some training, the generator synthesizes an image that is better than its previous output but still the discriminator identifies it as a fake. After some more training, the generator will synthesize a similar image of the real image which cannot be discriminated by the discriminator. Figures 9.1(a) and 9.1(b) illustrate this activity.

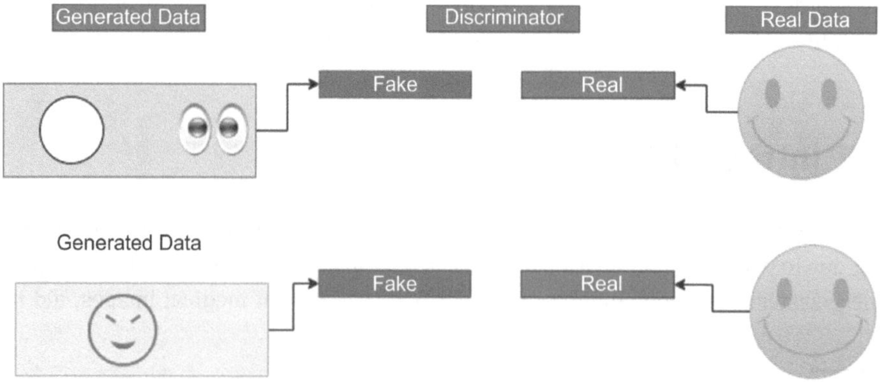


**Figure 9.1(a):** Working of a generator and discriminator.

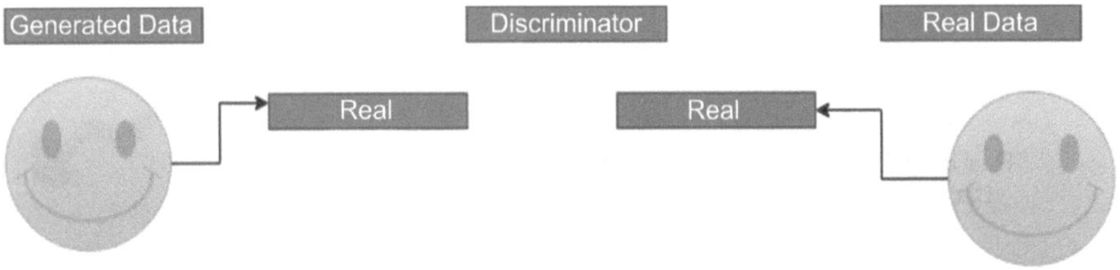


**Figure 9.1(b):** Generator fooling the discriminator after receiving proper training.

Figure 9.2 represents the general architecture of a GAN [7]. It comprises of two NNs viz., generator and discriminator. The boxes labeled sample represent data sources real and fake respectively. The real image data source that contains original images is used for conducting the positive training of the discriminator whereas the sample generated by the generator is injected into the discriminator directly for negative training.

The discriminator is connected to two loss functions. Whenever there is a misclassification, the discriminator loss function castigates the discriminator and similarly if the generator goes wrong in outwitting the discriminator, it will be castigated by the generator loss function. Both the generator and discriminator deploy backpropagation technique during their training. A random input such as noise is fed to the generator for synthesizing the images.

## 9.1.1 Training procedure of generator

1. The first step is to feed the generator with an irregular input such as noise.
2. Based upon the noise, the generator will synthesize an output
3. The generator's output is directly fed to the discriminator for classification.
4. The discriminator loss function returns the loss accumulated by the discriminator.
5. This loss is again backpropagated to both the discriminator and generator for obtaining the gradients.
6. The final step in the process is to modify the weights of the generator

## 9.1.2 Training procedure of discriminator

1. The initial step of a discriminator is to feed upon the images that generator synthesizes and classify them as original or phoney.
2. In the next step, the discriminator loss function computes the classification loss.
3. This loss is propagated back to the discriminator which is later used to adjust its weights.

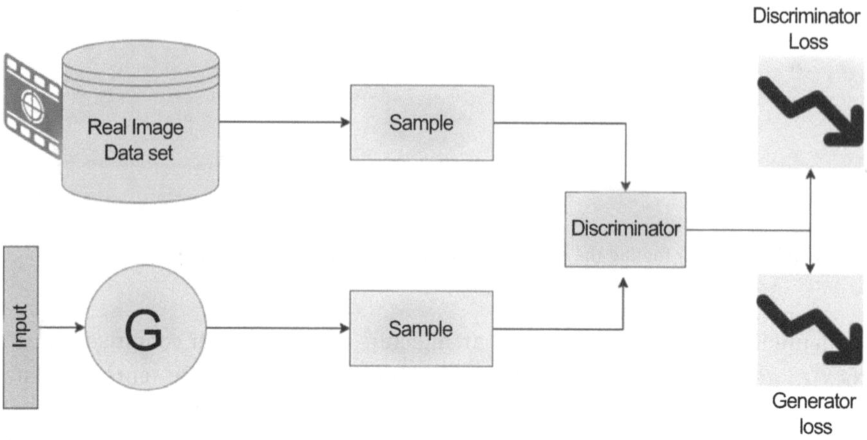


**Figure 9.2:** Architecture of a GAN.

The training of any GAN must face two issues:
1. The GANs must swindle two contrastingly trained networks.
2. GANs attain a stage where they become convergent and it is very complex to detect GAN convergence.

Generally, a GAN's training is scheduled in alternative timings. For example, if the discriminator is trained for one or more epochs, the generators training will not be scheduled during those epochs. This process iterates for a period of time. During the training of the discriminator, the generator is kept stable, and alternatively during the generator's training, the discriminator is kept stable. This alternate training mechanism is what powers the GANs deal with generative problems.

Convergence is a point in time where the discriminator's analysis becomes irrelevant [8]. As the training of the generator upgrades, the performance of the discriminator degrades. The discriminator predicts its outcome by making random assumptions. At this point, the GANs training should be stopped or else its performance will start to degrade. As explained earlier, the discriminator's output is propagated back and fed to the generator. If the generator is trained using the random feedback given by discriminator, the quality of the GAN disintegrates. This is called convergence of GAN.

### 9.1.3 Loss functions

In order to effectively train the discriminator and generator, loss functions have to be configured. There are two types of loss functions, namely, the discriminator loss and generator loss. The discriminator's loss can be modeled as a problem that can have two classes. Binary cross entropy loss function and Adam's stochastic gradient descent can be used to deal with discriminator loss. Generator does not have a direct loss. The discriminator tackles the loss function even for a generator.

# 9.2 Evaluation metrics of GANs

In the past few years, GANs are being implemented in most applications and each of it has got its own evaluation metric. This section of the chapter will discuss about the popular GAN evaluation metrics being used by researchers globally at present for their different tasks. This section of the chapter presents the various metrics available in the literature for evaluating the performance of GANs.

## 9.2.1 Quantitative metrics

### 9.2.1.1 Inception score

This metric was first introduced by T. Salimans et al. [9] in their research work. It automatically evaluates the quality of image-generative models. A higher inception score (IS) represents that the synthesized model can produce high quality and diverse images. An Inception V3 Network that is pretrained on ImageNet is used by IS. It also computes a statistic of the result generated by the network when solicited to generated images. Two key points were desired by the proposers of IS firstly, clear objects should be present in the images generated and secondly a high diversity of images should be synthesized by the generator. A large IS is expected if these two points are satisfied.

### 9.2.1.2 Average log-likelihood

One of the default evaluations metric for generative models over the years is log-likelihood method [10]. This measure the prospect of the fair data under the synthesized spread on "N" samples from the data. The general point to notice is that a model with high log-likelihood is expected to synthesize quality images. The Parzen windows are also applied to calculate the average log-likelihood. Annealed importance sampling was also used by some researchers in the past to calculate the average log-likelihood [11, 12].

#### 9.2.1.3 Coverage metric

This evaluation metric was proposed by Tolstkihn et al. [13]. The authors believed that coverage was better than likelihood, making it simpler to assess the gaps in effectiveness of the algorithms.

#### 9.2.1.4 Modified inception score

IS gives a greater estimate to models possessing a small entropy category conditional spread all over the synthesized image. Moreover, sometimes diversity in between samples of same class is expected. To characterize this diversity a novel technique that employed a cross-entropy style score was proposed [14]. Modified inception score (M-IS) may be imagined as a duplicate for estimating both intraclass sample deviation as well as image standard.

#### 9.2.1.5 Mode score

This metric represents the heterogeneity and visual quality of synthesized images [15]. Unlike IS, mode score considers the prior distribution of ground truth labels. But most researchers feel that IS and mode score (MS) are quite similar in nature.

#### 9.2.1.6 AM score

This metric is derived by making few changes in the IS. AM score can be used in cases where there is an uneven spread of data over the classes. The makers of this metric [16] considered KL divergence while formulating this metric. It is desired to have a small AM score for achieving better performance by a generator model. This score also works better when used with pretrained classifiers.

#### 9.2.1.7 Fréchet inception distance

Fréchet inception distance (FID) inputs a group of synthesized samples into a feature space specified by a particular layer of inception net or any convolution neural network [17]. The embedding layer is treated as a continuous multivariate Gaussian. Later, the mean and covariance of both synthesized and true data are compute. Finally, the Frechet distance is computed. A small value of FID indicates that there is very little difference between original data and synthetic data. With respect to discriminability, completeness and performance FID is a good metric. It detects intraclass mode dropping. It determines the manifoldness and quality of the synthesised images. Whenever new artifacts are added to images, FID score degrades. FID can tolerate noise in images and is constant with human judgments.

#### 9.2.1.8 Maximum mean discrepancy

This metric is used to represent the difference between two probability distributions using samples extracted independently from each other. A small maximum mean discrepancy (MMD) value indicates similarity between synthesized and real data. MMD is a two-sample testing technique [18].

#### 9.2.1.9 The Wasserstein critic

This metric provides an estimation of the Wasserstein distance between the two data distributions, real and synthesized. Practically speaking, the critic is a neural network that has clipped weights. This metric is also popularly known as the Earth mover's distance [19, 20] that is the smallest amount of weight to be displaced in between two distributions. Whenever the base distance is calculated in a conducive feature space, this metrics proves to be effective. The dimensionality and time complexity hamper the Wasserstein Critic evaluation metric.

#### 9.2.1.10 Birthday paradox test

This metric manually checks for a duplicate entry in a specimen of size "X" by approximating the assist volume of a discrete spread. The designers of this metrics [21] assure that the birthday paradox stays relevant even when the data follows a nonuniform distribution. Mode collapse in GANs can be identified by Birthday Paradox Test.

#### 9.2.1.11 Classifier two-sample tests

Whether the dual samples generated are identical or not is determined by two sample tests. Nearest-neighbor classifier was designed for computing classifier two-sample tests (C2ST). C2ST are better when compared to MMD [22].

#### 9.2.1.12 Classification performance

Unsupervised representation learning algorithms can be indirectly evaluated by using them as attribute selectors on tagged datascorpus and by calculating the efficiency of sequential models fixed upon the learned characteristics [23]. DC GANs and conditional GANs were evaluated for their quality using indirect techniques. "Semantic interpretability," "FCN score," "GAN quality index (GQI)" and "data augmentation utility" were introduced by researchers to measure the quality of GANs. Indirect evaluation techniques depend majorly on the type of classifier but are in most cases helpful to measure the quality of GANs.

#### 9.2.1.13 Number of statistically different bins

Whenever two groups of outputs from the similar spread are considered, the sampling noise and the count of samples that come in the range of a specific bin will be same. This metric was proposed to estimate the diversity in the synthesized samples and mode collapse [24]. Here, the count of statistically diverse bins are detected by using a uniform grid. The concept of Voronoi cells was introduced to find NDB. NDB is not tolerant and agnostic to multiple artifacts. Number of statistically different bins (NDB) is a good metric when compared with Birthday paradox test and MS-SSIM.

#### 9.2.1.14 Image retrieval performance

The metric starts by investigating the worst images synthesized by the generator [25]. A discriminatively trained CNN is used to represent both the images of a test dataset as well as synthesized images. A good image retrieval performance indicates a good GAN.

#### 9.2.1.15 Generative adversarial metric

The designers of this metric [26] analyzed two GANs by pitting them in a fight where the discriminators or generators of the GANS would be swapped. This metric estimates the relative efficiency of two GANs by estimating the similarity proportion of the two GANs. Generative multi-adversarial metric is a variation of generative adversarial metric (GAM) has also been proposed by researchers.

#### 9.2.1.16 Tournament win rate and skill rating

This evaluation metric was proposed by Olsson et al. [27]. The researchers utilized competitions between G's and D's for evaluating the performance of the GANs. Tournament win rate and skill rating are the two terms introduced by the researchers to keep in check the desired parameters. This metric became popular among GAN developers as it is simple and robust.

#### 9.2.1.17 Normalized relative discriminative score

As we know that the number of epochs required to discriminate real images from fake ones will be quite high, the NRDS was proposed [28]. Binary classifiers were used to distinguish real images from fake. A large value of normalized relative discriminative score (NRDS) indicates the efficiency of the GAN.

#### 9.2.1.18 Adversarial accuracy and adversarial divergence

This metric was proposed to measure the similarity of real data and generated data. Two measures are present in this metric namely the adversarial accuracy (AA) and adversarial divergence (AD). AA represents the classification accuracy attained by the two classifiers on a given validation set. On the other hand, the measure represents the KL divergence between the two data distributions [29]. It is desired to have a low adversarial divergence whereas a high adversarial accuracy. Major manual attempt is required in labeling the synthesized and real images due to which time taken also increases.

#### 9.2.1.19 Geometry score

This measure estimates the geometric properties of data that is present under manifold between the true and fake data. A simplicial complex is designed from the data by employing the proximity knowledge [30]. This score is complex and tedious to understand. Auxiliary networks are not used by geometry score and it is difficult to assess the visual characteristics of an image by using only this measure. Hence, it is suggested to use geometry score (GS) in conjunction with other evaluation metrics.

#### 9.2.1.20 Reconstruction error

It is to be noted that for a majority of generative models, the refabrication flaw over the instruction base is regularly and precisely optimized. Reconstruction error calculated on a test set can be used for evaluating generative models. Researchers proposed [31] reconstruction error for evaluating GANs which was a time consuming process. Feedforward networks were employed for this purpose.

#### 9.2.1.21 Image quality measures

Some researchers in the past used image-quality measures like structural similarity index measure (SSIM), peak signal-to-noise ratio and sharpness difference for evaluating the quality of a GAN. SSIM [32] cuts down on features that are irrelevant for manual impression. It differentiates relevant pixels and their neighborhoods in the given images using three parameters namely luminance (I), contrast (C) and structure (S). SSIM presumes a constant image sampling density and sighting range. The range of SSIM and its variants is between 0 and 1. The PSNR [33] score can be used to evaluate conditional GANs using paired data. During the image is generated by a GAN, its sharpness is affected. This can be estimated using sharpness difference. Thus, these three image-quality measures have been effectively used by researchers to evaluate the performances of the GANs.

#### 9.2.1.22 Low-level image statistics

Some researchers proposed [34] a way to analyze the performance of generative models. Moreover, natural images are non-Gaussian. Four measures were taken into account while estimating the quality of GANs namely the mean power spectrum, the count of bounded parts in a specified pixel region, the contrast spread and the spread of random filter responses, GANs can be controlled using these statistics. They can be used to monitor the optimality of discriminator and generator.

#### 9.2.1.23 Precision, recall and F1 score

The amount of overfitting of a GAN can be measured using the metrics precision, recall and F1 score [35]. The standard of the synthesized images are inspected by precision by default. On the other hand, the amount of reference distribution enveloped by the trained distribution is measured by recall. The disadvantage of these measures is that they work well on synthetic data but not on real images.

## 9.2.2 Qualitative metrics

Human beings are very good inspectors of images that GANs produce but they have major drawbacks. Human inspectors are expensive, biased, and cumbersome. Their ability is difficult to replicate and not always do they represent the effectiveness of models. Human inspectors are highly variant and inconsistent. Therefore, researchers have introduced qualitative measures like nearest neighbors, rapid screen categorization, rating and preference judgment, mode drop and mode collapse evaluation, investigating and visualizing the internals of networks. In this subsection, a brief introduction to the above mentioned measures is presented.

### 9.2.2.1 Nearest neighbors

Sometimes samples are projected next to their nearest neighbors [36]. This is done to detect overfitting. But this approach has drawbacks like they are intolerant to slightest of noises; the models passed by nearest neighbors are still susceptible to overfitting.

### 9.2.2.2 Rapid screen categorization

Humans are used in this measure to identify and distinguish synthesized images from real images [37]. This is a very good qualitative measure but does not work satisfactorily in uncontrolled environments. These tests are costly, biased and usually lead to overfitting of the model.

### 9.2.2.3 Rating and preference judgment

In this type of measure, people are asked to judge the models based on the quality of images generated by the GANs . Many approaches [38, 39] used this type of technique in their research but could not succeed completely as desired.

### 9.2.2.4 Evaluating mode drop and mode collapse

Whenever some of the modes of real images are ignored by generated images we call it as a mode drop. Whenever average of modes of original images is taken with respect to generated images, it is called mode collapse.

#### 9.2.2.5 Investigating and visualizing the internals of networks

Over the years, researchers have been exploring ways to gauge the working of GANs. Identifying the learning patterns of GANs, investigating their internal working mechanisms and exploring their limitations are one way of keeping track of the performance of GANs. Disentangled representations, space continuity and looking into the discriminator characteristics are some of the extra evaluation methods developed for generative models.

## 9.3 A survey on popular GANs

This section of the chapter presents a brief summary of popular classes of GANs available in the literature. Initially, the classification of GANs is presented. Later, the subsections take a deep dive into the various classes of GANs. A detailed illustration of popular GANs with neat diagrams is presented in this section.

### 9.3.1 Classification of GANs

A study performed by a group of researchers concluded that GANs can be classified into two categories, namely architecture optimization based GANs and objective function optimization based GANs [40]. The architecture optimization based GANs can be further classified as convolution-based GANs, condition-based GANs and autoencoder-based GANs. Deep convolutional generative adversarial networks (DCGAN) [41] is a popular example of convolution based GANs. Conditional generative adversarial network (CGANs) [42], InfoGAN [43] and auxiliary classifier generative adversarial network (ACGAN) [44] are the popular variants of condition-based GANs. Adversarial autoencoder (AAE) [45], bidirectional GAN (BiGAN) [46], adversarially learned inference (ALI) [47], AGE [48] and variational autoencoders generative adversial networks (VAE-GAN) [49] are some of the examples of autoencoder-based GANs.

Examples of objective function optimization based GANs include unrolled generative adversarial networks (unrolled GAN) [50], f-generative adversarial networks (f-GAN) [51], mode-regularized generative adversarial networks [52], Wasserstein generative adversarial networks–Lipschitz penalty (WGAN-LP) [53], Wasserstein generative adversarial networks–gradient penalty (WGAN-GP) [54], Wasserstein generative adversarial networks (WGAN) [55], least square generative adversarial networks (least square GAN) [56], loss-sensitive generative adversarial networks [57] and energy-based generative adversarial network (EBGAN) [58].

## 9.3.2 Architecture optimization based GANs

### 9.3.2.1 Convolution-based GANs

One the most popular supervised learning model is the convolution neural network (CNN) [59]. Traditionally, GANs use multi-layer perceptron (MLP) as generator and discriminator. But due to the superiority of CNN over MLP in feature extraction, MLP is replaced by CNN and termed as deep convolution generative adversial networks (DCGAN). In this model, the deconvolution layer substitutes the fully connected layer in the generator part. This modification increases the performance of the image synthesis jobs of the generator. Figure 9.3 illustrates the block diagram of a CNN.

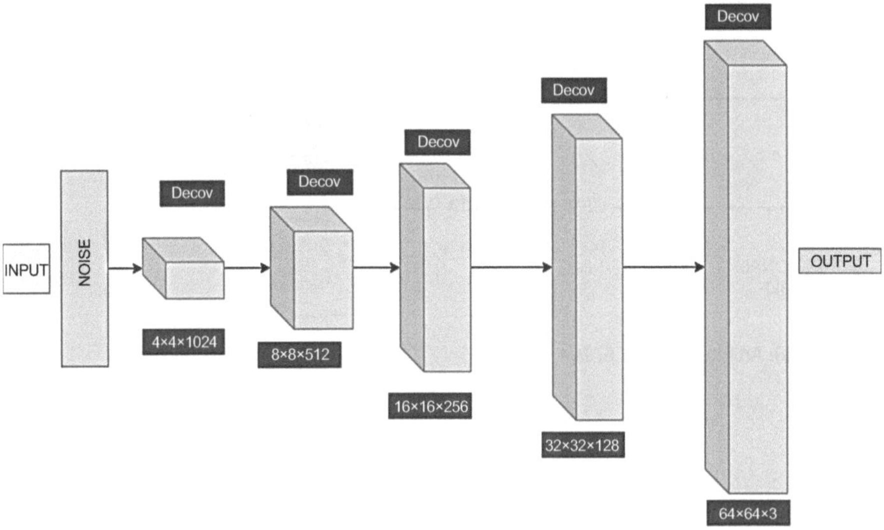


**Figure 9.3:** Block diagram of a CNN.

### 9.3.2.2 Condition-based GANs

Historically, the input for any GAN is a random noise vector. This noise sometimes crashes the training module of the generator. Therefore, a few researchers added a new variable called "condition" to both the generator and discriminator. This condition can be a variable or labels or text or any other formats of data. This new model of GAN was termed as conditional generative adversarial networks (CGANs) and is represented by Figure 9.4(a). Some researchers further enhanced the CGAN by adding mutual information. This novel GAN was termed as InfoGAN which made image generation more efficient. A separate network was embedded to the existing discriminator for synthesizing control variables. In InfoGan, the latent code is initially hidden which should be found out during the training routine of the generator. The mutual information is the deviated value between latent code and generated output. Figure 9.4(b) represents a model of InfoGAN. A group of researchers proposed a new type of GAN named ACGAN. Figure 9.4(c) illustrates the block diagram of ACGAN. In this model, instead of summing up the condition variable to the discriminator, a new classifier

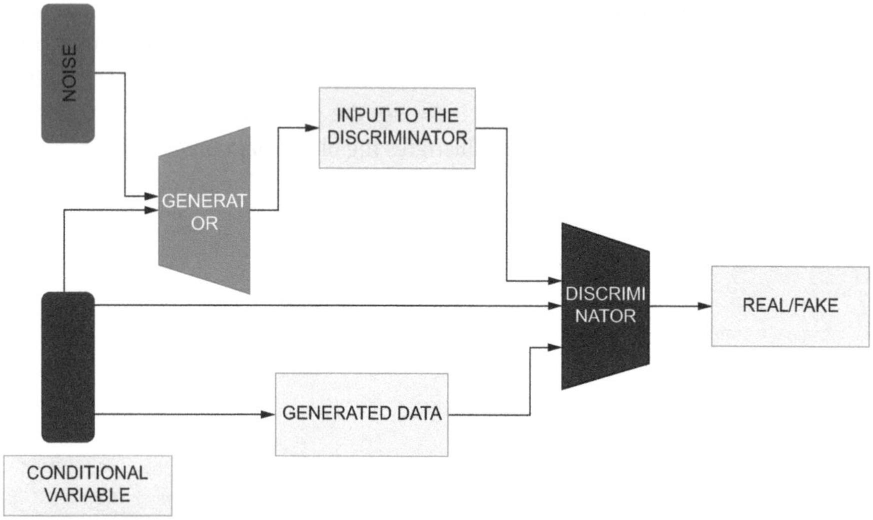


**Figure 9.4(a):** Architecture of CGAN.

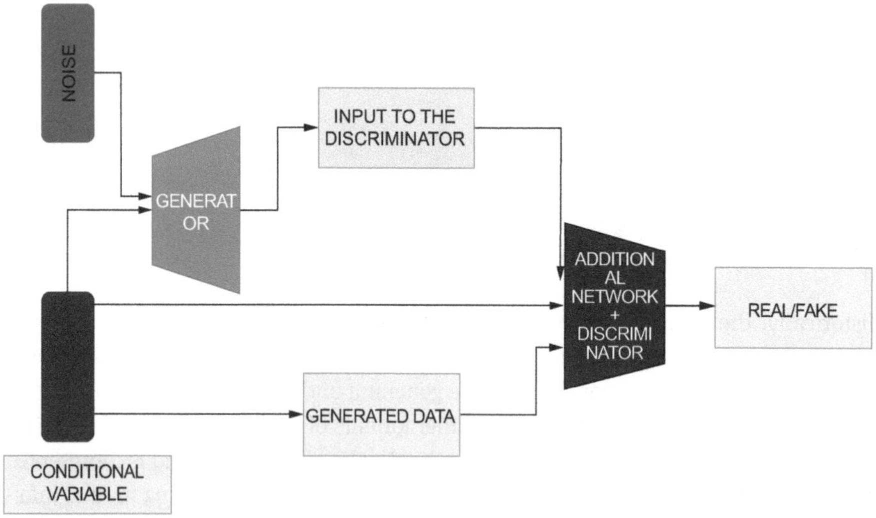


**Figure 9.4(b):** Architecture of InfoGAN.

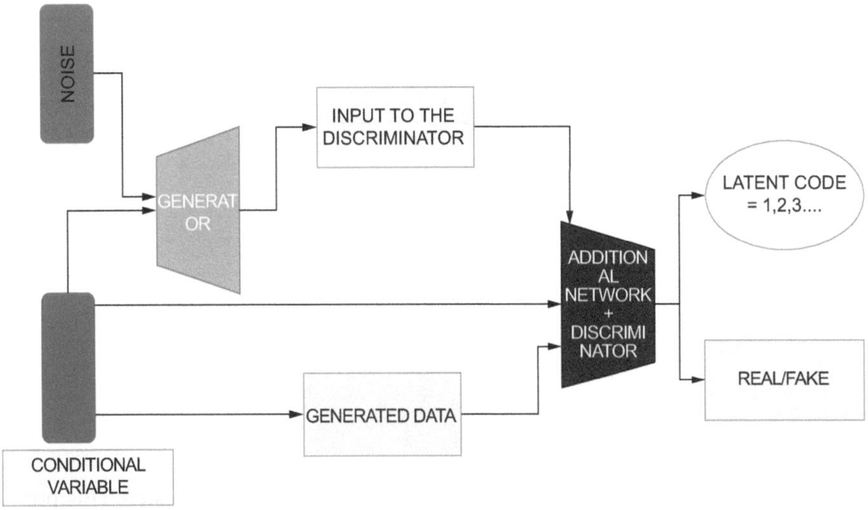


**Figure 9.4(c):** Block diagram of ACGAN.

is employed to display the probability above the class labels. In order to perform precise class estimation, the loss function is tuned accordingly.

#### 9.3.2.3 Autoencoder-based GANs

An autoencoder is a combination of two units namely the encoder and decoder. The encoder translates the input into a hidden layer. Generally, any of the multimedia data is given as input, whereas latent code is generated as output. Whenever output is generated dimensions of the input are reduced. The decoder accepts the latent code from hidden layer as input. This input is converted into output of the decoder once it is trained. By definition, autoencoder is a specially trained neural network for reformatting its input into output. Autoencoder is trained using unsupervised learning technique. Researchers have identified a number of deficiencies in the autoencoder-based GANs. Hence, they made enhancements according to their needs. One such improvement to autoencoder-based GAN is AAE. Here, the researchers fused adversial networks with autoencoders. In this model, the encoder distributes input before it is transferred to the hidden layer. This ensures effective translation of the input to output by the decoder. This model is illustrated by Figure 9.5.

A few researchers sported the idea of adding encoder to the GAN. But a few problems were identified during their operations which were addressed by implementing improvements to the existing models. One such model is the BiGAN. Traditionally, the generator of a GAN has the ability to grasp features in the latent space and extract modifications in the data distribution. But the generator does not have

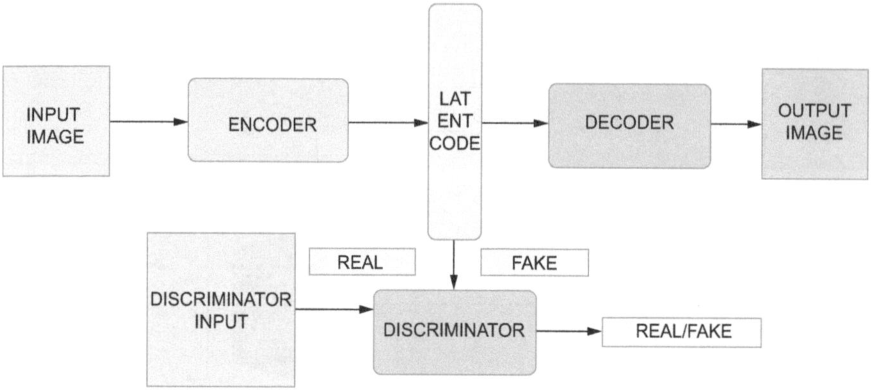


**Figure 9.5:** Architecture of AAE.

the ability to learn the mapping from data sample spread to latent space. This problem is solved in BiGAN. A BiGAN produces quality samples and has the ability to make valid inferences. An encoder is added to the existing GAN which inversely maps the data that was generated by the GAN in the data distribution reverse to the latent characteristic vector. The architecture of a BiGAN and ALI is illustrated in Figure 9.6(a). ALI was proposed where the encoder learns latent feature vector. In both BiGAN and ALI, the generator and encoder can learn in parallel. A new approach named AGE was proposed by a group of researchers wherein the adversarial network operates in between the generator and encoder. Here, discriminator is optional. Figure 9.6(b) represents the architecture of AGE. The task of generator in AIM is to minimize the deviation in between latent distribution and synthesized data distribution. The job of encoder is to increase the deviation in between latent distribution and generated data and minimize deviations in real data.

Some researchers supported the idea of fusing VAEs with GANs. One such example is VAE-GAN. In this model of GAN, the generator of a GAN replaces the decoder of VAE. The advantage of VAE is that they are immune to mode collapse but the quality of images they synthesize are not satisfactory. Here, the adversarial loss of GANs is summed up with the objective functions of VAEs. This increased the quality of images that a VAE generated without affecting their learning abilities. Thus, the architecture optimization based GANs can be explained briefly.

### 9.3.3 Objective function optimization based GANs

In order to stabilize the GANs, a number of approaches were introduced by researchers globally. One such approach is to optimize the objective function of the GAN. Unrolled GANs were proposed where the training methodology of the GAN is adjusted. A gradient based loss function i9s employed to enhance the operation of the generator.

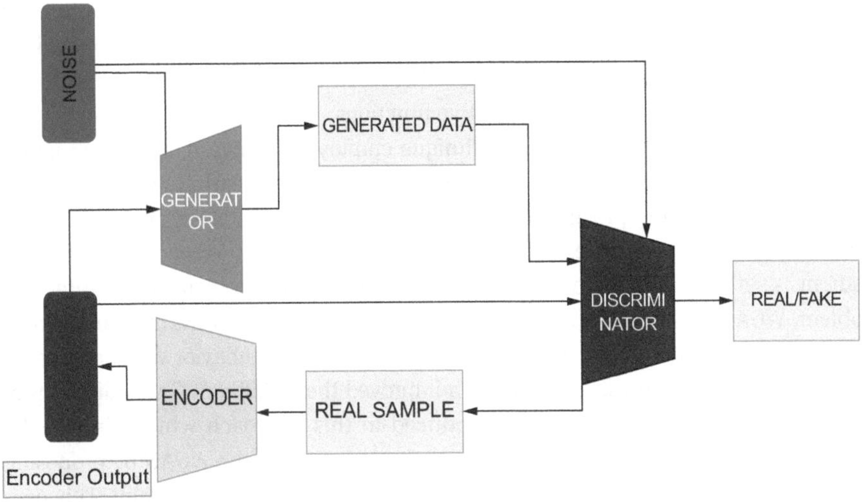


Figure 9.6(a): Architecture of ALI and BiGAN.

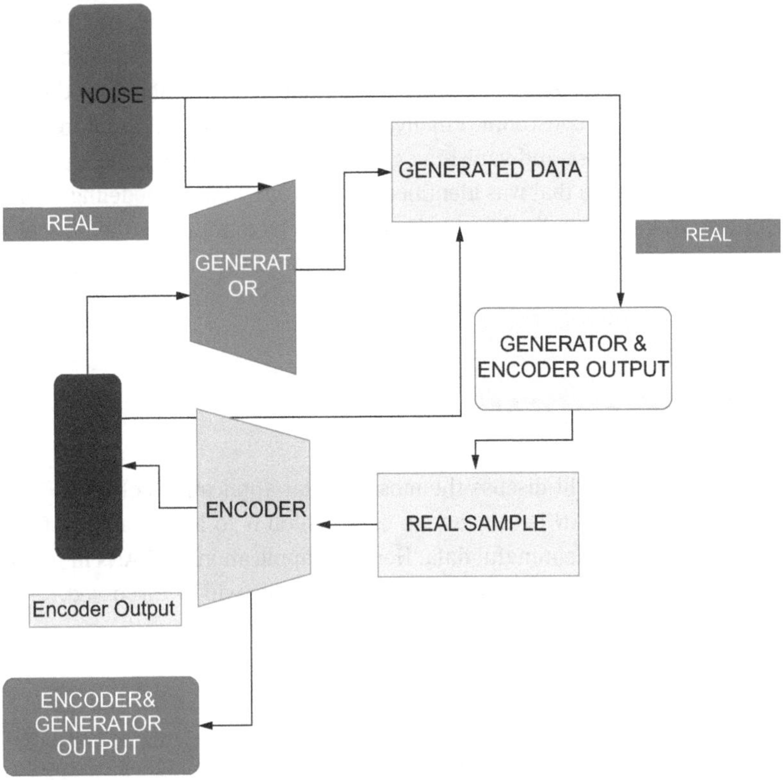


Figure 9.6(b): Architecture of AGE.

In the original GANs Jensen-Shannon (JS) deviation was minimized for minimizing the loss function of generator. F-divergence was also employed in GANs to stabilize them. In order to improve the stability of GANs, researchers introduced different types of divergences for formulating the objective functions.

Regularization is also a popular technique employed by investigators to enhance the stability of a GAN. A proposal was made wherein two regularizers were employed to improve the learning process of a GAN. If the deviation between actual data and synthesized data is negligible, then the divergence will be set to a constant and the gradient is set to zero. This initiates the vanishing-gradient problem. To counter this problem, WGAN was proposed by a group of investigators. In WGAN, the Earth-mover (EM) distance was introduced as it produced better gradient behavior when compared with other distance metrics. This approach improved the stability of GAN-training process. Lipschitz constraint was also introduced in this approach which clipped the weights of input vectors. Further enhancements were made to the WGAN by proposing a gradient penalty named WGAN-GP which enforces the Lipschitz constraint. This penalty ensured that the weights of inputs were optimally clipped and due to this the model achieved improved results when compared to WGAN. This penalty also stabilized the training methodology of various GANs without tuning any hyper parameters. Moreover, a new penalty was introduced namely WGAN-LP for enforcing the Lipschitz constraint. This model proved to be more efficient when compared to its predecessors WGAN and WGAN-GP. Some researchers proposed the Kantorovich-Rubinstein duality for generating the Lipschitz constraint. Finally, researchers succeeded in addressing the problems of mode collapse and vanishing gradient by implementing the Lipschitz constraint. One more problem that was identified in GANs was infinite modeling ability of GANs. This was addressed by limiting the loss function by introducing the Lipschitz constraint. Thus, the objective function optimization based GANs can be explained briefly.

## 9.4 Applications of GANs

This section of the chapter will discuss the most popular applications of GANs. The default application of any GAN is to generate data. Its job is to learn patterns from random noise and generate meaningful data. Here, the applications of GANs in prominent areas like computer vision, natural language processing will be discussed.

### 9.4.1 Computer vision

Currently, GANs are predominantly used for image translation, image superresolution, image synthesis and video generation in the field of computer vision. In this subsection, a detailed introduction of application of GANs in purple patches of computer vision is discussed.

#### 9.4.1.1 Image superresolution

Three advanced GANs namely superresolution generative adversarial networks (SRGAN) [60], Enhanced superresolution generative adversarial networks (ESRGAN) [61] were proposed by researchers for improving the resolution of images. Usually, a low-resolution image is given as input to the SRGAN. This scales up the resolution of the image by four times. The disadvantage of SRGAN is that there is noise associated with the generated image texture. To solve this problem, ESRGAN was proposed. This GAN was more enhanced than SRGAN as it had a robust architecture and also noise was minimized in the output images. The ESRGAN was further improved by adding a network unit named residual-in-residual dense block (RRDB). This was based on relativistic GAN [62]. The results proved that ESRGAN is a superior version of SRGAN. Thus, GANs can enhance resolution of images.

#### 9.4.1.2 Image translation

The process of converting an image from one dimension to another is termed as image translation. Pix2Pix [63], Pix2PixHD [64], CycleGAN [65], DiscoGAN [66], DualGAN [67] and StarGAN [68] are some of the examples of image-to-image translators. Pix2Pix is a CGAN that was employed for graphics and vision image to image translation tasks. An extension to Pix2Pix is Pix2PixHD which generates images of 2,048 × 1,024 resolution. Furthermore, the quality and definition of the generated images was much more enhanced when compared to Pix2Pix. An approach called cyclic consistency was employed by CycleGAN, DiscoGAN and DualGAN. These three GANs used encoder-decoder framework and can apply unpaired data for teaching the relationship from X space to Y space which was not possible in Pix2PixHD. The GANs discussed so far worked well for two-dimensional image translation but are not suitable for multidimensional image translation. To address this issue, StarGAN was introduced. StarGAN produced exemplary results in areas such as facial expression generation and facial characteristic transfer. Thus, GANs can be used for Image translation.

#### 9.4.1.3 Texture generation

Texture generation is an age-old problem in image processing. Markovian generative adversarial networks (MGAN) [69], spatial GAN(SGAN) [70] and periodic spatial

GAN (PSGAN) [71] are prominent examples of GANs that are used for texture synthesis. MGAN extracts the texture data from Markovian patches. This data is later used to synthesize images and videos quickly. SGAN implements an unsupervised learning strategy for synthesizing textures artificially. An enhancement to SGAN is the PSGAN which is more effective than its predecessors. PSGAN has the ability to learn periodic textures from images or large image data sets quickly. PSGAN efficiently manages noise while producing high resolution textures. Thus GANs can be used for texture generation.

#### 9.4.1.4 Face generation

A very important application of image processing is generation of faces. Researchers have used GANs to synthesize face of nonexistent people that mimic human faces. Popular GANs used for this purpose include two-pathway generative adversarial network (TP-GAN) [72], self-attention block with GAN (SAGAN) [73], Big-GAN [74] and MoCoGAN [75]. TP-GAN takes into account both local and global information to generate high resolution anterior face photos from a single flank image. The image generated can be used to synthesize multiple images comprising of varied positions and brightness. In order to improve the quality of images generated by TP-GAN, a few improvements were made and SAGAN was introduced. In this model, self-attention is integrated with GAN thus making discriminator more effective. SAGAN was further improved to Big-GAN which uses a "truncation trick" which improves the precision and diversity of the images synthesized. MoCoGAN synthesizes videos in an unsupervised manner. Thus, GANs can be used for face generation.

### 9.4.2 Natural language processing

GANs are also paving their way in the field of natural language processing (NLP) and speech processing. They are being used to develop automatic speech translators, voice recognizers, sentiment analyzers, text-to-text translators. One such GAN is the SeqGAN [76] which generates poetry, music and speech better than existing techniques. Policy gradient is used to train the generator. A novel form of GAN termed as RankGAN [77] is used to synthesize sentences. The discriminator is replaced using a unit called ranker. This replacement enhanced the quality of the GAN by multifold. A new GAN that employed the adversarial training method was used to synthesize open-domain dialogue [78].

### 9.4.3 Miscellaneous

GANs are used in other areas also like medical diagnosis [79], DNA sequencing, product customization and medical product manufacturing. AnoGAN [80] is a popular GAN that is used for finding anomalies in medical images. AnoGAN was trained on health data sets to detect the anomalies in lesions. GANs were used to perform DNA sequencing. Protein binding optimization was done using GANs [81]. GANs find their applications in the area of security also. Malware generation was also performed using GANs [82]. Medical product generation was also performed using GANs. This helped in customization of products to individual customers.

## 9.5 Issues in GANs

The major issues to be addressed in GANs are mode collapse, unstable training and evaluation metrics for GANs. There is always a lack of differentiation between images generated by GANs. How to attain differences in generated samples of images is still an issue associated with GANs. Attaining Nash equilibrium during its training is a major issue associated with GANs. For this first of all the training process of GANs has to be stabilized. Last but not least, the evaluation metrics for GANs are not yet standardized. As discussed earlier there are a number of evaluation metrics designed for GANs which will create ambiguity in the researchers.

## 9.6 Future scope in GANs

With the invention of advanced algorithms and network architectures GANs are expected to generate more precise text, images, speech and videos that cannot be distinguished by a human analyst. GANs have a lot of scope in the areas of NLP and information retrieval (IR). GANs can be integrated with the concepts of reinforcement learning for designing new GAN models. Embedded GANs are Policy Gradient GANs, latest examples of GANs. Actor critic algorithms can also be fused with GANs in future. GANs with their improved capabilities can be used in the field of network security against adversarial attacks. Stenographic techniques also can employ advanced GANs for hiding information. GANs can be used for unsupervised learning. In future, techniques involving semisupervised learning and GANs is also a good prospect for researchers. These are some of the directions in which GANs can be used in future.

## 9.7 Conclusions

The chapter crisply introduces GANs in the initial sections. It explains the architecture and working of a GAN with neat diagrams. Section 9.2 broadly discusses the various evaluation metrics of GANs available in the literature. Both qualitative and quantitative metrics were analyzed deeply. In Section 9.3, the classification of GANs was discussed with focus on various types of GANs. The architectures of various GANs were illustrated precisely. Afterwards in Section 9.4, the various applications of GANs were discussed in detail. In Section 9.5, the various issues surrounding GANs were presented. Mode collapse, unstable training and standardization of evaluation parameters were discussed briefly. Later in Section 9.6, the future scope of GANs was discussed. A clear cut direction of research in GANs was presented.

## References


[1] Hinton, G., Deng, L., Dahl, G.E., Mohamed, A., Jaitly, N., Senior, A., Vanhoucke, V., Nguyen, P., Sainath, T., Kingsbury, B. (2012a). Deep neural networks for acoustic modeling in speech recognition. IEEE Signal Processing Magazine, 29(6), 82–97.

[2] Krizhevsky, A., Sutskever, I., Hinton, G. (2012). ImageNet classification with deep convolutional neural networks. NIPS, 2012.

[3] Jarrett, K., Kavukcuoglu, K., Ranzato, M., LeCun, Y. (2009). What is the best multi-stage architecture for object recognition? In Proc. International Conference on Computer Vision (ICCV'09), pages 2146–2153. IEEE.

[4] Goodfellow, I.J., Pouget-Abadie, J., Mirza, M., Xu, B., Warde-Farley, D., Ozair, S., Courville, A.C., Bengio, Y. (2014). Generative adversarial nets. NIPS.

[5] van den Oord, A., Dieleman, S., Zen, H., Simonyan, K., Vinyals, O., Graves, A., Kalchbrenner, N., Senior, A.W., Kavukcuoglu, K. WaveNet: A generative model for raw audio. CoRR, abs/1609.03499, 2016a.

[6] Iizuka, S., Simo-Serra, E., Ishikawa, H. Globally and locally consistent image completion. ACM Transactions on Graphics, 36(4), 107:1–107:14.

[7] Barua, S., Ma, X., Erfani, S.M., Houle, M.E., Bailey, J. (2019). Quality evaluation of gans using cross local intrinsic dimensionality. arXiv Preprint arXiv:1905.00643.

[8] Kodali, N., Abernethy, J., Hays, J., Kira, Z. (2017). On convergence and stability of gans. arXiv Preprint arXiv:1705.07215.

[9] Salimans, T., Goodfellow, I., Zaremba, W., Cheung, V., Radford, A., Chen, X., Chen, X. (2016). Improved techniques for training gans. Advances in Neural Information Processing Systems, 29, 2234–2242.

[10] Tolstikhin, I.O., Gelly, S., Bousquet, O., Simon-Gabriel, C.-J., Schölkopf, B. (2017). Adagan: Boosting generative models. Advances in Neural Information Processing Systems, 5430–5439.

[11] Wu, Y., Burda, Y., Salakhutdinov, R., Grosse, R. On the quantitative analysisof decoder-based generative models. arXiv Preprint arXiv:1611.04273.

[12] Theis, L., Oord, A.V.D., Bethge, M. A note on the evaluation of generativemodels. arXiv Preprint arXiv:1511.01844.

[13] Tolstikhin, I.O., Gelly, S., Bousquet, O., Simon-Gabriel, C.-J., Schölkopf,Adagan:, B. (2017). Boosting generative models. Advances in Neural Information Processing Systems, 5430–5439.
[14] Gurumurthy, S., Sarvadevabhatla, R.K., Radhakrishnan, V.B. (2017). Deligan:Generative adversarial networks for diverse and limited data. The IEEEConference on Computer Vision and Pattern Recognition (CVPR), 1.
[15] Che, T., Li, Y., Jacob, A.P., Bengio, Y., Li, W. Mode regularized generative adversarial networks. arXiv Preprint arXiv:1612.02136.
[16] Zhou, Z., Cai, H., Rong, S., Song, Y., Ren, K., Zhang, W., Wang, J., Yu, Y., Activation maximization generative adversarial nets, International Conference on Learning Representations. URL https://openreview.net/forum?id=HyyP33gAZ
[17] Heusel, M., Ramsauer, H., Unterthiner, T., Nessler, B., Hochreiter, S. (2017). Gans trained by a two time-scale update rule converge to a local nash equilibrium. Advances in Neural Information Processing Systems, 6629–6640.
[18] Fortet, R., Mourier, E. (1953). Convergence de la répartition empirique vers la répartition théorique. In: Annales Scientifiques de L'école Normale Supérieure, vol. 70. Elsevier, 267–285.
[19] Arjovsky, M., Chintala, S., Bottou, L. Wasserstein gan. arXiv Preprint arXiv:1701.07875.
[20] Arora, S., Zhang, Y. Do gans actually learn the distribution? an empirical study. arXiv Preprint arXiv:1706.08224.
[21] Lehmann, E.L., Romano, J.P. (2006). Testing Statistical Hypotheses. Springer Science & Business Media.
[22] Ye, Y., Wang, L., Wu, Y., Chen, Y., Tian, Y., Liu, Z., Zhang, Z., Gan quality index (gqi) by gan-induced classifier.
[23] Santurkar, S., Schmidt, L., Madry, A. (2018). A classification-based study of covariate shift in gan distributions, in: International Conference on Machine Learning, 4487–4496.
[24] Richardson, E., Weiss, Y. On gans and gmms. arXiv Preprint arXiv:1805.12462.
[25] Wang, Y., Zhang, L., van de Weijer, J. Ensembles of generative adversarial networks. arXiv Preprint arXiv:1612.00991.
[26] Im, D.J., Kim, C.D., Jiang, H., Memisevic, R. Generating images with recurrent adversarial networks. arXiv Preprint arXiv:1602.05110.
[27] Olsson, C., Bhupatiraju, S., Brown, T., Odena, A., Goodfellow, I. Skill rating for generative models. arXiv Preprint arXiv:1808.04888.
[28] Zhang, Z., Song, Y., Qi, H. Decoupled learning for conditional adversarial networks. arXiv Preprint arXiv:1801.06790.
[29] Yang, J., Kannan, A., Batra, D., Parikh, D. Lr-gan: Layered recursive generative adversarial networks for image generation. arXiv Preprint arXiv:1703.01560.
[30] Khrulkov, V., Oseledets, I. Geometry score: A method for comparing generative adversarial networks. arXiv Preprint arXiv:1802.02664.
[31] Ledig, C., Theis, L., Huszar, F., Caballero, J., Cunningham, A., Acosta, A., Aitken, A., Tejani, A., Totz, J., Wang, Z., et al. Photo-realistic single image super-resolution using a generative adversarial network. arXiv Preprint.
[32] Wang, Z., Bovik, A.C., Sheikh, H.R., Simoncelli, E.P. (2004). Image quality assessment: From error visibility to structural similarity. IEEE Transactions on Image Processing, 13(4), 600–612.
[33] Regmi, K., Borji, A. Cross-view image synthesis using conditional gans. CVPR.
[34] Zeng, Y., Lu, H., Borji, A. Statistics of deep generated images. arXiv Preprint arXiv:1708.02688.
[35] Lucic, M., Kurach, K., Michalski, M., Gelly, S., Bousquet, O. Are gans created equal? a large-scale study. arXiv Preprint arXiv:1711.10337.

[36] Theis, L., Oord, A.V.D., Bethge, M. A note on the evaluation of generative models. arXiv Preprint arXiv:1511.01844.
[37] Denton, E.L., Chintala, S., Fergus, R., et al. (2015). Deep generative image models using a laplacian pyramid of adversarial networks. Advances in Neural Information Processing Systems, 1486–1494.
[38] Lu, Y., Wu, S., Tai, Y.-W., Tang, C.-K. Sketch-to-image generation using deep contextual completion. arXiv Preprint arXiv:1711.08972.
[39] Xiao, C., Li, B., Zhu, J.-Y., He, W., Liu, M., Song, D. Generating adversarial examples with adversarial networks. arXiv Preprint arXiv:1801.02610.
[40] Pan, Z., Yu, W., Yi, X., Khan, A., Yuan, F., Zheng, Y. (2019). Recent progress on generative adversarial networks (GANs): A survey. IEEE Access, 7, 36322–36333.
[41] Radford, A., Metz, L., Chintala, S. (2016). Unsupervised representation learning with deep convolutional generative adversarial networks, in International Conference on Learning Representations.
[42] Mirza, M., Osindero, S. (2014 Nov). Conditional generative adversarial nets. arXiv E-prints, P. arXiv:1411.1784.
[43] Chen, X., Duan, Y., Houthooft, R., Schulman, J., Sutskever, I., Abbeel, P. (2016). Infogan: Interpretable representation learning by information maximizing generative adversarial nets. Advances in Neural Information Processing Systems, 29, 2172–2180.
[44] Odena, A., Olah, C., Shlens, J. (2017 Aug). Conditional image synthesis with auxiliary classifier GANs, in Proceedings of the 34th International Conference on Machine Learning, vol. 70, 2642–2651.
[45] Makhzani, A., Shlens, J., Jaitly, N., Goodfellow, I. (2016). Adversarial autoencoders, in International Conference on Learning Representations.
[46] Donahue, J., Krahenb́ĺzhl, P., Darrell, T. (2017). Adversarial feature learning, in International Conference on Learning Representations.
[47] Dumoulin, V., Belghazi, I., Poole, B., Mastropietro, O., Lamb, A., Arjovsky, M., Courville, A. (2017). Adversarially learned inference, in International Conference on Learning Representations.
[48] Ulyanov, D., Vedaldi, A., Lempitsky, V. (2018). It takes (only) two: Adversarial generator-encoder networks, in AAAI Conference on Artificial Intelligence.
[49] Larsen, A.B.L., S?nderby, S.K., Larochelle, H., Winther, O. (2016 Jun). Autoencoding beyond pixels using a learned similarity metric, in Proceedings of The 33rd International Conference on Machine Learning, 1558–1566.
[50] Metz, L., Poole, B., Pfau, D., Sohl-Dickstein, J. (2017). Unrolled generative adversarial networks, in International Conference on Learning Representations.
[51] Nowozin, S., Cseke, B., Tomioka, R. (2016). f-gan: Training generative neural samplers using variational divergence minimization. In: Lee, D.D., Sugiyama, M., Luxburg, U.V., Guyon, I., Garnett, R., eds. Advances in Neural Information Processing Systems 29. Curran Associates, Inc., 271–279.
[52] Che, T., Li, Y., Jacob, A.P., Bengio, Y., Li, W. (2017). Mode regularized generative adversarial networks, in International Conference on Learning Representations.
[53] Petzka, H., Fischer, A., Lukovnicov, D. (2018). On the regularization of wasserstein gans, in International Conference on Learning Representations.
[54] Gulrajani, I., Ahmed, F., Arjovsky, M., Dumoulin, V., Courville, A.C. (2017). Improved training of wasserstein gans. Advances in Neural Information Processing Systems, 30, 5767–5777.
[55] Arjovsky, M., Chintala, S., Bottou, L. (2017 Aug). Wasserstein generative adversarial networks, in Proceedings of the 34th International Conference on Machine Learning, vol. 70, 214–223.

[56] Mao, X., Li, Q., Xie, H., Lau, R.Y.K., Wang, Z., Smolley, S.P. (2018). On the effectiveness of least squares generative adversarial networks. IEEE Transactions on Pattern Analysis and Machine Intelligence.
[57] Qi, G.-J. (2017 Jan). Loss-sensitive generative adversarial networks on Lipschitz densities. arXiv E-prints, arXiv:1701.06264.
[58] Zhao, J., Mathieu, M., LeCun, Y. (2017). Energy-based generative adversarial network, in International Conference on Learning Representations.
[59] LeCun, Y., Boser, B., Denker, J.S., Henderson, D., Howard, R.E., Hubbard, W., Jackel, L.D. (1989 Dec). Backpropagation applied to handwritten zip code recognition. Neural Computation, 1(4), 541–551.
[60] Ledig, C., Theis, L., Husz'l ́cr, F., Caballero, J., Cunningham, A., Acosta, A., Aitken, A., Tejani, A., Totz, J., Wang, Z., Shi, W. (2017 July). Photo-realistic single image super-resolution using a generative adversarial network, in 2017 IEEE Conference on Computer Vision and Pattern Recognition (CVPR), 105–114.
[61] Wang, X., Yu, K., Wu, S., Gu, J., Liu, Y., Dong, C., Qiao, Y., Loy, C.C. (2018 September). Esrgan: Enhanced super-resolution generative adversarial networks, in The European Conference on Computer Vision Workshops (ECCVW).
[62] Jolicoeur-Martineau, A. (2018 Jul). The relativistic discriminator: A key element missing from standard GAN. arXiv E-prints, P. arXiv:1807.00734.
[63] Isola, P., Zhu, J., Zhou, T., Efros, A.A. (2017 July). Image-to-image translation with conditional adversarial networks, in 2017 IEEE Conference on Computer Vision and Pattern Recognition (CVPR), 5967–5976.
[64] Wang, T.-C., Liu, M.-Y., Zhu, J.-Y., Tao, A., Kautz, J., Catanzaro, B. (2018 June). High-resolution image synthesis and semantic manipulation with conditional gans, in The IEEE Conference on Computer Vision and Pattern Recognition (CVPR).
[65] Zhu, J.-Y., Park, T., Isola, P., Efros, A.A. (2017 Oct). Unpaired image-to-image translation using cycle-consistent adversarial networks, in The IEEE International Conference on Computer Vision (ICCV).
[66] Kim, T., Cha, M., Kim, H., Lee, J.K., Kim, J. (2017 Aug). Learning to discover cross-domain relations with generative adversarial networks, in Proceedings of the 34th International Conference on Machine Learning, 1857–1865.
[67] Yi, Z., Zhang, H., Tan, P., Gong, M. (2017 Oct). Dualgan: Unsupervised dual learning for image-to-image translation, in The IEEE International Conference on Computer Vision (ICCV).
[68] Choi, Y., Choi, M., Kim, M., Ha, J., Kim, S., Choo, J. (2018 June). Stargan: Unified generative adversarial networks for multi-domain image-to-image translation, in 2018 IEEE/CVF Conference on Computer Vision and Pattern Recognition, 8789–8797.
[69] Li, C., Wand, M. (2016). Precomputed real-time texture synthesis with markovian generative adversarial networks, in European Conference on Computer Vision, 702–716.
[70] Li, C., Wand, M. (2016). Precomputed real-time texture synthesis with markovian generative adversarial networks, in European Conference on Computer Vision, 702–716.
[71] Jetchev, N., Bergmann, U., Vollgraf, R., Texture synthesis with spatial generative adversarial networks, in Advances in Neural Information Processing Systems, Conference Proceedings.
[72] Huang, R., Zhang, S., Li, T., He, R. (2017, Oct). Beyond face rotation: Global and local perception gan for photorealistic and identity preserving frontal view synthesis, in The IEEE International Conference on Computer Vision (ICCV).
[73] Zhang, H., Goodfellow, I., Metaxas, D., Odena, A. (2018 May). Self-attention generative adversarial networks. arXiv E-prints, P. arXiv:1805.08318.
[74] Brock, A., Donahue, J., Simonyan, K. (2019). Large scale GAN training for high fidelity natural image synthesis, in International Conference on Learning Representations.

[75] Tulyakov, S., Liu, M., Yang, X., Kautz, J. (2018 June). Mocogan: Decomposing motion and content for video generation, in 2018 IEEE/CVF Conference on Computer Vision and Pattern Recognition, 1526–1535.
[76] Yu, L., Zhang, W., Wang, J., Yu, Y. (2017). Seqgan: Sequence generative adversarial nets with policy gradient. in AAAI Conference on Artificial Intelligence, 2852–2858.
[77] Lin, K., Li, D., He, X., Zhang, Z., Sun, M.-T. (2017). Adversarial ranking for language generation. Advances in Neural Information Processing Systems, 30, 3155–3165.
[78] Li, J., Monroe, W., Shi, T., Jean, S., Ritter, A., Jurafsky, D. (2017). Adversarial learning for neural dialogue generation, in Proceedings of the 2017 Conference on Empirical Methods in Natural Language Processing, 2157–2169.
[79] Deshpande, N.M., Gite, S., Aluvalu, R. (2021). A review of microscopic analysis of blood cells for disease detection with AI perspective. PeerJ Computer Science, 7, e460.
[80] Schlegl, T., Seeböck, P., Waldstein, S.M., Schmidt-Erfurth, U., Langs, G. (2017). Unsupervised anomaly detection with generative adversarial networks to guide marker discovery. Information Processing in Medical Imaging, 146–157.
[81] Killoran, N., Lee, L.J., Delong, A., Duvenaud, D., Frey, B.J. (2017). Generating and designing dna with deep generative models. Advances in Neural Information Processing Systems.
[82] Hu, W., Tan, Y. (2017 Feb). Generating adversarial malware examples for black-box attacks based on GAN. arXiv E-prints, P. arXiv:1702.05983.

Rutuja D. Bhandari*, Shraddha Phansalkar

# 10 Computer vision-based approach for alphabetic recognition and Indian Sign Language recognition system

**Abstract:** Purpose: Sign language is a communication media for hearing impaired people. Although sign language can be effectively used for communication between two disabled persons, the challenge is higher when they need to communicate with the outside world which rarely has the understanding toward sign language.

The aim of this work is to develop computer vision based Indian Sign Language (ISL) fingerspelling identification system which will help in recognizing alphabetic signs as well as emergency words in the sign language without any additional equipment. It will thus improve the existing communication gap between the disabled and abled people.

Methodology: We surveyed the literature using Prisma methodology. The study shows that deep learning techniques like convolutional neural network (CNN) can promise better accuracy than any other neural network architecture with feature extraction. We have also used data aggregation between two data sets and prepared one dataset of around 50 K images.

Our idea is to implement a CNN which can be employed for real-time applications to detect hand gestures. The input of this system can be then forwarded to long short-term memory (LSTM) model (sequence to sequence learning algorithm) which will encode the gesture and then decode it into text format.

Findings: Many research works have been proposed on American Sign Language (ASL) recognition. The method used for ASL cannot be used for ISL, as it differs from Indian Sign Language (ISL) recognition. As in ISL, two hands are used for communicating, whereas ASL uses one hand for communication. This will lead to overlapping of hands making ISL unclear and difficult to understand the features. The major problem of all related research works was lack of available dataset. There are very few works done on creation of such datasets.

Values: A system which will help in automatic recognition of sign language and will help people with hearing impaired to communicate with the real world.

Originality: Similar systems have been developed on only alphabet recognition with sign language. We also aim to leverage this dataset by augmenting with frequently

***Corresponding author: Rutuja D. Bhandari,** MIT ADT University, MIT SOE, Pune, India, e-mail: rutuja2199@gmail.com
**Shraddha Phansalkar,** MIT ADT University, MIT SOE, Pune, India, e-mail: shradda.phansalkar@mituniversity.edu.in

https://doi.org/10.1515/9783110756722-010

used emergency words and gestures that can help hearing impaired people to convey important messages in emergency as well as routine activities.

**Keywords:** Indian Sign Language, fingerspelling recognition system, emergency words, prism methodology, deep learning, convolutional neural network, long short-term memory

# 10.1 Introduction

## 10.1.1 Significance and rationale

A person who is hearing impaired is often left out in this rapidly changing society we are living in. Over 430 million people, that is, 5% of the global population are suffering from hearing loss, according to the World Health Organization (WHO & April 2021). In India roughly 6.3% of the population are suffering from hearing impairment including men, women and children (WHO) [1]. Sign language is used by hearing impaired persons to communicate with each other, but it is usually not understood easily by normal people having no deaf or dumb impairment.

Most of this population are still dependent on an intermediary or a translator who is capable of translating hand sign gestures to them. An alternate option can be written communication, but it does not seem very useful as the deaf people are normally not gifted with the talent to read and write spoken languages properly. Therefore, this project aims at identifying alphabets, digits and some emergency words in Indian Sign Language (ISL) without involving any expert and helps personnel who use sign language in emergency situations.

Communication has proven to be the most basic need for everyday survival in society. Sign languages are used by deaf communities to communicate with the society [2]. Sign language has come out to be the main mode of communication for the people suffering with deaf impairment, but in some situations sign language can be used by normal people as well. The language is also found very useful to communicate with the people having autism, down syndrome, apraxia of speech and autism.

A sign language is composed of symbols shown by various postures of hand, its actions, orientations accompanied by facial expressions [3]. Our basic aim is to find a solution to this problem by using high end methods and technologies like deep learning and computer vision.

There are numerous sign languages used around the globe for communication which are developed through various groups of people communicating with each other, just like spoken languages. In many countries where the same spoken language is used, same sign language is not used. For example, there are three varieties of English language, namely: Australian sign language (Auslan), American sign language (ASL) and British sign language (BSL).

ISL is used by the deaf community in India [4]. Sign Language Recognition has been well researched in ASL but unfortunately few works have been carried out on ISL Recognition. Lack of datasets, feature extraction problems and variation in language may be the major barrios which is the reason behind less work done on ISL. We are trying to solve the problem that mainly deals with the communication gap between the hearing impairment people and society (normal people). Our project focuses on feature extraction and classification techniques which will help both the society as well as hearing impairment people in India to communicate with each other digitally and without help of a sign language interpreter (a person who knows sign language).

Second, conveying a word with letters may not always be faster and few emergency or common words can be shown and added to the existing system to enrich the sign language. This will help make the communication faster and effective during emergencies.

### 10.1.2 Motivation

As mentioned earlier that deaf and dumb people use sign language to communicate among themselves as well as with abled people. However, this is complicated to grasp for a common man. Thus, we are helping these disabled people to communicate with today's world by creating a system which will help them to communicate in some places like in banks (for words such as open account, withdraw money, amount value, etc.), hospitals, police stations, post offices for frequent as well as in situations like introducing themselves in particular places, giving numbers to someone, use words like "help," "stop," "name," "yes" and "no."

We aim to use this system in ISL as there has been less work done on it. It will prove to be beneficial for the people following ISL. Effective addition to this project can include the words used in emergency situations which will not only make the people having disabilities communicate faster with others, but also deliver a lift in the development of autonomous systems used in understanding and supporting them.

### 10.1.3 Challenges

ISL Recognition poses a challenge. The American Sign Language leads the Indian Counterpart, as there are not many standard datasets available, thus hampering the research in this field. Recognizing the Indian sign language system is complicated because of the involvement of both the hands and also the overlapping of the hands [5]. It requires combinations of images for understanding information with different hand poses, hand movements, face expression, etc. [6].

Sign language also has thousands of words composed of a number of letters and these words have similar hand poses. In this work, we will be using a subset of words which are chosen on the basis of popularity and are an important part of sign language recognition. Thus, the resultant dataset is a set that includes alphabets, numbers and highly frequent emergency words for interacting with the outside world.

The development in the technology of deep learning and computer vision can be applied to recognize sign language of hearing impaired into letters, words that a normal person can interpret. The Computer Vision techniques are popularly used to detect images and classify them into categories with appropriate and automated feature extraction and greater accuracy [7]. For the presented work, we employ deep learning technique of convolutional neural network (CNN) for classifying the hand gestures into multiple classes of alphabets, words and numbers. This will be the necessary step for further sequential processing with recurrent neural network (RNN) technique or LSTM to display the output text sequentially with proper syntax and semantics [8, 9]. Dataset is used in the form of videos and images. In our project we have used the image dataset.

The representation of our proposed system is shown in Figure 10.1 with different phases as data acquisition, gesture detection, feature extraction, classification and sequential processing. Image dataset is provided as an input in our system and output will be in English text or in digits. The images are collected by merging two different datasets. The model was trained using convolutional layers.

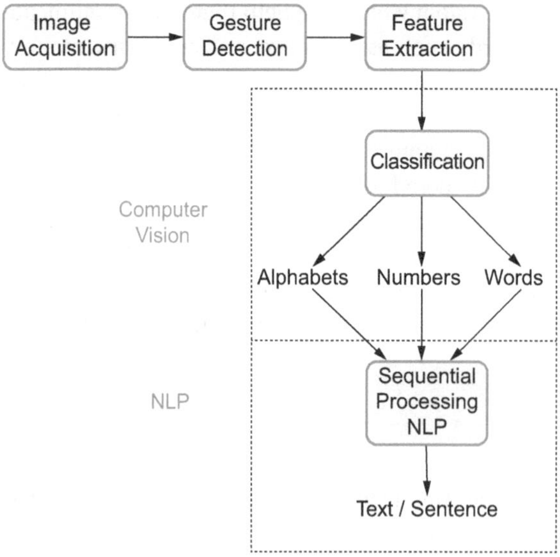


**Figure 10.1:** Proposed Indian Sign Language Recognition & Interpretation System.

In our proposed system, the first step is to collect data from available sources. We have collected a dataset from the available sources we found on the internet (Kaggle, GitHub). Expensive cameras or sensors have been used by many researchers to capture gestures. For our system we fortunately are able to collect data from two different sources (Kaggle and GitHub). We have collected digits and alphabets image dataset from Kaggle and emergency words from GitHub. And finally, we merge them together to train the dataset. The images which we have found are of the same size. Hence, there is no need for data preprocessing and data augmentation for the dataset as it has been found to be a well- balanced dataset.

CNN layers are applied in our system for feature extraction and classification. Hand gestures are recognized by using a classifier. Feature extraction is performed using CNN layers which extract the important information from the data and the raw data is reduced for further processing. Image classification is also done using CNN layers in which the image is taken as input and an output is produced which will give us the probability from which particular class the image belongs. Along with classification sequential processing which is Natural Language Processing (NLP) is done to keep the track of each sequence as an output. The architecture of recurrent neural network (RNN) known as LSTM is used to keep track of the sequence of each image so the final output given by the system will have some meaning. Finally, the output generated by the system is in the form of text or digits.

### 10.1.4 Objective

The works presented in this paper aims to contribute in following manner:
1. We have carried out an extensive survey of different work articles on classification models used for hand gesture recognition.
2. Digits from 1 to 9 are also embedded in our database as it will also be easy for deaf people to communicate in a few places like banks or police stations.
3. Few emergency words are augmented in our dataset as, always it's not possible to use alphabets then combine them together and form a word during emergencies.
4. Generally, hand gesture are identified by using static or dynamic recognition. In our system we will be focusing on dynamic sign recognition.
5. We are developing a deep learning-based model for classification of hand gestures for emergency words, letters and digits.

This work has been divided into following sections:
1. Section 10.1 deals with the introduction which is basically the information and challenges on Indian Sign Language (ISL) and about our project's objectives.
2. Section 10.2 explains the surveys carried on hand gestures and sign recognition with the use of various technologies and algorithms and research breaches.
3. The workflow and architecture of the proposed system is explained in Section 10.3
4. Research analysis has been done in the fourth section.
5. In the last section, conclusions and future work are explained.

## 10.2 Literature survey

For Sign Language Recognition many explorations are done to detect signs through images and videos using different methods and algorithms. We have surveyed significant papers from 2010 to 2021 and application of machine learning and deep learning techniques are chosen as the focus of the study. Dataset is used in the form of videos and images. We have used the image dataset.

The Literature review of the paper has been carried out by reviewing various papers and articles but we have shortlisted and mentioned twelve main papers (both research and review). We followed the guidelines of Preferred Reporting Items for Systematic Reviews and Meta Analyses (PRISMA) flow diagram and created one as shown in Figure 10.2.

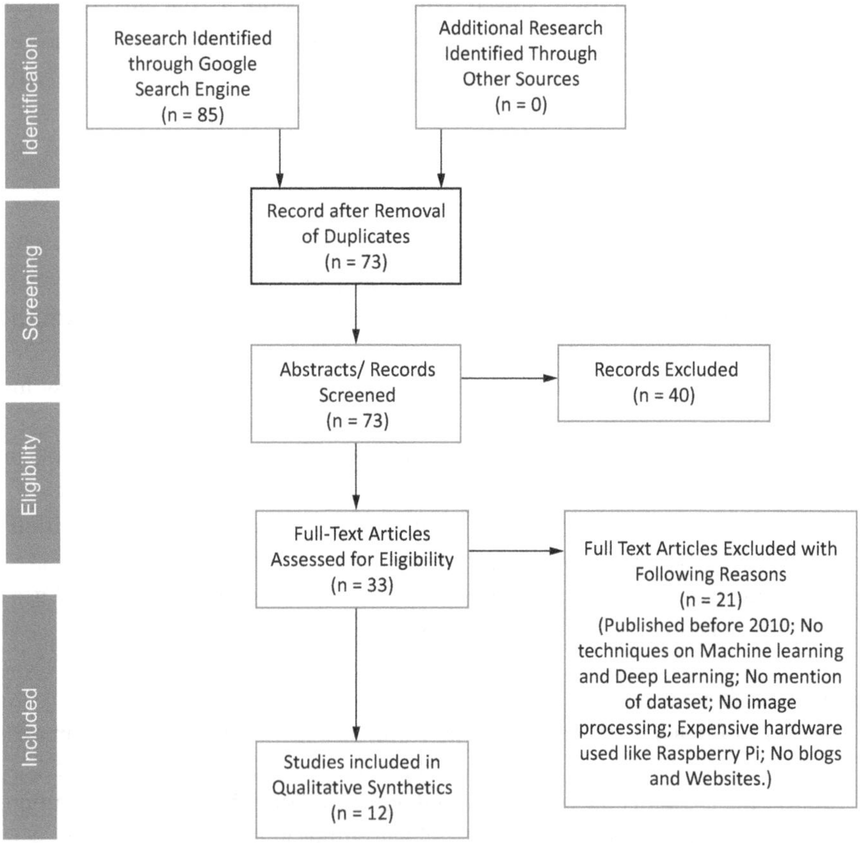


**Figure 10.2:** Preferred Reporting Items for Systematic Reviews and Meta Analyses (PRISMA) flow diagram for systematic review detailing searches, number of abstract/records screened and full text articles retrieved where n = number of papers referred.

## 10.2.1 Eligibility criteria and information sources

As our primary focus were deaf and dumb people, therefore, projects related to computer vision and people having disabilities are considered. Because of the recentness in this field, there are many industry based applications and projects going on. Therefore, we have also considered many different web sources for this paper. This significance of topics led to finding applications of computer vision in the health care sector primarily focussing on people with hearing impairments. Many research papers, web sources and articles have been referred to related to computer vision and deep learning.

This project only focuses on people suffering with hearing impairment in society and people having no disabilities are not considered. Therefore, related phases have been searched and topped ranked papers in this area have been taken into attention. Table 10.1 shows the Inclusion and Exclusion criteria for the same.

**Table 10.1:** Inclusion and Exclusion Criteria for the presented study.

| Topic of Study | Inclusion Criteria | Exclusion Criteria |
|---|---|---|
| Publication type | Publication in or after 2010 | Publication before 2010 |
| Paper type | Journal, Conferences or Workshop papers | Blogs or Websites |
| Paper Written in | English Language | Any other Language than English |
| Study type | For Deaf and Dumb People | Any other category |
| Studies Focused on | Software Engineering Education | Hardware Engineering Education |
| Hardware type | Cameras or Gloves | Other expensive hardware like Raspberry Pi |
| Technique type | Machine learning and deep learning | Any other Techniques except machine learning and deep learning |
| Outcomes | Gesture Recognition | Any other |

Keywords used in the search for the survey were "*Gesture recognitions*" AND "*sign detection*" AND "*CNN Technique*" used for detection etc. We carry out a survey of these papers and have summarized the findings. The significant papers and their highlights are shown in Table 10.2.

The paper [10] mainly focuses on classification and recognition of ISL by using dynamic gestures in real time. They used a consistent image background while capturing the video. Usually eliminating the background is very complicated and it reduces the classification accuracy. They recorded the video using a Sony handy came with 2.5 megapixel resolution. Their dataset consists of 11 specific ISL words. They applied two classifiers and found out that K nearest neighbour gives a good classification result

Table 10.2: Survey of machine learning (ML) and deep learning (DL) techniques in the sign language and gesture recognition.

| Citation No. | Year | Learning Techniques (ML / DL) | Feature extraction techniques | Classifier Techniques | labels | Dataset (Indian/US) | Video / Image Dataset |
|---|---|---|---|---|---|---|---|
| [10] | 2010 | Machine Learning | – | Euclidean distances and K nearest neighbor | 11 Words | Indian | Video |
| [11] | 2012 | Machine Learning | PCA | – | 25 Alphabets and 9 Digits | Indian | Image |
| [12] | 2015 | Machine Learning | Scale invariant feature transform (SIFT) | Support Vector Machines | Alphabets | Indian | Video |
| [13] | 2016 | Deep learning | PCANet | – | Alphabets | Arabic | Image |
| [14] | 2018 | Machine learning deep learning | Continuously adaptive mean shift (CAMShift) | CNN, AdaBoost | 10 Digits | | Video |
| [15] | 2020 | Deep learning | – | CNN | Alphabets and numbers | American | Video |
| [16] | 2020 | Deep learning | HSV (Hue,Saturation,Value) | CNN | Alphabets | American | Image |
| [17] | 2020 | Deep learning | – | CNN | Alphabets and Words | Indian | Image |
| [18] | 2020 | Deep learning | – | CNN | Alphabets and 2 classes (Delete & space) | American | Image |
| [9] | 2021 | Deep learning | 3D-CNN | LSTM | Alphabets | Indian | Image |
| [19] | 2021 | Deep learning | – | CNN | Alphabets | American | Image |
| [20] | 2021 | Deep learning | – | CNN | Digits (0–9) | Bhutanese | Image & Videos |

but not satisfactory. Further they proposed to apply a Support Vector Machine (SVM) classifier and applied statistical Hidden Markov Model (HMM) to improve classification performance.

ISL Recognition is done in [11]. They used 3 Mega pixel cameras for data acquisition, where they maintained the image intensity throughout the experiment by using table lamps having uniform light intensity. The person used to wear red and blue coloured gloves for two different hands. They collected a total of 510 images and for each sign they took 15 images and stored them in a database. Their system is made up of alphabets and numbers. They have carried segmentation using coloured gloves and recognition of hand gestures was carried out using Principal Component Analysis (PCA). The live data frames were recognized using PCA. The system was successfully able to recognize the signs made and gave the accuracy of 94%. They also use fingertip algorithms to make a database based on the number of finger tips. They get problems in the signs in which both the hands overlap each other and signs which have motion in them. They further added that the recognition can be done in future using neural networks.

A system [12] works on ISL recognition. They collected dataset from a school of deaf "Jyothi Badhir Vidhyalaya" located in Bithoor, Kanpur. A video of around 6–11 s was taken for each alphabet from 8 different students with different skin tones using a 30fps camera. They try to take videos with background as light as possible to make segmentation easier by tilting and moving the camera so they can get images from different positions too. They observed that similar looking gestures are there for alphabets like (E, F), (U, V, W), (M, N) and better classification of clusters may result in better accuracy.

This paper [13], is proposed for fingerspelling sign recognition using Arabic language in which the image is captured from a Soft Kinect camera. They observed that by applying features extraction, depth images yield better accuracy than for intensity images. Hence, the blending of both these techniques for good performance. They tested the dataset which was collected from 3 different users and achieved an accuracy of 99.5%.

In system [14] 10 hand gestures were adopted in the experiment. 1600 pictures were taken for training each gesture and 400 pictures for test each type was taken. The system was unaffected by the hand gestures but was not able to execute due to the objects having the same skin colour like face. So, the restrictions were on the system's background. LeNet-5 network is used to recognize hand gestures and realize the classification of 10 digits. The LeNet-5 includes layers like convolutional layer, pooling layer, fully connected layer and softmax layer.

In [15], the signs made by the deaf person are recorded using the camera. The video is then split into frames and each frame containing a sign is converted into a binary image using the thresholding technique. They used CNNs as a classification technique. The data from the binary image is passed through dense and flatten layers, and maximum pooling layers, before being compared with the training data.

The output label that presents the highest accuracy is displayed on the liquid crystal display (LCD) screen, thus successfully converting sign languages into standard text based languages.

A system [16] focuses on ASL with 2000 images. There were 10 classes of different alphabet signs. Two datasets have been made by 2 different signers. They found out different technologies like artificial neural network (ANN), LSTM and 3D CNN most of them require extra computing power. So, they used CNN to categorize 10 alphabetical American sign gestures and yielded an accuracy of 98% and their system requires low computing power.

This paper [17] deals with static signs for sign language recognition. Total 35000 gesture images of about 100 class signs classes which includes 23 alphabets of English language, The digits from zero to 10 and other 67 commonly used words are collected. The efficiency had been evaluated on 50 CNN models. The model was tested with various optimizers like Stochastic Gradient Descent (SGD), Adam, Adagrad, RMSprop and others. They trained their model using AlexNet and Adam which gives less accuracy so they reduced the number of layers and found that accuracy increased. Using Adam as an optimizer yielded with best classification accuracy as 99.17% and 98.80%. The work found out that Stochastic gradient descent (SGD) performed better with classification accuracy of 99.90% in training and 98.70% in validation respectively.

This paper [18] is focused on fingerspelling in ASL in its static form. A total of 28 classes were collected by them, which includes 26 alphabets and two classes of space and delete. For each image they took varied background and lightnings and different angles of view for each sign. The total number of datasets used were 2200 for each class. For training a total 43,120 was used and 18,480 for the validation of datasets. For the test 2816 images were kept. They applied CNN as a classification technique. Softmax activation function was used by them for hyper parameter tuning. The maximum accuracy obtained for training was 98.53% and for validation 98.84%. The lowest accuracy achieved was 98.6506% for testing.

A study [9] was done on ISL recognition. It deals with a vision based Approach. The real time image will be captured from the camera and uploaded in the system then the image hand gesture will be converted into its meaning. The speaker button is also added to convert from text to speech. The system poses no threat of hardware malfunction and yields better accuracy and efficient results.

This paper [19] aims to produce a system that recognizes hand gestures and signs. Their dataset consists of a total 24 alphabets except J and Z from ASL alphabets. The size of images was taken as 28 × 28 pixel images, with background subtraction. They implemented the project by using several ways and methods like k-nearest neighbors (KNN) algorithm, Logistic Regression, Naïve Bayes Classification, Support vector machine (SVM) and CNN. The method which gives them better accuracy was CNN compared to other methods they tried out. They used CNN to recognize the alphabets. The layers used by them were Conv2D and Maximum Pooling followed by fully connected layers. They also used an activation function in their last layer. Softmax as an activation function was used to give output. They take images as an input and are able to convert sign language into text format as an output of the system.

The paper [20] presents the Bhutanese Sign Language (BSL) using CNN. They prepared the dataset with images and video frames. They collected 20,000 RGB images of 10 digits from Thai students. They captured images and videos from different angles, different backgrounds, different positions and lighting. In their system features were extracted using CNN algorithm. They use different convolutional layers, filters of different sizes, fully connected dense layers, rectified linear unit (RELU) and SoftMax activation functions. They get the training and testing accuracies as 99.94% and 97.62%, respectively, for their proposed model.

## 10.2.2 Discussion

After going through all the above papers, we come to know that most of them are focussing on machine learning and nowadays on deep learning. In [10] they found out K nearest neighbor to be the good classifier but further they planned to use SVM to improve classifier performance. In [12] they used SVM for classification and found out that for similar looking gestures like (E, F), (U, V, W), (M, N), (2 with V) and (3 with W) they are not getting good accuracy after classification. In [13] they only used PCANet for feature extraction. They observed that by using only feature extraction, depth images give better accuracy than intensity images and by combining both together result in good performance. In [14] both the techniques (feature extraction and classification) are used to see how they both will give accuracy when combined together. They use CAMShift and CNN, now the model was not affected by similar hand postures but was not able to execute properly due to objects having similar skin color. Here, CNN gives good accuracy and solves the problem of similar looking postures but only for 10 digits.

In [15] they use only CNN for both alphabets and digits dataset and they get a good accuracy with different layers. The problem we found was that during an emergency the disabled person used all the alphabets to give out a final word. So, taking this paper as a reference we add some emergency words in our dataset.

Further research [16] was also done using LSTM, 3DCNN and ANN and found that most of them require extra computing power. LSTM was used here to produce output sequentially. In [18] they used a larger dataset of alphabets and two classes (space and delete), they achieved great accuracy using CNN. So, the only use of CNN was to give great accuracy and there is no use of other extraction techniques like PCANet and CAMShift. Also, by referring to research paper [20] it was clear that using 3DCNN as a classifier will use extra computing power.

By going through all the research papers, it was clear that ISL recognition can be done only by using CNN with a dataset containing alphabets, digits and some extra classes of words.

### 10.2.3 Long short-term memory: framework for text classification and prediction from gestures

Our proposed framework aims to use a sequence model problem to deal with text classification. For this we read various articles, blogs, journals and papers which tells us about CNN used with LSTM) which will give us the most accurate output with meaningful sentences. We divide our project in two phases. Phase 1 consists of implementation of CNNs (done in this paper) and Phase 2 gives us information about the NLP model which is used for sequence model processing. In this paper, we mainly focus on Phase 1 of our project and implement it with great accuracy. For NLP we have made a short plan which is explained below.

If we talk about solving sequence model problems, NLP is the best architecture on which one can think of. There are various sequence to sequence problems which includes Machine Translation which is used to translate a series of words from one language to another, Text Classification is another sequential problem in which a sentence should be formed sequentially having a meaningful sentence, and Video Captioning also uses sequential modelling in which the subtitles are created automatically for each frame. From all the above sequential problems we are mainly focussing on Text Classification; as in our system, the final output will be in form of text or digit, which need to be sequentially arranged and create some meaning to a sentence.

In Text Classification there are many classification algorithms like SVM, Decision Trees and hidden Markov model (HMM) used for classification but we have mainly focussed on LSTM for various reasons. One of the reasons we are using LSTM for classification is that LSTM memorizes the important information. Another good reason we found about LSTM is that it stores the data sequentially. If we look toward other classification techniques, we can see that they are predicting output but the words joined together are not forming a meaningful sentence. In LSTM we will get actual output which has actual meaning.

We have used CNN in our project but there is a drawback of using only CNN which is CNN cannot accept sequences of images. That's why we are using sequential processing NLP which allows us to extract important information and also stores the data or images sequentially so that the output will be the sequence of words one can understand. In our system we will be using an architecture of RNN which is LSTM. LSTM will help us to store memory for long term as well as for short term. Proposed ISL Recognition System which consists of CNN–LSTM architecture is shown in Figure 10.1. Here the CNN is used for feature extraction and classification whereas LSTM will be used for sequence prediction.

After extracting the features from CNN layers, we can use a network of layers of LSTM which will be followed by other classification layers which is in our case fully connected layers and softmax layers. The linearly sequential model consists of encoder and decoder. The encoder will encode the image pixels and the decoder will then decode it for image gestures. The encoder will understand the sequence it will get from CNN layers and then send it to the decoder. The decoder will produce its own sequence that will be represented as an output. The module will have two states: the cell state and hidden state. The cell state will store the information which will be further used by the model. Long-term memory is used for gesture identification which is stored by state cell and short-term memory is stored by hidden layer which is used for maintaining the proper sequence.

### 10.2.4 Research gaps

By going through various papers, surveys and web sources, we have come across various research gaps. The research gaps from some of the sources are mentioned below:

1. Most of the works focus on the universal ASL. ISL lags behind due to lack of ISL datasets.
2. There are also similar sign gestures for some alphabets and digits. E.g., V and 2 have the same sign and W and 3 also have the same sign.
3. By including a greater number of training samples, the classification and recognition results can be improved.
4. Many research papers have focused on only alphabets and digits and lack attention to emergency words.
5. Literature shows that on an average the standard dataset was of size 2,000 for each class.

## 10.3 Proposed system

In this section, methods and the materials which are used to achieve the gesture recognition have been explained. The different modules of the system architecture are shown in Section 10.3.1. The algorithms and layers used in detecting gestures are explained under Section 10.3.2.

### 10.3.1 System architecture

ISL Fingerspelling Recognition was built using the CNN algorithm using RGB images. It consists of a total of 42 classes.

Extracting features through the frames and predicting the hand gestures have been done using a CNN model. The CNN architecture contains some convolution layers, each having a pooling layer and activation function. It also contains a set of fully connected layers. As soon as one image moves across the network, its size reduces. This happens because of maximum pooling. The prediction of class probabilities has been given in the last layer.

Figure 10.3 represents the ISL system architecture which is divided into three phases. The first phase is the input layer which takes input as an image and passes it to the next layer. The second phase is the model layer of CNN and the last layer consists of an output layer where output is displayed in the form of text or digits.

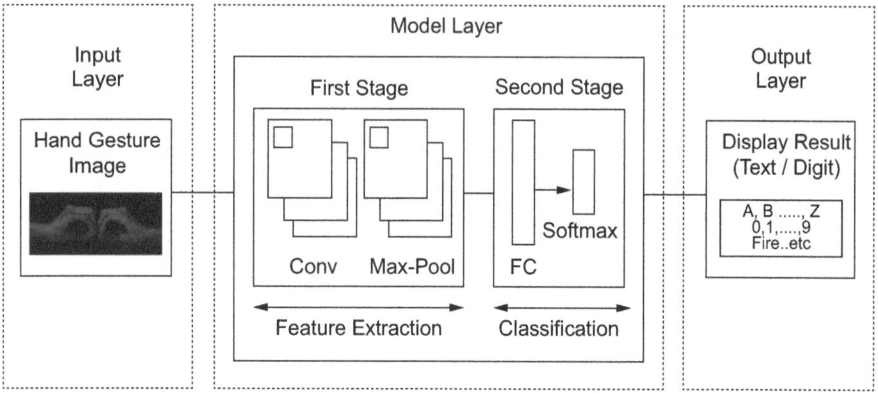


**Figure 10.3:** System Architecture. (Output is the alphabet/number in text).

#### 10.3.1.1 Input layer

In this system the dataset consists of different RGB images for different sign gestures. The dataset consists of 50,436 images which is used to train the CNN. These images are further classified into 42 different classes. Each class consists of 1,200 sign images. Our classes have 26 alphabets of English (Figure 10.5), 9 digits ranging from 1 to 9 (Figure 10.4) and 7 emergency words (Figure 10.6) which includes "help," "house," "name," "no," "pay," "stop" and "yes." We split our dataset in the ratio of 70:30 for training and validation, respectively.

The emergency words added in our system will lower the efforts taken by the disabled person to communicate with someone. Every time the disabled person needs to show a separate sign for each alphabet so that the system will generate the whole word. So, it was taking a lot of time to communicate with each other even during emergency situations. So, by adding emergency words in our system will help disabled people to directly show a sign of a particular word (used during an emergency) and the person from whom he is asking help will be able to help him as soon as possible.

Figure 10.7 represents the histogram representation of the number of sample images in digits, alphabets and emergency words. Here, we can see that the blue represents the 9 classes of digits which have total 10,800 sample images, the

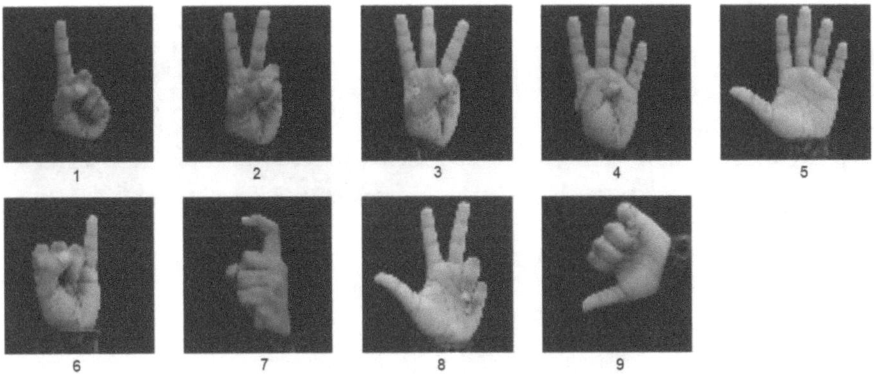


Figure 10.4: Indian Sign Language images for digits (1 to 9).

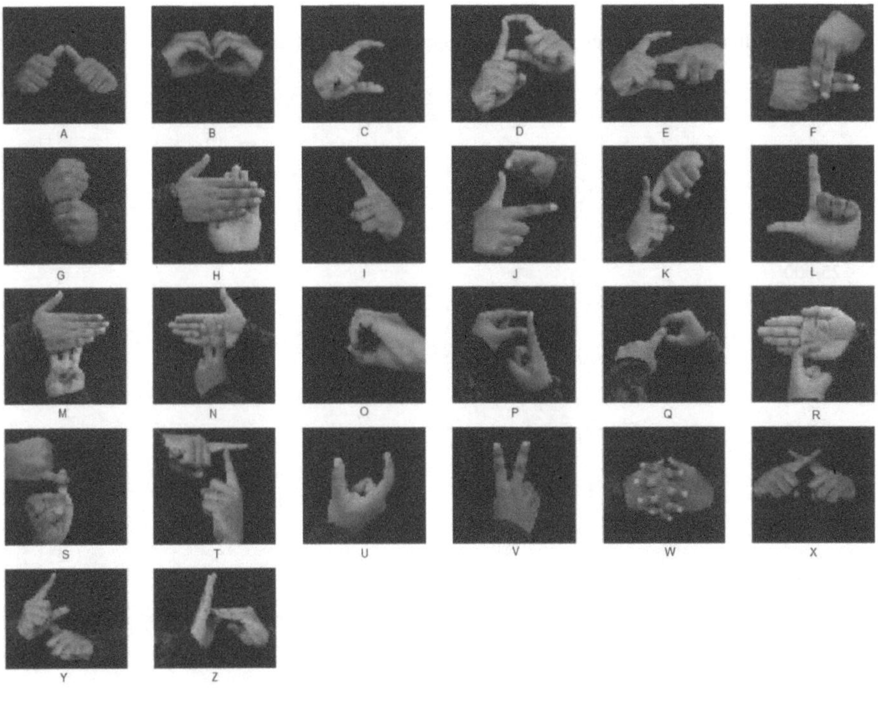


Figure 10.5: Indian Sign Language images for alphabets (A to Z).

orange represents the 26 classes of alphabets having 31,200 sample images, and finally the gray color represents the emergency words which consists of 7 classes and a total of 8,400 images.

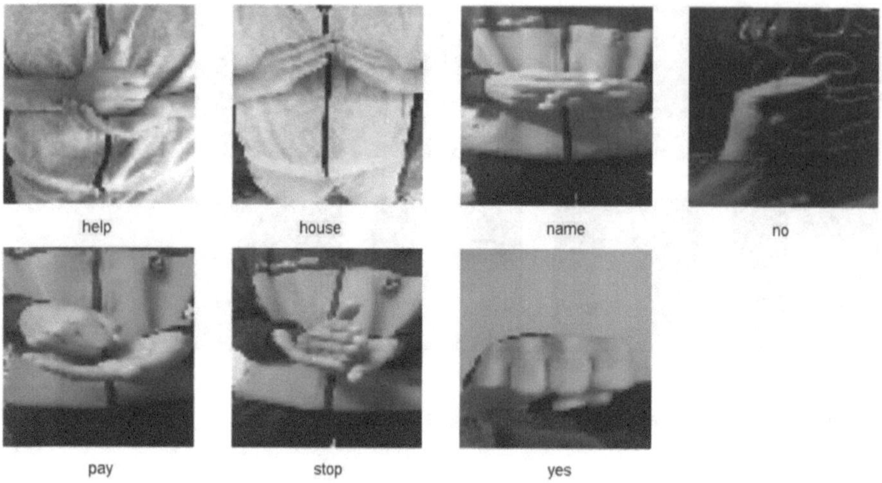


**Figure 10.6:** Indian Sign Language images for emergency words (7 words).

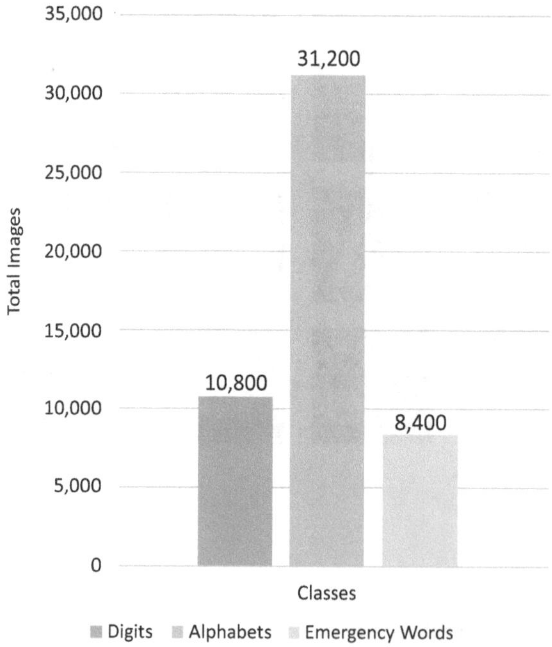


**Figure 10.7:** Histogram showing total sample images of all classes.

#### 10.3.1.2 Model layer

CNN or ConvNet is a type of deep learning technology. CNN is classically used to recognize patterns. CNN recognizes patterns in images and also is used for computer vision, NLP, signal processing and other purposes as well. If we talk about CNN architecture, it is similar to the connectivity patterns (neurons) in the human brain. CNNs have filters which are also known as kernels (in a form of matrix) that are used to detect the features from input data, features such as edges.

We can say that a CNN model is a combination of two components: Feature Extraction and Classification. CNN also performs four main operations:
- Convolution layer
- RELU activation function for learning.
- Pooling layer (max classifier) for downsampling.
- Fully connected dense layer for classification

From the above operations the convolution and pooling layer is used to perform feature extraction. For example, in our system feature extraction is used to detect hand gestures. Fully connected layer is used to perform classification. In our system classification is done by using fully connected dense layers and softmax layers. We referred the paper "Application of Machine Learning Algorithms for facial expression analysis" [21] to know about extraction processes.

Our architecture uses convolution layers to classify the output. In CNN, the main building blocks are convolutional layers. CNN is used for image classification, in our case to detect the gestures. Our architecture consists of different layers of CNN layers which includes convolutional layers, pooling layers (max pooling), RELU, fully connected dense layer and softmax layer (output) for image processing, classification and segmentation. Architecture of the summarized model is shown in Table 10.3.

It has feature maps and hidden units as shown in Figure 10.8, where the feature map is the output of convolution layer and maximum pooling layer. Hidden units have fully connected layers and softmax layers.

Convolutional layer receives images as input and filter are used to learn features from the images. Pooling layer performs operations on the input image to compress the image using max pooling. Now the matrix is flattened and is fed to a fully connected dense layer. Softmax (activation function) works as the final layer of our system and is used to classify the output.

#### 10.3.1.3 Output layer

Output layer forms a communication between user and system. System will display the output on screen in the form of text or digits. Output layer is the final hidden layer which is softmax. Softmax is the activation function which is used as a classifier. This layer receives input from the last hidden layer, performs the calculations and produces the final result. The output layer consists of 42 layers (number of neurons) which is equal to the number of classes in our dataset.

## 10.3.2 Workflow model

For a good training of deep neural network, a proper powerful GPU hardware is required. This project will also require GPU hardware for training. But we decided to reduce the hardware load by using an online platform which is Google Collab and collecting the dataset from two different sources then we merge them together.

The workflow model represents the consecutive series of tasks. This section explains feature extraction and classification.

### 10.3.2.1 Feature extraction

This section explains the dimensionality reduction in which a preliminary set of data is converted to manageable groups for processing. In the end, reduction of data is helpful in building the model with less machine's effort and this eventually increases the learning speed and generalization of steps in the whole process.

In our system we used CNN or the ConvNets to get the sign or gesture as a text. A neural network which consists of one or more convolutional layers are mainly used for image processing, classification and segmentation.

In our model we use following CNN artifacts:

1. Convolutional layer
2. Pooling layer (max classifier)
3. Hidden layer with RELU
4. Fully connected layer
5. Classification layer using softmax activation

The convolution layer takes the input as an image and the last layer, which is a fully connected dense layer, is connected with the output layer (softmax). We had also used dropout layers after every maximum pooling layer to regularize the training. The activation function softmax is used after the fully connected layers to give us output.

The first layer of our model is the convolutional layer (input layer) which detects various features such as fingers and environment from the image. The convolutional operation is applied on the entire image using matrix and kernel of $3 \times 3$ with RELU activation function. Convolutional operation is the mathematical operation applied on the original image. Filter is applied of a particular size of $3 \times 3$ on the input image. The output is fed to the maximum pooling layer to learn several other features of the image.

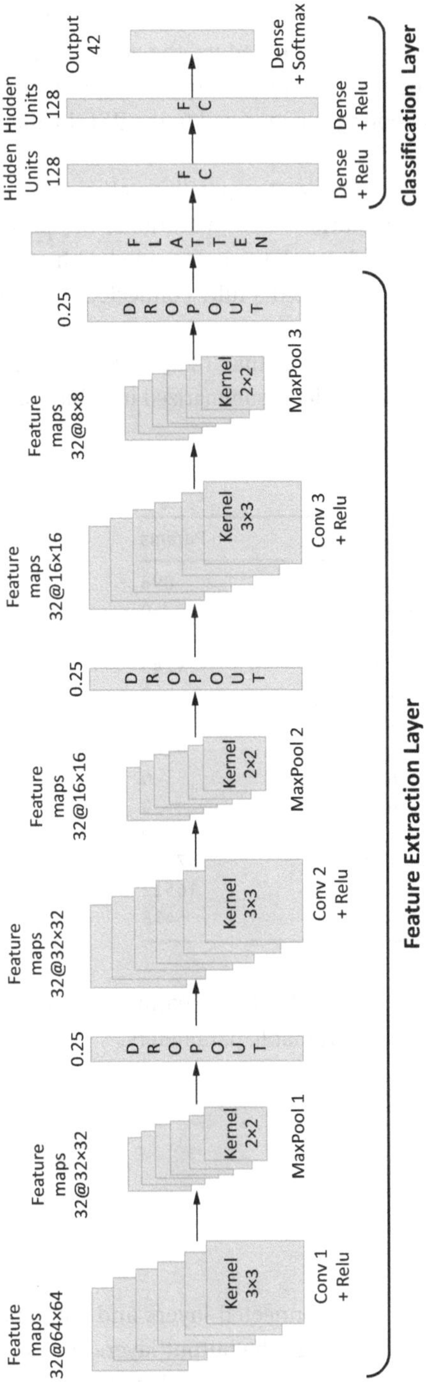


**Figure 10.8:** CNN layers.

Pooling layers (max classifier) reduce the size or dimension of the convoluted image. Pool size is selected as 2 00D7 2, which is used to reduce the actual size by scanning each pixel of the image. In our model we use maximum pooling in which the largest element is taken. The maximum pooling is connected to the RELU which is also known as a positive function.

An RELU, which is an activation function, enables the positive value and for negative value it proceeds zero. After all convolutions and operations, the processed image is then sent to the successive layer. Every layer is cascaded with each other which carries arbitrary values called weights called a fully connected layered network.

Our CNN input layer is convolutional layers and other than that, we have a softmax which is an output layer. The developed CNN model is summarized in Table 10.3.

**Table 10.3:** Model summary.

| Types of layers (layer no.) | Output shape | Params |
|---|---|---|
| Convolutional layer (Layer 1) | (None, 62, 62, 32) | 896 |
| Maximum pooling layer (Layer 1) | (None, 31, 31, 32) | 0 |
| Dropout layer (Layer 1) | (None, 31, 31, 32) | 0 |
| Convolutional layer (Layer 2) | (None, 29, 29, 32) | 9248 |
| Maximum pooling layer (Layer 2) | (None, 14, 14, 32) | 0 |
| Dropout layer (Layer 2) | (None, 14, 14, 32) | 0 |
| Convolutional layer (Layer 3) | (None, 12, 12, 32) | 9248 |
| Maximum pooling layer (Layer 3) | (None, 6, 6, 32) | 0 |
| Dropout layer (Layer 3) | (None, 6, 6, 32) | 0 |
| Flattened layer (Layer 1) | (None, 1152) | 0 |
| Fully connected dense layer (Layer 1) | (None, 128) | 147584 |
| Fully connected dense layer (Layer 2) | (None, 128) | 16512 |
| Fully connected dense layer (Layer 3) | (None, 42) | 5418 |

In the Table 10.3 mentioned above, there are a total of 188,906 parameters out of which 188,906 are trainable parameters and 0 non-trainable parameters.

(Total prams: 188,906

Trainable params: 188,906

Non Trainable params: 0)

#### 10.3.2.2 Classification

In our system classification layers consist of fully connected layers and softmax layers. Fully connected dense layer is positioned before the output layer. The system will first flatten the entire image and then feed in the FC layer where the mathematical operation will occur. The fully connected dense layers (Layers 1 and 2) as shown in summary model in Table 10.3 are layers which are fully connected layers. These dense layers are added with RELU.

The final layer of our model is softmax. As this is a multi-classification problem, we used softmax in our output layer as a classifier. Our system saves the learning model which is generally used by the system to predict new image inputs.

## 10.4 Result

Our model includes 42 classes of 9 digits, 26 alphabets and 7 emergency words. Which are taken from two different sources. Digits and alphabets are taken from Kaggle and emergency words from GitHub then we merge them together. We found 1,200 images for each set of classes and there was no need for data augmentation as it was a well-balanced dataset.

The implementation of our model is shown in Figure 10.3, we used Google Collab for execution of our model. In our dataset we are having a mixed dataset of images containing both plain images as well as images having other objects in the background. We used a total of 35,306 images to train, 7,565 images to validate and 7,565 images to test the CNN. For our model we split the dataset in the ratio of 70:15:15 for training, validation and testing. Our model was successfully able to give us an accuracy of over 99.80% and validation accuracy of 100%.

For better accuracy we started training our model with the minimum epochs of 10 which gives us accuracy of 99.70% and validation accuracy of 99.95%, then we increase our epochs to 20 to which we observe the slight increase in our accuracy to 99.77%. Then, with the epochs of 30 we successfully get an accuracy of 99.80%.

Figure 10.9 displays the training, testing and validation accuracy with different epochs. Here we can observe that at 30 epoch the training, testing as well as validation accuracy goes highest which shows that as we increase the number of epochs the classification accuracy of our model also increases. Here, after 30 epochs we also tried for 50 epochs but the accuracy remained the same which was around 99%.

Accuracy curve shows the curve for both the training and validation dataset. The accuracy curve in Figure 10.10 shows accuracy versus epoch graph where the accuracy of 30 epochs for training and validation dataset is plotted.

The graph shown in Figure 10.11 is of loss versus epoch which is a plot of loss on the training and validation dataset where blue shows the training data and orange shows the validation set for 30 epochs.

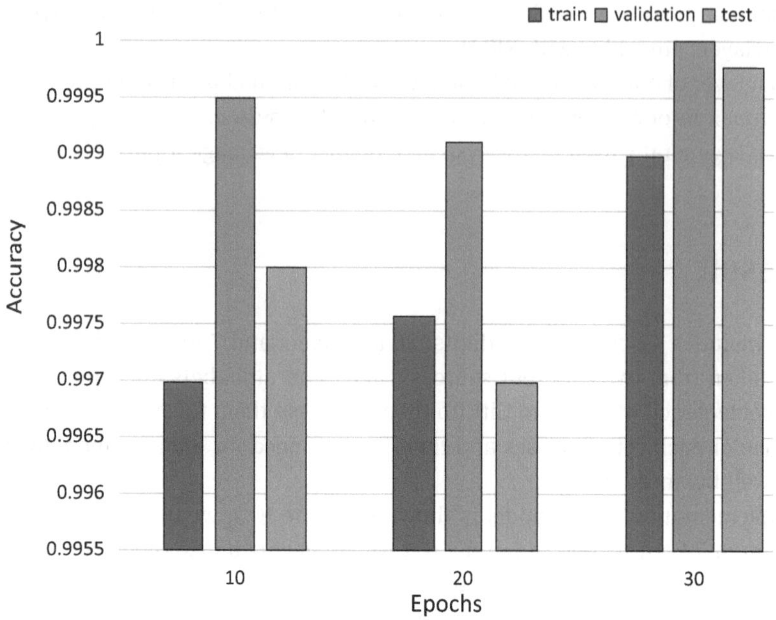


**Figure 10.9:** Loss curve for training and validation dataset (30 epoch).

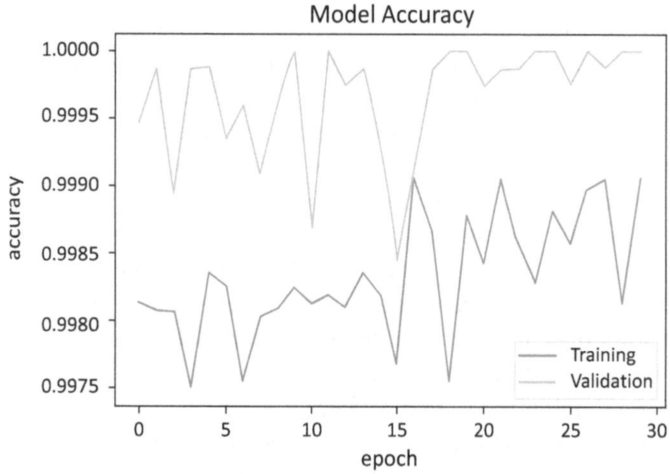


**Figure 10.10:** Accuracy curve for training and validation dataset (30 epochs).

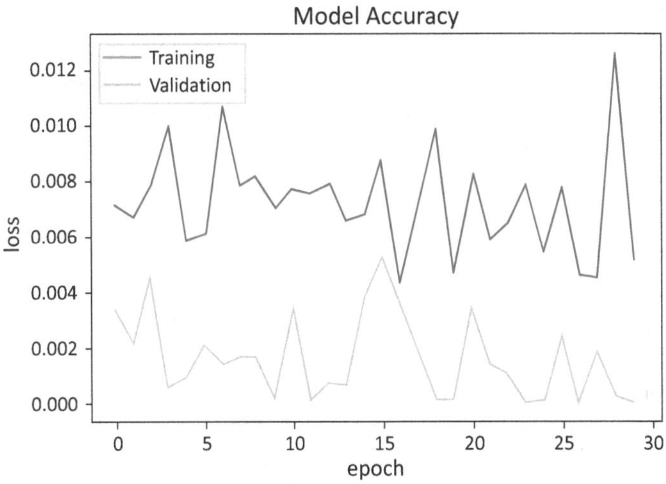


**Figure 10.11:** Training and Testing accuracy curves on the model.

## 10.5 Conclusion

There are various approaches to design a Sign Language Recognition for impaired people (deaf and dumb), most of the works are focussed on ASL and lesser work was found on the ISL. So, we decided to go ahead with ISL. It will offer help and give support to the people with disabilities to communicate in a better way with the societies in India and in the sign language they know (ISL). Our project will help deaf and dumb people to communicate with society digitally, so it will be reliable and efficient.

ISL has some challenges that need to be resolved. Here it requires both hands for communication and overlapping of hands also make ISL more unclear and sign language more complex. To know more about the ISLs, challenges ISL have and technologies used in different papers to solve the complexities, we had gone through various articles, journals and papers. After going through various papers, journals, conferences and blogs we finally decided to go ahead with deep learning techniques which have developed a lot in image segmentation and classification.

Our dataset consists of 42 classes which includes 26 classes of alphabets, 9 classes of numbers/digits and 7 emergency words like "help," "yes," "stop" and "home." We decided to include emergency words in our dataset because during emergency situations or places where the signers (a person who knows sign language) will not be available, the disabled people will be able to communicate quicker with the people around them.

The presented work is the Phase 1 of a system which establishes effective communication between hearing impaired individuals and the outside world. The proposed system is Phase 1 development of a complete system which is aimed to interpret a sign language, convert it into text. The text however is done by character recognition, digit recognition. It is a vision-based approach and employs deep learning-based approach. The dataset presented performs well in accuracy with CNN classifier. Our system is modeled with convolutional layers, followed by pooling layer (max classifier), RELU, fully connected dense layer and softmax layer for classification. Feature extraction and classification was done using these layers as shown in Figure 10.8. Softmax is the activation function which receives input from the last hidden layer and is used as a classifier to perform calculations and produce output.

In Phase 2 of our project, we will be working on sequence models that sequentially take input and output in the form of image, speech, video or text. RNN is popularly known for sequence modeling and is very efficient. NLP is an architecture of RNN which will be used in our project to generate the output sequentially. In our project long short-term memory will be used for text classification which is explained above in Section 10.2.3. LSTM will help us to improve the model's performance.

In Part 1 of our project we successfully trained, tested and validated the dataset. The system was able to classify all the 42 classes (26 alphabets, 9 digits and 7 emergency words) from our dataset which was collected from two different sources and was able to achieve an accuracy of 99.80%. Currently we tried to solve the problem using deep learning methodologies. We will work further in this project to improve our performance and thus the module can be replaced with a more accurate one to achieve better accuracy.

## 10.6 Future work

**Larger Dataset**: We were looking forward to improving our model by adding more words and increasing the size of the dataset with more vocabulary which are most commonly used in our everyday life. By adding more words which are commonly used it will be easy for the disabled person to directly show the sign of a particular word instead of showing many alphabets and form a word.

**Speech Recognition**: We would also like to add speech recognition so it will also be helpful for blind people. Blind people will use the signs they learned and the person who is trying to communicate will use the system to understand him and maybe help him in a way he/she can. Including speech recognition in our system will not only help blind person and normal people but also used by deaf and dumb people by putting the input in the system and system will directly convey their emotions or speech to the person they are trying to communicate with. So, here we can see that speech recognition is not only used by blind communities who use sign language but also by all the people who use sign language as their basic language to communicate.

**Better Classifiers:** For that we can use LSTM for speech recognition and also for text processing. To communicate in a better way, we will also add text to the sign module so that the user will be able to respond.

Long short-term memory may also be used in a system where auto spell or auto correct is used. Let's say a disabled person wants to communicate with someone, he needs to show a sign for all alphabets that come in a word. This will take a long time to understand what they are trying to tell or want to share. By using a system with auto spell or auto correct, the person need not to show all the alphabets one by one after showing some the system will automatically display the remaining word. This system will perform in the same way google does when we type something in the search bar.

## References

[1] https://nhm.gov.in/index1.php?lang=1&level=2&sublinkid=1051&lid=606#:~:text=Hearing%20loss%20is%20the%20most,at%206.3%25%20in%20Indian%20population.

[2] https://tulsaworld.com/archive/what-is-wrong-with-the-use-of-these-terms-deaf-mute-deaf-and-dumb-or/article_7c50d337-9ceb-5278-b59c-ea201f01ef46.html

[3] Goldin-Meadow, S., Brentari, D. (2017). Gesture, sign, and language: The coming of age of sign language and gesture studies. The Behavioral and Brain Sciences, 40, e46. 10.1017/S0140525X15001247.

[4] Sinha, S. (2017). Indian sign language: An linguistic analysis of its grammar.

[5] Rojasara, D., Chitaliya, N.G. (2013). Indian sign language recognition – a survey. International Journal of Engineering Research & Technology (IJERT), 02(10).

[6] Savant, R., Nasriwala, J. (2019). Indian sign language recognition system: Approaches and challenges.

[7] Yim, J., Ju, J., Jung, H., Kim, J. (2015). Image classification using convolutional neural networks with multi-stage feature. In: Kim, J.H., Yang, W., Jo, J., Sincak, P., Myung, H., eds. Robot Intelligence Technology and Applications 3. Advances in Intelligent Systems and Computing, vol. 345. Cham, Springer. https://doi.org/10.1007/978-3-319-16841-8_52.

[8] https://machinelearningmastery.com/gentle-introduction-long-short-term-memory-networks-experts/

[9] Chavan, A., Ghorpade-Aher, J., Bhat, A., Raj, A., Mishra, S. (2021). Indian sign language interpreter for deaf and mute people. International Journal Of Creative Research Thoughts (IJCRT), 2320–2882.

[10] Nandy, A., Prasad, J., Chakraborty, P., Nandi, G., Mondal, S. (2010). Classification of Indian sign language in real time. International Journal on Computer Engineering and Information Technology (IJCEIT), 10, 52–57.

[11] Deora, D., Bajaj, N. (2012). Indian sign language recognition, 2012 1st International Conference on Emerging Technology Trends in Electronics, Communication & Networking, 1–5, doi: 10.1109/ET2ECN.2012.6470093.

[12] Jain, S., Raja, K.V.S., Mukherjee, A., Indian sign language character recognition 2015 Indian Institute of Technology, Kanpur, 2015, Course Project-CS365A.

[13] Aly, S., Osman, B., Aly, W., Saber, M. (2016). Arabic sign language fingerspelling recognition from depth and intensity images, 2016 12th International Computer Engineering Conference (ICENCO), 99–104, doi: 10.1109/ICENCO.2016.7856452.
[14] Sun, J., Ji, T., Zhang, S., Yang, J., Ji, G. (2018). Research on the hand gesture recognition based on deep learning, 2018 12th International Symposium on Antennas, Propagation and EM Theory (ISAPE), 1–4, doi: 10.1109/ISAPE.2018.8634348.
[15] Mannava, V., Gnanasagar, V. (2020). Portable sign language translator for emergency response teams, 6, 2395–2566.
[16] Hurroo, M., Elham, M. (2020). Sign language recognition system using convolutional neural network and computer vision. International Journal of Engineering Research & Technology (IJERT), 09(12).
[17] Wadhawan, A., Kumar, P. (2020). Deep learning-based sign language recognition system for static signs. Neural Computing & Application, 32, 7957–7968. https://doi.org/10.1007/s00521-019-04691-y.
[18] Khalissi, A., Kadhim, R.A., Khamess, M. (2020). A real-time american sign language recognition system using convolutional neural network for real datasets. TEM Journal, 9. 10.18421/TEM93-14.
[19] Bhavana, D., Kishore Kumar, K., Chandra, M.B., Bhargav, P.V.S.K., Sanjana, D.J., Gopi, G.M. (2021). Hand sign recognition using CNN. International Journal of Performance Analysis in Sport, 17(3), 314–321. 10.23940/ijpe.21.03.p7.314321.
[20] Wangchuk, K., Riyamongkol, P., Waranusast, R. (2021). Real-time Bhutanese Sign Language digits recognition system using Convolutional Neural Network. ICT Express, 7(2), 215–220, ISSN 2405-9595. https://doi.org/10.1016/j.icte.2020.08.002.
[21] Uma Maheswari, V., Aluvalu, R., Chennam, K.K. (2021). Application of machine learning algorithms for facial expression analysis. Machine Learning for Sustainable Development, 9, 77.

Megharani Patil, Yash Gupta, Vatsal Kadakia, Harsh Doshi

# 11 Automated healthcare system using gamification and AI-based chatbot

**Abstract:** Due to the current global pandemic in the form of Corona virus, healthcare industry is under immense pressure to meet the ends of patient's needs. Doctors and nurses are working relentlessly to treat and help the patients in best possible way and still they are facing issues in terms of time management, technical resources, healthcare infrastructure, support staff as well as healthcare personnel. This type of viral pandemic has created a very high demand to use our resources and manpower to a great extent in order to just cure this virus. Due to this most of the medical professionals have been busy finding solutions and vaccines to treat the virus hence unknowingly decreasing the attention toward the treatment of patients suffering from various other existing diseases. Owing to the current technical advancements in the field of artificial intelligence and machine learning, we, as engineers, can help the healthcare industry to a significant level. In this paper, we aim to create an automated Healthcare System using gamification and AI-based chatbot which can prove to be a boon to medical professionals to assist them in treating the patients as well as justifying the robust technological advancements to be of great use. Chatbot is a software application designed to simulate conversation with human users through interactive and tailored content. It is often described as the most trending and promising expressions of interaction between humans and machines using artificial intelligence and natural language processing. Furthermore, we will add a module of gamification to make the website more interactive and user friendly such as healthcare quizzes and/or health score generators. Through this study, we aim to create a fully functional chatbot which will help the patients/users to know about the disease by simply entering the symptoms they possess. Additionally, they can find nearby hospitals and book appointments. This paper presents the study and usage of AI-based chatbot in healthcare industry for basic diagnostic evaluation and suggesting robust measures for the patients/users.

**Keywords:** chatbot, AI, gamification, NLP, healthcare

## 11.1 Introduction

Starting with a simple yet powerful quote: "Health is Wealth," which in itself is quite self explanatory stating that without proper and regular management of one's health everything falls apart. It is tragic to know that how a single virus can fail some of the best healthcare services around the globe. There are a lot of large-scale

https://doi.org/10.1515/9783110756722-011

health institutions like Hospitals to small-scale Clinics but most patients find it difficult to choose and find best suitable for them. Hence the proposed idea of Automated Healthcare System using AI-based chatbot helps the users/patients to find small clinics to major hospitals around their preferred location. Apart from that they can gain information of other various diseases and small problems right from a small wound to cancer from our chatbot named **MEDIXA.** The following Figures 11.1 and 11.2 help us to understand the growth of chatbots and increasing usage of the chatbots in different countries.

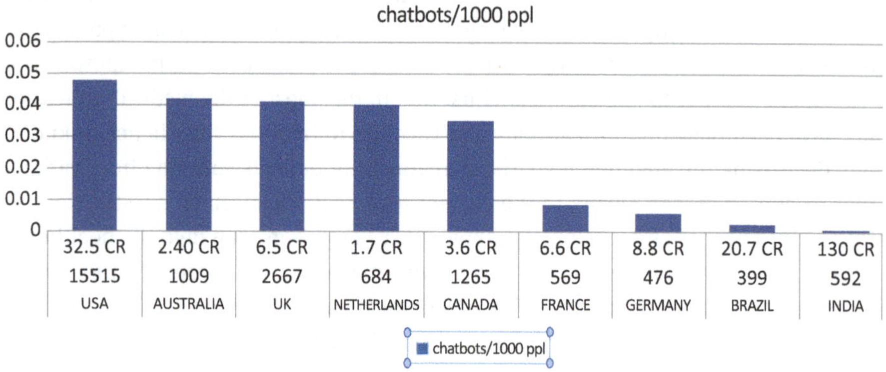


Figure 11.1: Chatbot usage per 1,000 people.

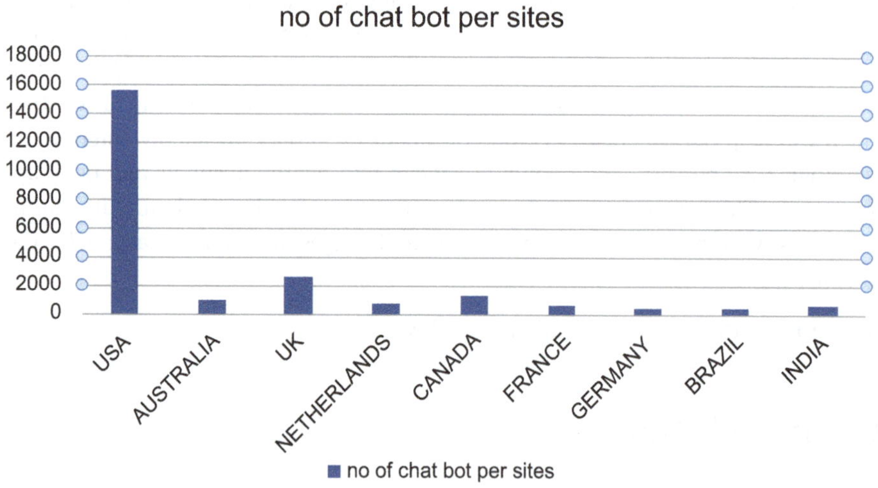


Figure 11.2: Number of chatbot used per sites in different countries.

We are now doing more than ever in the field of artificial intelligence in various sectors of our society. Right now healthcare sector is in the utmost need for it. According to IBM, “Artificial intelligence leverages computers and machines to mimic the problem-solving and decision-making capabilities of the human mind.” One of the reliable and helpful application of AI is a chatbot. A chatbot is a conversational agent which simulates real- time interactions with patients/users. It is a program that conducts conversation via auditory or textual methods. This proposed chatbot is based on textual method. The dataset which is going to used will be proficient in answering most of the queries of the users/patients. Dataset is the Json file with keys such as intention – what patient is expecting answers from the chatbot, responses and tags (specific disease). After that, data pre-processing will be carried out using natural language processing (NLP) which provides data for fitting it into different deep learning (DL) predefined algorithms.

The proposed model will be integrated into a website with better user interface (UI) design which will be filled with various other helpful and easy-to-access information about healthcare and services such as health quizzes and pop-ups which will justify the gamification module of our website hence making the title Automated Healthcare System using gamification and AI-based chatbot. Besides this, the main motto behind preparation of this proposed system is to help and inspire the forthcoming young batch of engineers who are interested to take a dive into the fields of AI, machine learning, deep learning and NLP. The entire working and use of technology is shared in detail in the following segments to help them understand the basics of the same.

## 11.2 Related work

We were truly motivated to build a chatbot by researching and referring various other existing chatbot system designs. Researching other papers and existing works always helps to build a better proposed system. Hence we have also jotted down a literature survey of our own recognizing the best few papers we encountered and presented a short summary of each paper to get the gist of their designed systems. It’s important to study the related works to help realize the need of additions to one’s own research/project. Like a chatbot named as “INFERMEDICA” with 81 + million heath checkups, introduce Symptomate which is a machine learning-based technology for detecting symptoms with chatbot. It is speech-to-text model or voice to text model, which is cross platform application which can be used in mobile phones and websites. After assessing user’s health condition based on symptoms and provide actionable recommendation. After conducting over 3 million diagnostic interviews company doubled its monthly recurring revenues in 2018. Other chatbot such as “sensely” which is computerized assistant which can assess patient checkups using images, videos, text and audio. Sensely interprets the user's symptoms and recommends a diagnosis, based on the gathered data as well as the information fed to its required algorithm. Another chatbot, "Florence" is basically a "personal nurse" in the color blue and is integrated with facebook messenger, kik and skype. Generally it notifies the user when and how to take their pills or drugs through text-to-text model. The chatbot notifies based on the various inputs provided by user such as name of medicine, time to take it and number of times in a day. The chatbot then processes the response through data saved in backend architecture of the system and gives timely reminder. These kind of application is useful for every other human being because one can sometime forget to take the pills on right time.

### 11.2.1 Literature review

Agrawal et al. [1] build a text-to-text Medical Diagnosis Bot that helps patient into discussion about their issues and gives a solution for their diagnosis based on their symptoms and their past records. But their algorithm accuracy and diagnosis accuracy was very low. "Diabot" a medical chatbot that was seem to be predictive using a term called "essemble learning" which is a meta algorithm that combines a bundle of weaker model and averages them to produce one final accurate model but the accuracy was lower than random forest model and concluding that ensemble model provides less accuracy [2]. A mobile-based app for health care using artificial intelligence called "Doc-bot" to predict the users diseases and provide them the solution to cure it [3]. A small application was made to guess patients diseases using their symptoms. First user has to create a account on this app by filling some basic details like username, password and a unique id is shared by this app. However in blank space the user has to fill the required information like age and sex to fix the diseases of the algorithm fixed here. Once the diseases is found many review questions can be asked to clarify the efficacy of the medicine of that diseases. Ghosha et al. [4], to handle complex question answering tasks, employed a linked medical knowledge graph (developed internally) to explore the associations between all the potential medical entities recognized by the user input. The selected entities were then ranked by their strength of associations with user selected entities. The ranking of entities was also facilitated by a frequent occurrence of entities attached to the original extracted entities. Finally, they used this mechanism to select the top-ranked symptom that is sent back using a natural language template response to the user. But this procedure took a very long time to identify the question of the user. In a chatbot, Krishnendu and Rarhi [5] aims to present a design for a virtual doctor that provides diagnosis and remedies based on the symptoms provided to the system. The chatbot they proposed were extracting the disease from the symptoms provided by the user and based upon the criticalness of the disease the chatbot identifies whether it is a major or minor disease. If the disease is major one then the algorithm will propose the patient to book an appointment with the doctor from the database, but these chatbot can only handle the user-specified symptoms, if it is present in the database. Divya et al. [6] build using artificial intelligent predicts the disease based upon symptoms provided by the patient and provides some information about the disease before consulting the doctor. Its a text-to-text-based algorithmic process which means it's a text-to-text classification model. Chatbot is also efficient to predict all the kinds of disease as proposed in the paper, but there is no virtual doctor for them to talk on a video call face to face. Anil Kumar et al. [7] proposed a system that is is based on symptom mapping, and the chatbot finds out the solution of the patients diseases by providing them the way to cure and drugs to consume by scanning their symptoms and if the diseases are major it directs to a page where you can have an appointment with doctor. It is a database of all doctors according to their qualification from where the data will be fetched and will be sent to users according to their profile. There are three modules in this chatbot: collection of users symptoms, matching with the chatbot database and curing it. In order to achieve accurate diagnosis finite graph is used. String searching algorithm is used to extract symptoms; however most of the times the user need to say exact symptoms in order to get the desired diagnosis; otherwise the conversation will go into a loop and start the convo all over again. The dataset and mapping are accurate but simple and small at the same time. Datasets are trained well, and use of NLP is pinpointed [8]. The key feature of this paper is that it explains how AI, machine learning and deep learning can change the course of medical history. Past methods to provide a medical diagnosis and recommendations is not living up to the mark; therefore AI can be used here. Due to sudden rise and advances in AI for image analysis, radiology and pathology images will be taken care off with machines. Patient communication and capture of medical-related critical notes are some of usage of AI in speech recognition which is increasing day by day. Ensuring daily adoption in healthcare sector is the main aim of artificial intelligence and it should be seen as the greatest challenge that the technology is useful or not in practice. AI system must be approved by regulators and also must be integrated with EHR system and must be updated on a timely basis. Akram Mustafa and Mostafa Rahimi Azghadi [9], in a paper "Automated Machine Learning for Healthcare and Clinical Notes Analysis," aim to spread a comprehensive research on AutoML application in healthcare system for critical notes analysis. The purpose of paper to bring out ML and integrate it in daily life for better healthcare outcomes. But the evaluation methods used here doesn't produce a significant and reliable result due to the need of more computational power. The paper "Application of Machine Learning Algorithms for Facial Expression Analysis" focuses on how facial classification is important for medical and education domain where non-verbal is possible [10]. The papers mainly integrate various machine learning and deep learning methods for accurate classification on various streams of data like images, videos, text with keeping in mind the crucial parameter like age and gender.

# 11.3 Proposed work

## 11.3.1 Methodology

The main purpose of healthcare data sets is to identify the data elements to be collected for each patient and to provide uniform definitions for common terms. There are different datasets used for this paper for different modules such as disease prediction, Q&A (Question & Answer) for general questions about different health test parameters. The datasets were solely prepared and extracted from different website where they conducted different surveys and interviews with health professionals as well as patients. The dataset we found out is very rich in maintaining ethics related to health culture as well as accurately predicting diseases. Introducing gamification module to our website is altogether more interesting because it helps to create a sense of active participation of the users/patients which will be in the form of health score quizzes. Health score quizzes will pop up to the users throughout the year on numerous health awareness days, for example: World Heart Day on 29 September. Refer Figure 11.3 for flow of work.

Besides this the tools required for the project are as follows:

1) Python
2) Natural Language Tool Kit[NLTK]
3) Numpy and Pandas for data cleaning
4) Matplotlib for data visualization
5) Sklearn for model building
6) Jupyter notebook, visual studio code
7) Python flask for http server
8) HTML/CSS/Javascript for UI
9) Django Framework for backend

### 11.3.2 Work flow

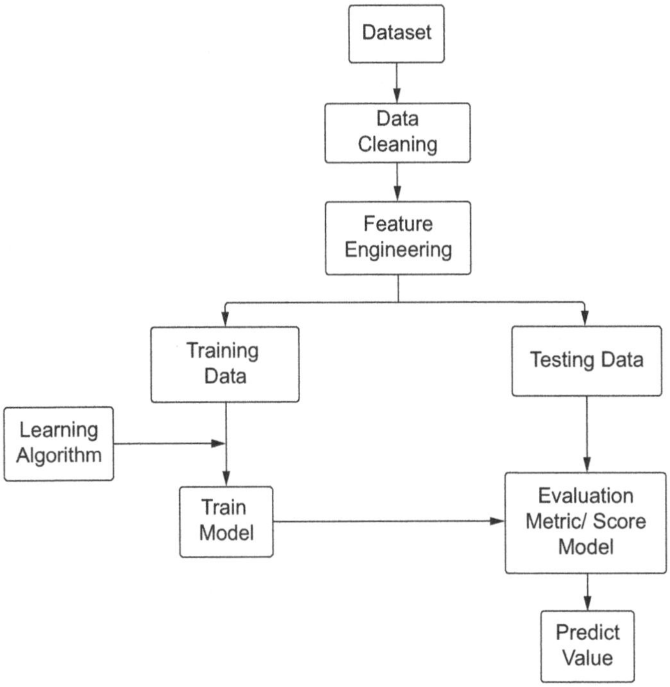


**Figure 11.3:** Flow of work.

### 11.3.3 The dataset

First dataset we used is basically "Question&Answer"-type conversation between MEDIXA and the user which helps to know what the user wants and how the user wants to solve it. The dataset is the json file with different keys such as INTENT, which are different ways of asking questions to chatbot or in simple words the intention of patient; RESPONSE, which stores the responses of particular questions asked by the patient related to particular disease; TAG-Tag represents the disease name as shown in Table 11.1. The dataset contains 80 different tags thus capable of answering questions of 80 different diseases. This dataset is used for answering basic queries of patient's related to common/basic remedies of a disease and detailed information of parameters used in various health tests such as blood test, urine test and vitamin test. The dataset contains 10–12 questions of a particular tag thus giving the model sufficient amount of data to correctly predict the intention of users. Another is the open source dataset which can be acquired on Kaggle website. The dataset is used to detect disease based on the symptoms provided by the users.

The symptoms are marked as binary values, that is, 0 and 1. If a bunch of symptoms maps to a particular disease then it is marked as 1 else marked as 0. This dataset has 132 parameters on which 42 different types of diseases can be predicted. Since different users can have different symptoms for a particular same disease,then we have different rows for the same prognosis to reduce the incorrectness in predicting the disease.

**Table 11.1:** Columns of dataset.

| Column name | Description |
|---|---|
| Intent | Queries asked by the users. |
| Tag | Corresponding disease with respect to Intent. |
| Response | Answer provided by the chatbot. |

## 11.3.4 Text pre-processing using NLP(data handling)

For data or text to be predictable and analyzable for different task we need to pre-process it. Pre-processing text is a method to transform text so that it can be ready to use with model for better accuracy. To pre-process data or text there are different techniques mentioned below.

### 11.3.4.1 Lower casing and removal of stop-words

One of the simplest and most efficient way of text pre-processing to convert all text to the lowercase. Words like "Pain" and "pain" mean the same but are represented as different words in the vector space which results in increase the dimension, so here lowercasing is used.

Here is an example of solving sparsity issue through lowercasing, where same word with different meaning are mapped to single lowercased word. Table 11.2 represents Lower casing.

**Table 11.2:** Lower casing.

| Raw | Lowercased |
|---|---|
| Fever | Fever |
| fever | |
| FEVER | |

In natural language processing, stop words are generally called as "useless" data, that is, they are not useful for the model. A stop word is a commonly used word (such as "the," "a," "an" and "in") which are not at all useful for classification model and can create a noise which will effectively decrease the accuracy model.

#### 11.3.4.2 Lemmatization

Doing things properly with the use of vocabulary through grouping different inflected forms of words to its base root mode with same meaning is known as lemmatization. Lemmatization extracts the correct lemma of each word. Lemmatization is the slower process but it is useful for the word which is important for analysis. Its is a Dictionary-based approach and produce a better accuracy if used. Lemmatization transforms the word without chopping things off and the table below uses a different word, for example, "Itching,Itchy,itches" mapped to its lemma known as "itch." Table 11.3 represents the lemmatization process.

**Table 11.3:** Lemmatization.

| Original word | Lemmatized word |
|---|---|
| Itching<br>itchy<br>itches | Itch |
| Sting<br>stingers<br>stings | Sting |

#### 11.3.4.3 Tokenization

Tokenization algorithm uses tokenization technique mostly common. It splits a sentence into individual words generally based on comma. Tokens are the different words in the following sentence. It is the first technique used in NLP for pre-processing text data. Word tokenization is splitting text into words whereas sentence tokenization is splitting text into sentences. Spaces and punctuations are ignored while tokenizing the text data. Example if input sentence is – "I am having a stomach pain," then word tokenization is ["I," "am," "having," "a," "stomach," "pain"].

## 11.4 Design and analysis

### 11.4.1 Appointment scheduling

Since taking a physical appointment is tired some work in the unusual COVID-19 pandemic time through visiting different doctors and clinics. So we have designed a module

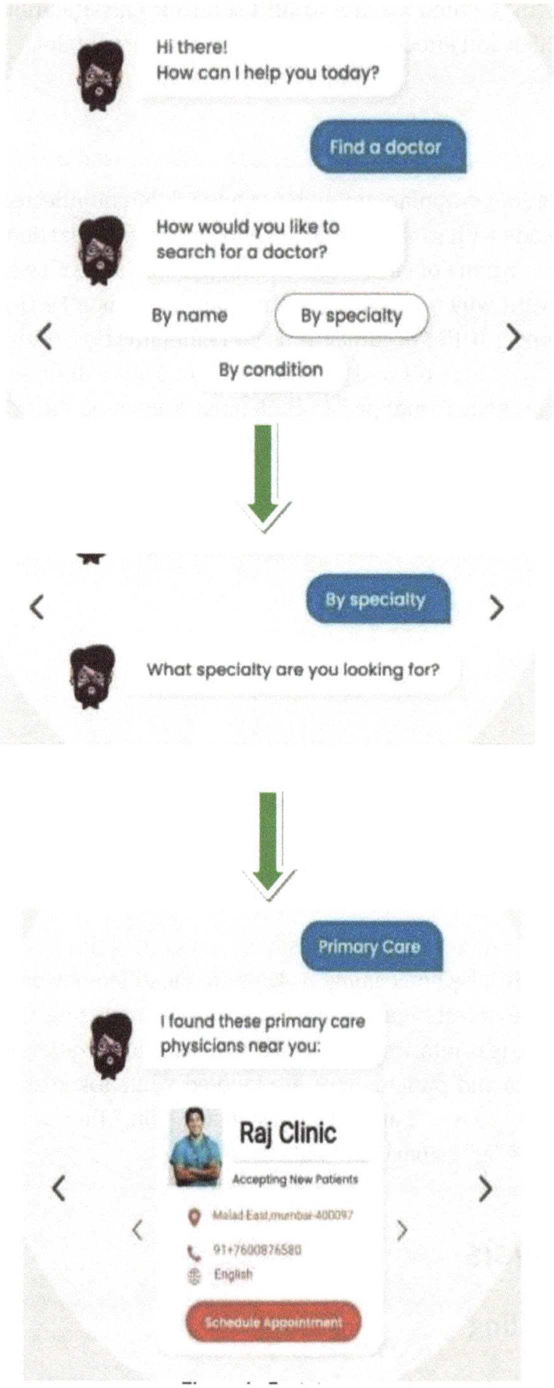


**Figure 11.4:** Prototype.

for making a patient appointment on his/her desired location. We have integrated appointment module into our chatbot so that we can provide a appointment best suited to patients through direct interaction with a chatbot. The frontend model is build through HTML/CSS and JavaScript whereas for backend we have used Django. Given figure below gives proper flow of making appointment. Refer figure 11.4 for the prototype.

### 11.4.2 Gamification module

The major role of gamification module is to make the UI more interactive for the users/patients. Since our chatbot will be deployed on our website it is important to have a user-friendly website. Gamification helps to create active participation and a hint of positivity among the people. Technology should always be used by taking into consideration not only the core technical aspects but also the social welfare of the people.

Our gamification module will include general quizzes based on different health-related occasions throughout the year. For example, World Heart Day is on 29 September every year. So a health score generator quiz will pop on that particular day and the users can generate a score based on their answers and they will come to know certain unknown facts about the conditions of Heart. Quizzes will be handled by our frontend using HTML & JavaScript and the scores generated will be stored in the backend using Django Framework.

## 11.5 Algorithms used

### 11.5.1 Naïve Bayes algorithm

Naïve Bayes model is fast, simple classification algorithm used for a very high-dimensional dataset. The classifier is based upon Bayesian classification method, which can be shown by the following formula:

$$P(X|Y) = \frac{P(Y|X)P(X)}{P(Y)}$$

Using Bayes theorem, we can find the probability of X happening, given that Y has occurred. Here, Y is the feature and X is the label. The assumption made here is that the predictors/features are independent. Other features are not affected because of one feature, which is why it is known as naïve method.

Naïve bayes algorithm is yet very intuitive but a very powerful classification algorithm. To predict the purpose of classification, the algorithm uses likelihood probability. In simple term likelihood is known as reverse probability. Each word from the question asked by the user is considered as features and tag or answer with respect our question are considered as labels. Now every input word is compared with all words there in the dataset and based on above Bayesian formula the different labels are associated with input question. Those label who has the highest probability are considered and in this way the algorithm works. For example let's say if the patient asked question as "What do if i am having mild fever?," so now every word will form a feature vector space. Every word will be compared with different words in a bag and based upon that the class labels are assigned with certain probabilistic measure.

Class labels assigned are "mild fever" with probability of 95.67% and an other class label is assigned as "cough" with 92.78%, then class label with highest probability is considered that is "mild fever."

## 11.5.2 Random forest classifier

Random forest classifier algorithm is a supervised learning algorithm which is used for classification as well as regression. As a name suggests it is a forest where there are trees which carries correct labels within them. So now random forest classifier algorithm select the best tree from forest based on voting. Generally it is considered that if there are more trees in the forest then the accuracy of the algorithm increases, a real life fact. The main advantage of using random forest classifier is saving the model from overfitting because it itself takes the average of the prediction. As given in the diagram below, random subsets are decision tree which consist of root node, leaf node, leaves and correct prediction of each decision tree is stored on its leaf and after through majority voting correct labels are predicted. Refer figure 11.5 for Random Forest Implementation.

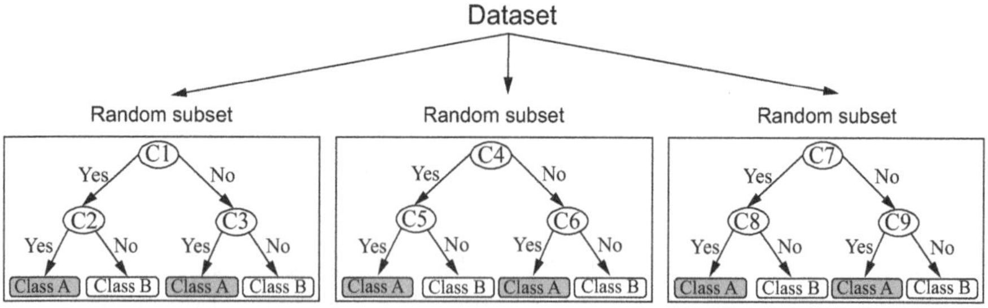


**Figure 11.5:** RF implementation.

## 11.6 Evaluation metrics

To build an effective machine learning model, certain evaluation metrics are used. Evaluation helps to get the useful insights whether the labels are correctly or incorrectly predicted. There are various metrics for evaluating a model such as mean-squared error (MSE), area under curve (AOC) and confusion matrix which is used mostly for classification. In confusion matrix the correct and incorrect predicted labels are segregated or broken with each class. Confusion matrix also gives the type of error the model is making. Figure 11.6 illustrates how the confusion matrix looks like with its different terms such as FP, TP, FN and TN. Refer Figure 11.6 for classification metrics. Table 11.4 also represents various evaluation metrics.

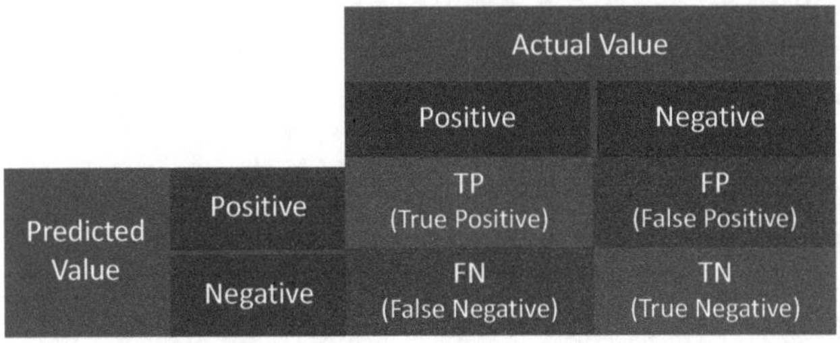


- True Positive (TP) : Observation is positive, and is predicted to be positive.
- False Negative (FN) : Observation is positive, but is predicted negative.
- True Negative (TN) : Observation is negative, and is predicted to be negative.
- False Positive (FP) : Observation is negative, but is predicted positive.

**Figure 11.6:** Classification metrics.

**Table 11.4:** Different evaluation metrics.

| Metric | Description | Formula |
|---|---|---|
| Accuracy | Accuracy is one metric which gives ratio of right prediction to the number of classes. | $ACC = \frac{TP + TN}{TP + TN + FP + FN}$ |
| Recall | Recall gives the fraction of positive labels that are predicted positively. | $\text{Recall} = \frac{TP}{TP + FN}$ |
| Precision | Precision gives the fraction of correct patterns out of all positive patterns in a correct class. | $\text{Precision} = \frac{TP}{TP + FP}$ |
| $F_1$ score | It is defined as the harmonic mean of the model's precision and recall. | $F_1 = \frac{2}{\frac{1}{\text{Recall}} + \frac{1}{\text{Precision}}}$ |

## 11.7 Result of prediction

The dataset used for disease prediction is divided into 70% of the training dataset and 30% of the testing dataset. The Random Forest classifier algorithm was used which gave 90% accuracy as a result. Thus RF classifier provided better results than expected. Better results were achieved due to the large and clean volume of dataset. The following figure shows a sample of diseases taken into consideration and based on that evaluation metrics were generated. Kindly Refer Figure 11.7 for results.

| | precision | recall | f1-score | support |
|---|---|---|---|---|
| (vertigo) Paroymsal Positional Vertigo | 1.00 | 1.00 | 1.00 | 34 |
| AIDS | 1.00 | 1.00 | 1.00 | 33 |
| Acne | 1.00 | 1.00 | 1.00 | 31 |
| Alcoholic hepatitis | 1.00 | 1.00 | 1.00 | 50 |
| Allergy | 1.00 | 1.00 | 1.00 | 39 |
| Arthritis | 1.00 | 1.00 | 1.00 | 34 |
| Bronchial Asthma | 1.00 | 1.00 | 1.00 | 41 |
| Cervical spondylosis | 1.00 | 1.00 | 1.00 | 38 |
| Chicken pox | 1.00 | 1.00 | 1.00 | 38 |
| Chronic cholestasis | 1.00 | 1.00 | 1.00 | 33 |
| Common Cold | 1.00 | 1.00 | 1.00 | 30 |
| Dengue | 1.00 | 1.00 | 1.00 | 23 |
| Diabetes | 1.00 | 1.00 | 1.00 | 39 |
| Dimorphic hemmorhoids(piles) | 1.00 | 1.00 | 1.00 | 30 |
| Drug Reaction | 1.00 | 1.00 | 1.00 | 44 |
| Fungal infection | 1.00 | 1.00 | 1.00 | 30 |
| GERD | 1.00 | 1.00 | 1.00 | 43 |
| Gastroenteritis | 1.00 | 1.00 | 1.00 | 35 |
| Heart attack | 1.00 | 1.00 | 1.00 | 28 |
| Hepatitis B | 1.00 | 1.00 | 1.00 | 35 |
| Hepatitis C | 1.00 | 1.00 | 1.00 | 36 |
| Hepatitis D | 1.00 | 1.00 | 1.00 | 36 |
| Hepatitis E | 1.00 | 1.00 | 1.00 | 47 |
| Hypertension | 1.00 | 1.00 | 1.00 | 32 |

**Figure 11.7:** Result.

## 11.8 Conclusion and future scope

Above 60% of users are involved in different categories of chatbot say whether its related to fashion, healthcare, streaming, etc. which is useful for business as stated by mindtree.chatbots.org, an online site that keeps up with the number and types of chatbots around the world has reported around 1,400 chatbots currently in production. A big reason for the spike in popularity is the addition of artificial intelligence to chat applications provides a short and get every effective answer to why are chatbot application are becoming more and more popular addition to any website or mobile application. Artificial intelligent robots are programed to fulfil the intention of questions asked by the user while answering them quickly. This is the sole reason why artificial intelligence is used with chatbots, so that it can instantiate any response robustly based upon questions asked by the user. Chatbot integration in any website or mobile application gives the user a rewarding experience and user sometime feels it as a human-friendly way to conduct any task. We plan to improve our bot "MEDIXA" to a great level which will help us to master the research and development in the fields of AI, ML, DL and NLP. We will introduce new modules into our project such as prescribing medicines, interaction in native language, so that our model can be reached out to each and everyone who are less aware of technology and can give upperhand on their treatment. Along with that, we have tried to contribute in the healthcare sector to help the healthcare industry as much as possible due to the current pandemic situation.

## References

[1] Agrawal, M., Cheng, J., Tran, C., What's Up, Doc? A Medical Diagnosis Bot.
[2] Bali, M. Diabot: A Predictive Medical Chatbot using Ensemble Learning.
[3] Tripathy, A.K., Carvalho, R., Pawaskar, K., Yadav, S. Mobile based healthcare management using artificial intelligence.
[4] https://towardsdatascience.com/the-5-classification-evaluation-metrics-you-must-know-aa97784ff226
[5] https://towardsdatascience.com/naive-bayes-classifier-81d512f50a7c
[6] https://towardsdatascience.com/understanding-random-forest-58381e0602d2
[7] https://www.analyticsvidhya.com/blog/2021/06/text-preprocessing-in-nlp-with-python-codes/
[8] Aggarwal, V., Jain, A., Khatter, H., Gupta, K. (2019 August). Evolution of Chatbots for Smart Assistance.
[9] Sharma, V., Goyal, M., Malik, D. (2017 April). An intelligent behaviour shown by chatbot system, 3(4). ISSN:2454-4116.
[10] Mu, X., Shen, X., Kirby, J. (2017). Support vector machine classifier based on approximate entropy metric for chatbot text-based communication. International Journal of Artificial Intelligence, 15(2), 1–16, ISSN: 0974-0635.

Kiran L. N. Eranki*

# 12 Prediction of MRI tumor condition through image segmentation using deep neural network approach

**Abstract:** Need for medical imaging has immensely increased to diagnose functionality of organs and tissues. Application of magnetic resonance imaging (MRI) or computerized tomography scan (CT scan) has automated the process of disease prediction which otherwise would need manual intervention of experienced physicians. Critical transplants and tumor diagnosis of kidney, heart and liver could be very challenging without the application of high-resolution medical imaging. Semantic image segmentation helps to identify features within the image and label them to demarcate the background from organs and tumors formed on them. Current proposed model is built by applying image segmentation using deep learning models by training 3D image information obtained from KITS-19 using CNN U-Net architecture to segregate essential tumor details from the kidney background images. Considering the evaluation of medical condition based on imagery is very challenging due to lack of pre-processing methods to handle unbalanced nature of class distribution which is common in medical imaging. In the current research, we apply state-of-the-art convolution neural network and deep learning models to predict the tumor in MRI imagery. We performed evaluation of a subset of 60 CT scans with slice thickness values obtained from KiTS19 dataset. We also compared the image segmentation model with BRATS'2013 challenging dataset also to validate the prediction of tumor in brain MRI scan images. Performance evaluation of kidney and tumor features obtained through threefold cross validation comparing with ground truth gave us dice coefficient score of 0.96 while tumor segmentation prediction was close to 0.80. We believe the performance can be increased through higher GPU memory requirements, which was a limitation to our existing GPU hardware. Nevertheless, the possibility of applying improved deep learning model based on existing results and can be a promising step forward in medical imaging segmentation.

**Acknowledgments:** The authors would like to thank the Kidney Tumor Image Segmentation 2019 support team for their courteous support provided to improve the study. And also to the support group of brain tumor image datasets used in this article were obtained from the MICCAI 2014 and 2015 Challenges on Multi-Modal Brain Tumor Segmentation. The challenge datasets used from both portal are available to researchers and release under creative-commons license for research and educational purpose. All the datasets provided contain fully anonymized images from the Cancer Imaging Atlas Archive and the BRATS 2014 challenge.

***Corresponding author: Kiran L. N. Eranki,** CSE and AI, SR University, Warangal,
E-mail: eranki.kiran@sru.edu.in

https://doi.org/10.1515/9783110756722-012

**Keywords:** Magnetic Resonance Imaging (MRI), Medical Image Segmentation, Convolutional Neural Networks (CNN), U-Net Framework

## 12.1 Introduction

Medical imaging-based diagnosis and therapy are becoming popular considering the remote tele-medicine health care practices. As medical imaging devices accessibility and availability keep growing as more and more technology advances occur. Magnetic resonance imaging (MRI) helps to identify tumor in the segmented images [1]. Sliced images still require to be validated by physicians, especially as application of medical imagery has become lot popular choice for experts as it reduces the diagnosis time and validation process, and automated segmentation of MRI images would be helpful. Image split into low-dimensional segments has gained gravity in computer science considering the benefits it add to disease diagnosis and prognosis [2]. Semantic segmentation is widely used to detect notable features and label the segments of the sample into classes – kidney, brain or tumor within the MRI images. Although we have multiple segmentation algorithms to solve segmented imaging problems, convolution neural networks and deep learning model outperform others [3]. With growing need of medical imaging, advanced computation networks are required to process medical images. CNN-based models provide efficient results [4, 5]. Along with prediction, one can also learn about the similarity between ground truth revealed by an expert compared to our model [6–8]. Recent studies all indicate the CNN model performance to compete with the revelations of an expert radiologist [9, 10].

CNN also work well with 3D images for pre-processing, data augmentation and metrics specific to the datasets along improved performance and accuracy of prediction [11] as different medical conditions provide varied performance across different data sets. However, comparison among these models provides a challenging task due to their varying sizes, localization and distinct features among the data sets. In order to compare segmentation models used in medical diagnosis one consideration would be data augmentation and performance metrics to implement using CNN architectures. Frameworks for medical imaging generate complex image formats which requires different I/O interfaces along with pre-processing of the data sets mostly providing highly unbalanced class distributions followed in standard medical imaging techniques. Among the recent medical imaging platforms NiftyNet [12], BRATS [13] along with challenging data sets provided. Since 2012, the Perelman School of Medicine at the University of Pennsylvania, Center for Biomedical Image Computing & Analytics (CBICA) have been organizing online competition, the Multimodal Brain Tumor Segmentation Challenge (BRATS) [13, 14]. Similarly, Kidney Imaging and Tumor Segmentation Challenge (KiTS) [15]. However, other concern would be most of these pipelines are not designed for frame-works as they lack easy customization for building interfaces and format of datasets also applies different extraction methods used in image segmentation problems. Current work uses a state-of-art-framework build using convolutional neural network and deep learning model to predict tumors among the MRI datasets applied in the study. Cascaded Image Segmentation Convolutional neural network (CIScnn) model proposed in the current study has been adopted from existing frameworks applied by the researchers used on challenging datasets. Cancer detection in preliminary stage is most often unnoticed which can cost the live of an individual. Early detection can control the growth and damage done by carcinogens to individuals. Medical terminology of cancer exists in three forms namely – giomas, meningiomas and pituitary tumors [16]. Cancer can exist in either benign or malignant forms, among which benign cancer usually prevails as meningiomas while giomas exist as malignant form which can spread to various other organs causing major damage to the vital organs of the body. On the other hand, benign forms usually do not spread to other organs as these originate from brain tissue as tumors while meningiomas arise from membrane surrounding the brain tissue and central nervous system. Exception among the three forms is pituitary tumors which are benign, however can cause damage as they sit inside the skull causing massive friction among the nerves which is more harmful compared to other forms [5]. Based on the forms of cancer noticed in the subjects, radiologists look for features to predict the tumor condition and suggest the necessary diagnostic treatment followed by medication or surgery as required. Most commonly used data to conduct such diagnosis is done using MRI of varying resolutions. However such data is very limited and subjective to the condition of patient diagnosed and highly dependent on the experience of the radiologist [16]. Ground truth to be thoroughly verified before concluding a brain tumor to be benign or malignant form, as it also needs further biopsy tests to validate the same. Most often surgery will not the first choice rather determining the cause of the tumor, its origin and stage at which its prevailing in the body. All these require deep analysis of the vitals gathered from lab tests and build the diagnostic necessary for tumor segmentation and classification of MRI images as required [2].

Recent developments in medical imaging have improved the performance of prediction by using artificial intelligence and machine learning, applied to disease diagnosis or prognosis, drug therapy and other health-care ailments as a tool for many nephrologists, neurologists, radiologists and medical practitioners using MRI imaging. Different features from image segmentation and classification are obtained using MRI images which help radiologists validate their medical predictions. Center for Biomedical Image Computing & Analytics (CBICA) has been running an online competition since 2012 which focuses on the Multimodal Brain Tumor Segmentation Challenge (BRATS) [10]. Datasets are publicly downloadable after the completion of the competition. Although the size of the dataset is very small providing about 300 images which often contains details of tumor level such as low or high gioma tumor, shown on the axis panel [6]. Apart from classification and segmentation of tumor images, data augmentation is also required in most of the medical imaging challenging datasets.

Researchers have used smaller sample sizes of 60–100 images to classify tumor into possible four classes – no tumor, giomas, meningioma, sarcoma or metastasis [7]. Using deep learning algorithms to perform image segmentation provides 96% accuracy which also depends on the datasets quality, ground truth provided by the radiologists. Accuracy of image segmentation provides another challenge as giomas and intratumoral lesions provide an indication for levels of severity or surgical procedure to be followed. Nevertheless, the same if manually performed would be time-consuming and also adds up the intra or inter-rater errors while making a prediction to characterize the tumor. Physicians apply their experiential understanding to perform the evaluation which can be unpredictable at times increasing the severity of the condition [5]. In addition to shape, size and structure of the tumor, tumor mass or outgrowth also has an effect on the surrounding nerve tissues which cannot be predicted in MRI images due to varying intensity range applied while scanning the patent and also depends on the scanner used for diagnosis [8]. For all the above reasons cited a semi-automatic or automatic data segmentation procedure is required with improve performance and accuracy [17]. Brain tumor segmentation is carried through different methods which can be parametric or non-parametric probabilistic models. Almost all models use a likelihood function to determine the feasibility of tumor or not. Most often, the abnormalities associated with tumor growth can be a hassle to prediction, which also influences the segmentation time required to process the images. In addition to the tumor shape and location of nerve lesions noticed in the imagery can also be helpful in diagnosis process which all adds up to the performance and accuracy of the tumor prediction [18]. Tumor prediction has many challenges to address before implementation of radiomic metrics into oncology workflow. These include harmonization of features to be extracted from the datasets and consensus in subsequent prediction models. Tumor volume is calculated by voxel size and number of voxels within the tumor as it accounts for feature normalization [19]. Several segmentation algorithms have been implemented to determine the voxel within the neighborhood of the tumor. Studies conducted by Zhao et al. [20] show the application of Markov random fields-based segmentation to estimate the supervoxels that gave better accuracy through histogram-based likelihood function. Similarly Menze et al. [14] used generative models to apply prior knowledge of experts to probabilistic model to demarcate voxels within the tumor. Some of these voxel clusters may be wrongly classified due to uneven anatomy of the voxel locations. In order to overcome mis-classification improved classifiers such as Support Vector Machines [21] and Random Forests [22] are popular in feature extraction and multi-class classification problems. Many feature extraction methods have been adapted by researchers in the past which include supervised classifiers and semi-supervised classifier approaches. Similarly several features have been considered in the literature such as fractal-based texture [23], gradients [24] and brain symmetry [25] that are popular. Some studies are performed using random forest classifier to improve image segmentation. While other studies used hierarchical segmentation based on spatial adaptive random forests classifier. Other studies have also used semi-supervised random forest classifier to perform tumor segmentation [22]. In addition to RF classifier, neural network model with auto-encoder models is also widely used in image segmentation. Deep learning models provide automated feature extraction through representation learning from the datasets which also provides better performance and accuracy as compared to earlier models. CNNs have been used in several image pattern recognition and facial recognition challenges and provided promising results. In the current context MRI tumor recognition is a complex task to perform using conventional and manual methods, as a result would need an automated process to identify and characterize the tumor with better performance and accuracy. Studies have shown enhanced convolutional neural networks (CNN) models with varying levels of pooling, stride followed by fully connected layer with softmax has proved to be resourceful in most cases [26]. Studies have also used CNNs for speech and text narrative problems as well to predict the emotional state of the subjects involved in conversation [27]. Studies have shown the usage of 3D and 2D filters to refine the classification process of tumor prediction [28]. Application of 3D filter helps to visualize the benefit of 3D projections of the medical images. However, it does add up to the cost of performance and computational time, GPU requirements. While 2D filters still continue to be a choice in recent studies as well. In addition to filters, researchers have also implemented refinements to convolutional neural networks through two-way path flow network to multi-fold cascaded architecture to enhance the performance and accuracy of prediction [9]. Application of regularization methods such as dropout, optimization through adaptive moment estimation (ADAM), RMSProp, AdaGrad to control gradient descent swing and vanishing gradient and also controlling the over-fitting and under-fitting model issues [29]. Extraction of patches on the axial plane of voxels identified in the tumor neighborhood using a trained CNN has also been done by few researchers with multiple layers of maxpooling followed by fully connected layer running with softmax and random forest classifier [30]. Studies have also shown demarcated tumor regions into multiple segments and each segment is assigned as binary sub-tasks to classify the region while training the CNN model [23].

## 12.2 Research methodology

The open-source Python library has been used to implement Cascaded Image Segmentation CNN (CIScnn) framework with pipelines to extract the dataset comprising MRI images from the MICCAI 2019 [31] and apply to convolutional neural networks and deep learning models. CIScnn implementation includes the following stages from data extraction, pre-processing and data augmentation, convolutional neural network for feature extraction, classification and performance metrics as illustrated in Figure 12.1.

- Data gathered from challenging datasets, extract 3D medical image segmentation of MRI dataset in .png format.
- Perform data pre-processing, create I/O extract pipelines and data augmentation to extract features/patches from MRI images
- Provide the patch wise extracts to CNN model by splitting into training, testing datasets.
- Train the refined convolutional neural network with datasets and evaluate the performance metrics using testing datasets.
- Visualize the performance metrics and refine the loss function to improve the accuracy.
- Apply threefold cross validation to improve the prediction of features from the MRI images
- Refine the hyper-parameters to fine tune the performance of the CNN model.

### 12.2.1 Dataset pre-processing

MRI data extracted from challenging dataset carries from different forms of bias field distortions which have to be cleansed before utilizing them for further data preprocessing. Issues related to tissue intensity variations across all the MRI image repository have to be regularized by applying intensity normalization [32]. As it helps to regulate the intensity variation, if any noticed among the images caused due to variations of intensity distribution. These variation could occur due to in-built scanner as a manufacturing defect or depending on the rate of exposure, time of exposure and levels of intensity given to the subject while generating the MRI images. In this process intensity landmarks are used to address the variations. Histograms of each sequence of image are gathered and homogeneity across all sequences is maintained. During the training phase normalization of intensity is performed which improves the visibility of patches among the images and further patch pre-processing is performed to refine the quality of the MRI images obtained from the challenging datasets. Mean and standard deviation are applied to training samples.

### 12.2.2 Refined convolutional neural network

CNN were used to achieve some breakthrough results and win well-known contests [5, 31, 33]. Application of convolutional neural network helps to extract the feature maps from the convolutional layers using the kernels applied in the CNN model. These kernels help to connect one layer to another layer through weights assigned to them. These weights are acquired from each node of the convolutional layer during the training phase computed by back propagation. Kernels also help to improve characterization of features extracted from the datsets by sharing them across all the feature maps extracted. As a result, convolutional neural networks run with fewer weights while training as compared to fully connected dense layers while testing the model making it easier to train and less prone to model fitting issues. In addition to this, as same kernel is convoluted across all the images, features with similar characteristics can be detected and information related to the neighborhood of the image can help in contextual feature extraction [34]. Current model that uses nonlinear activation function was applied to each neural unit to control loss function. Usually the stacked convolution layers make the features more abstract as compared to the linear models. However, the image features such as edges and low intensity spots could be enhanced through this process. We now discuss the important phases of the convolutional neural network and their role in prediction of tumor from the MRI datasets provided as training and test datasets to validate the CIScnn model.

#### 12.2.2.1 Initialization

We have applied Xavier initialization [35] to implement converge, which will ensure that the gradient is maintained at controlled value, to compute the loss function. Otherwise the back-propagation could create a vanishing gradient descent problem, not common across CNN models which can effect the performance metrics and accuracy results generated from the model. This phase is carried forward by applying the necessary activation function in the next phase of the model framework.

#### 12.2.2.2 Activation function

In this phase, we have applied non-linearly activation function for transforming the data. Rectifier linear units (ReLU)was found to achieve better results than the more classical sigmoid($\sigma$), or hyperbolic tangent functions($\rho$), and speed up training [36] as defined in equation (12.1).

$$f\ (x) = \max(o,\ x) \tag{12.1}$$

However, one of the drawbacks of ReLU function is all negative values generated would be normalized to zero(0) which can be a concern as it ignores consequent weights adjusted to control the flow of the gradient [36]. In order to overcome this limitation we have applied an improved form of ReLU referred as leaky rectifier linear unit which helps to improve gradient flow by considering a small slope on the negative part of the function as shown in equation (12.2).

$$f\ (x) = \max(0,\ x)\ +\ \alpha \min(0,\ x) \tag{12.2}$$

where $\alpha$ used in the expression is the leakyness parameter. In the CNN architecture last phase would be a fully connected layer run with softmax function. Each phase of the network combines features extracted from each layer to identify the changes in the image segments. Nevertheless, this approach of pooling the features to compact representation will consistently reduce computational load on training and improve performance. This phase is often carried using max-pooling or average-pooling methods [37].

#### 12.2.2.3 Regularization

In order to control model over-fitting issues, when too much of training is provided to the model, it tends to memorize the features and fails to perform well. We overcome this issue by applying dropout or batch normalization methods to control the model fitting issues. These methods help to regularize the gradient descent across the layers to reduce the model over-fitting. Dropout method ($L^{[l]}_{\text{dropout}}$) discards random set of nodes and considers the rest for parameter tuning. This approach helps to control model over fitting and under fitting aspects. As shown in equations (12.3)–(12.7) there are two components mainly cross-entropy cost and L2 regularization cost $\left(\frac{1}{m}\frac{\lambda}{2}\sum_i^m\sum_j^m\sum_k^m W^{[l]2}{}_{i,k}\right)$ which will determine the overall regularization cost of each unit within the network [38].

$$J_{\text{reg}} = \frac{1}{m}\sum_{i=1}^{m}\left(\left(y^i \log\left(a^{[L](i)}\right) + (1-y^i)\log(1-a^{[L](i)}\right)\right) \tag{12.3}$$

$$J_{\text{reg}} = \frac{1}{m}\sum_{i=1}^{m}\left(\left(y^i \log\left(a^{[L](i)}\right) + (1-y^i)\log\left(1-a^{[L](i)}\right)\right)\right) + \frac{1}{m}\frac{\lambda}{2}\sum_{i}^{m}\sum_{j}^{m}\sum_{k}^{m} W^{[l]2}{}_{i,k} \tag{12.4}$$

$$Z^{[l]} = W^{[l]} + L^{[l-1]} + b^{[l]} \tag{12.5}$$

$$L^{[l]} = g^l\left(Z^{[l]}\right) \tag{12.6}$$

$$L^{[l]}_{\text{dropout}} = D^{[l]}\left(L^{[l]}\right) \tag{12.7}$$

where $g^l$ is the activation function applied and $D^{[l]}(L^{[l]})$ will provide the dropout factor to regularize the CNN model.

#### 12.2.2.4 Data augmentation

Augmentation is done to enhance the sample size to reduce the effect of model over-fitting issues. From the MRI images patch extract was done and pre-processing of patches was done. All patches were obtained from voxels surrounding the tumor and restricted the augmentation to rotation operations only while there are other forms of augmentation such as usage of kernels and applying cross-folds to image are popular. One reason to augment would be to reduce the mis-classification of patches while generating from rotation operations. Followed by we also discuss about the loss function computed to determine the performance metrics of the model.

#### 12.2.2.5 Loss function

Based on the image segmentation features extract at each layer, we also compute the loss function which will determine the performance of model both during training and testing phases. And has to be minimized during training. We used the categorical cross-entropy to compute the loss function for the CNN model as shown in equation (12.8).

$$E_{\text{categorical}} = -\Sigma_{t\in\text{voxels}}\Sigma_{c\in\text{classes}} C_{t,c}\log(\hat{c}_{t,c}) \tag{12.8}$$

where $C_{t,c}\log(\hat{c}_{t,c})$ represents the probabilistic predictions obtained after the softmax has been applied in the final fully connected layer of the CNN model. Followed by in the next section we discuss the data acquisition process to obtain MRI datasets from Neuroimaging Informatics Technology Initiative (NifTI) challenge portal [39].

### 12.2.3 NifTI data acquisition process

#### 12.2.3.1 Input–output pipeline

Python-based data I/O pipeline has been provided in the Neuroimaging Informatics Technology Initiative portal [39]. We have used the same framework to connect to the improved CNN model to implement the prediction of tumors in the MRI images. Datasets provided will be in compressed format and I/O pipeline is used to extract MRI images using informatics tools provided in the NifTI portal. Procedure for extraction is common across almost in all forms of medical imaging challenges where datasets will be in 3D image format along with metadata related to the thickness and volume of MRI slices. We have integrated the I/O pipeline to the CNN architecture to secure all the essential metadata obtained along with the 3D images extracted from the source. Some of the pipeline methods provided in the framework are not conducive to extract MRI images. As we also need to store the MRI thickness, volume of the slices details to build the I/O pre-processing pipeline. While extracting the images feature information should be intact with the image to synthesize a complete MRI image structure. Pytorch and Numpy packages were used in the I/O pipeline process to extract the MRI images along with the metadata information.

#### 12.2.3.2 MRI image pre-processing

Image pixel intensity can vary depending on the signal intensity ranges of the scanning device used by the radiologists. Performance of the image segmentation methods gets affected by the medical scanning instrument used as factors such as calibration of the device, hardware/software malfunctioning or technical glitches and human errors while operating the scanning device [33]. While machine learning algorithms on the other hand give better performance on feature extraction and image segmentation, nevertheless, in order to achieve consistent signal intensity while performing the image segmentation, normalization of pixel intensity is required. As it will improve the quality of signal intensity and normalize the range values to a uniform scale. Different normalization methods are available in literature such as Z-score normalization, batch normalization methods to homogenize the values across the image matrix. This procedure also helps to improve the performance metrics and reduce error or noise generated from the MRI images.

#### 12.2.3.3 Image re-structuring and re-sampling

Image can be restructured using pixel intensity clipping method. Through this method we can selectively identify the intensity pixel values within a range and eliminate the pixels outside the range to segregate identical organs or tissues even if the image was generated from different scanning devices. Clips of organ-specific pixel intensity can be obtained and analyzed further to classify the features into tumor or non-tumor region of the images. Another approach is image re-sampling which performs a modification to width and height of the images and re-generate a new image from existing samples. In addition to increasing the sample size, we also see another benefit of the sample as re-sampling also reduces the computational GPU memory required to process the images by downsizing the pixel sizes or voxel spacing in the image. As a result, the cost of training and prediction of the tumor within the image also improves. Most of the CNN models require a consistent pixel structure across all the images subjected for training and testing datasets which can be achieved through re-sampling methods.

Cascaded image segmentation CNN model implemented for the study can handle multi-class image segmentation datasets. Datasets can be pre-processed by using the methods explained earlier and processed datasets are classified into multiple classes which require the usage of encoding mechanism which dynamically labels the image segments into multiple classes [42]. Prediction of tumor based on class labels using the CNN model require mathematical representation of datasets into classes using binary values for each image segment within the images as shown in Figure 12.1.

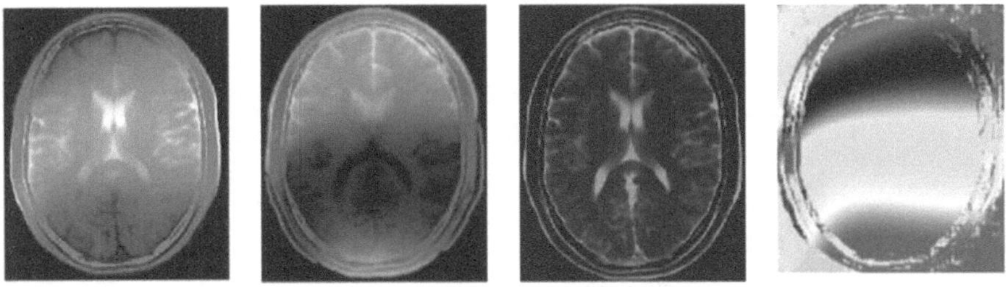

(a) MRI Brain scan showing real image

(b) MRI brain scan showing imaginary pixel projections

(c) MRI brain magnitude scan used for diagnosis

(d) MRI brain phase scan are used for susceptible distortion checks

**Figure 12.1:** Clipped sample BraTS2013 MRI images [40, 41] showing brain functionality flow and susceptibility of induced distortion in the scanned imagery.

## 12.2.4 MRI patch extraction and pre-processing

### 12.2.4.1 Phase 1: removal of blank patches

Firstly, we solve the problem of unbalanced image segments comprising blank or no information possibly considered as outlier or noise generated from the datasets pre-processing. As these segments add no value to the existing set of segments considered for classification. And, moreover, such segments could also bring overhead to computation time and hence to be excluded while identifying the MRI patches within the image segments [37]. We perform this phase through CNN model which needs to be fine-tuned to provide accurate classification of segments based on the patch details obtained. Otherwise with scanning the datasets using the model they would consider such blank segments as background metadata which neither has any information to process nor can be classified to the background class. Therefore such image segments have to be removed in order to avoid unbalanced model fitting issues.

### 12.2.4.2 Phase 2: managing patch batches

Second phase takes care of the formation of batches by stocking the MRI patch images together. So that all the images within the batch can be processed in one go during the training phase. Batch normalization approach is applied to ensure that all the datasets put for training are uniformly distributed for smooth transition and less computation time to complete the training process. CNN model adds the necessary weights to each batch processed depending upon the memory, step-size defined in the learning rate and number of epochs to be carried to complete the phase. All these hyper-parameters determine the performance metrics and cross-

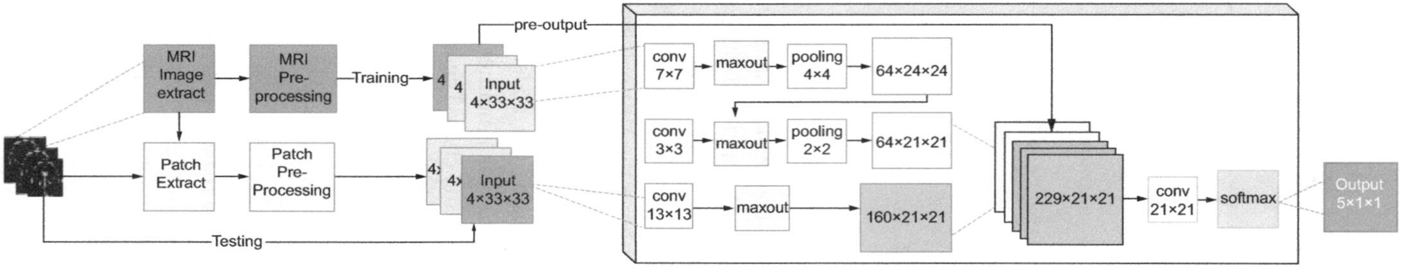


**Figure 12.2:** Cascaded Image Segmentation Convolutional Neural Network architecture flow diagram showing MRI image challenging datasets obtained from MICCAI 2019 [31] and applied to CNN implementation phases of data acquistion, augmentation and evaluation of performance metrics.

entropy loss function value obtained after each iteration of the model. Every batch with patch information can be either stored in the disk to reduce computational time or if executed in cloud can be processed using inbuilt memory or cache which can be expensive to perform unless the size of the MRI images is very large and can't be stored in disk, one can opt for cloud-based computing platform. During the training phase we also randomize the batches to avoid over-fitting or under-fitting issues. By shuffling the batch generated from the patch information randomly the variance among the obtained batch datasets would be reduced consistently and also contributes to improved accuracy and low error. Although the batch shuffling is performed but patch images within the batch will remain intact without any change in the sequence as generated by the model during patch pre-processing. However, only the randomization happens at the batch level data only without effecting the underlying patch images within the batches. As it is essential to retain the patch details intact to generate a coherent MRI structure as discussed in the earlier section. Batch normalization and shuffling process would require good amount of GPU and memory to have achieve faster computational time and run more number of iterations. Our model has been tested for limited set of GPU computational time with small sample of datasets used for testing which is one of the limitations to existing architecture. However, we believe that similar CNN model architecture with more GPU and memory can outperform our results and also can handle much larger medical datasets. In order to avoid GPU bottlenecks we have used in-built storage accessibility of datasets to improve the computational time. However, is also a limitation considering the data size beyond the disk capacity would require more storage. We found cloud computing-based platforms such as Amazon S3 buckets or Google cloud computing environments to run the CNN model with larger datasets would be more suitable rather than standalone disk capacities.

#### 12.2.4.3 Phase 3: analysis of MRI patches

In this phase analysis of the image patches obtained through batches will be carried. Depending on the availability of GPU and memory of the experimental study, we have re-positioned the image patches to a standard 128 × 128 × 128 slices to comfortable fit into the CNN model and deep learning architecture. By doing this, we are also ensuring that the essential metadata is also retained to form the complete MRI structure and be able to perform through analysis of the patch data obtained through training data sets. Although the conventional 3D MRI image matrix would be quite large to fit completely consists of 480 × 512 × 512 due to huge GPU requirements. Batches of large size not only hinder the computational time but also have storage constraints as well. Because more disk space would be needed to process the MRI repositories which is more suitable for a cloud computing-based environment rather than a conventional disk storage facility. Nevertheless the existing CNN model framework works well with the image size considered for the study. Research facilities with larger GPUs having RAM capacities of 16–32 GB can accommodate MRI images of large dimensions without any further transformations, re-sampling or rotation of image axis. As discussed in the earlier sections, different forms of image pre-processing would be necessary to fit the data into CNN and deep learning models of lesser GPU and memory available to train and test the model.

## 12.3 Deep learning model implementation

### 12.3.1 Model architecture

Medical imaging requires robust image processing tools to effectively predict the medical condition of the patients. In the conventional approach experience of the physicians plays a pivotal role in diagnosis and prognosis of the medical condition. Research in medical imaging has seen more usage of convolutional neural networks in recent years considering their performance metrics and accuracy close to expert radiologists [37]. However, every machine learning model has its own limitations while considering for the research. Some of them would be more suitable for complex image processing tasks applying mathematical representations but fail to address computational and storage requirements. Many open-source neural network model built using Keras [43] and Tensorflow [28] are popular and also showed promising results in medical diagnosis. Pre-trained models also serve well with preconfigured settings related to learning rate, number of epochs, regularization to be performed either using optimisers, droput or batch normalization. Some of the existing models are used in the architecture from segmented U-net [43] and refined to fine-tune the hyper-parameters used in the model. Model architecture flow diagram is shown in Figures 12.2 and 12.5. Nevertheless, the benefit of adapting from open-source neural network models is the fexibility to switch over between the refinements to the already existing model parameters, which will help to improve the accuracy and reduce error controlling model fittings issues.

### 12.3.2 Model training and testing

Deep learning model with pre-processed datasets has been trained using the data I/O pipeline discussed in earlier sections. After the necessary hyper-parameters have been set to run the model. Model has also been fine-tuned to simultaneously perform re-sampling, shuffling of batches as the model continues to operate phase by phase during the training stage. Followed by model is also tested with validation segments by subjecting the datasets to different evaluation metrics to asses the quality of the image segmentation and multi-class classification performance of the model. Open-source architecture allows to share and extend the pipelines within the CIScnn model to integrate API calls to perform any additional task on the MRI images if required as part of model refinements.

In addition to the above-discussed, we also perform batch-wise data augmentation during the training stage which requires the batch size and shuffling parameters to be set before the model is put for training. Similarly, number of epochs and step-size of the learning rate also have to be defined before the model is executed. We have considered a epoch($\rho$) of 800 iterations across the CNN model and learning rate ($\alpha$) of 0.0001(1E-4) for the model selection and training phase. We also have implemented regularization using dropout with *keep_prob* of 0.8 size to ensure that 80% of each segment has a chance to keep in the iteration. Similarly, learning rate ($\alpha$) has also been optimized by considering the weights assigned after each epoch run by the neural network. The model also computes the cross-entropy loss and *f*1-score for every iteration executed by the model while performing batch shuffling, data augmentation as well. So that, we can consider optimal number of epochs based on the loss and accuracy values obtained after every iteration or epoch. These parameters also determine the number of batches put for pre-processing which also shows the volume of data utilized for training based on the size requirements model storage and memory can also be determined. As this would be helpful for researchers running with low GPU and memory requirements. Keras framework allows to obtain the model performance insights while training the model. Using this facility we can obtain the performance metrics related to recall and precision to make a justification of the hype-parameters selected for the model and any further refinements to be done to the existing parameters of the model.

### 12.3.3 Model evaluation metrics

In the current study we have applied Dice similarity index [44], Jaccard's coefficient [4], Tversky loss [4] along with cross-entropy, accuracy metrics using *f*1-score, precision, recall and loss function are incorporated into the cascaded image segmentation convolutional neural network model. Each of the listed metrics have pros and cons to consider for model evaluation. Dice similarity index is used to assess similarity between ground truth and obtained image segments of the medical images. However, one drawback with this metric is it penalizes the false positives which would be required to compute the overall predictions of performance of image segmentation. Hence we considered class-wise prediction rather than simple counts as it helps to address the class imbalance issues of the model as well.

Similarly by applying Jaccard's coefficient we would be able to address both under and over segmentation issues of the model considering the smaller sample sizes re-sampling has been opted during pre-processing phase as discussed in earlier sections. So Jaccard's index would be able to address this issue. Followed by Tversky loss addresses imbalance issues among the datasets provided for training and ensures a better performance while classifying the image batches into multi-class segments. Apart the mentioned indexes discussed we also have applied standard evaluation and performance metrics such as *f*1-score, precision and recall to assess the quality of datasets and accuracy of the prediction done using the CIScnn architecture.

## 12.3.4 Model prediction and analysis of results

Model prediction phase of the neural network looks at the probability estimates of the pixels identified in the multi-class labels. Activation function chosen for the training and testing phase also affects the step-size chosen for the selection of the model. In addition to the activation function, argmax feature also helps to improve the segmentation constraints. Multi-class labeling requires uniform selection of sample with distinct characteristics which is a concern with distorted or non-uniform intensity-scaled MRI images. We have applied measure to control the intensity pixel normalization and batch management to reduce model fitting issues.

In addition to the above-discussed, we also have applied overlapped patch-wise analysis to improve the prediction of tumors within the MRI images. Overlapping of patches within the batch images will improve the accuracy of patch prediction formed from the stock of patch images extracted from data source. CNN models usually suffer from edge prediction issues while training the datasets, overlapping the patch images through data augmentation will improve the edge prediction of the model. By generating a stock of patch images into a batch and performing shuffling of batches with the patch images to control the model over-fitting issues. After which mean of each predicted class in every overlapped patch is computed using chosen activation function of the model. The image matrix obtained from the data I/O pipeline used in the pre-processing phase of the CIScnn model generated the images in NifTI image formats. However the images are further subjected to pre-processing to extract MRI images as .png file formats which can be easily provided as data repository for convolutional neural networks to perform Model Training, testing and evaluation. However, image files obtained for challenging datasets source also provide metadata as well which captures information related to MRI thickness, volume of voxels, etc. Nevertheless, our approach uses all the metadata along with generated MRI images to re-structure the complete profile of the images segmented.

### 12.3.5 Implication of auto evaluation techniques

Auto evaluation techniques will help to improve the prediction process of the convolutional neural networks using cross-entropy, 10-fold-cross-validation, regularization methods. All these methods will ensure that ration of 80:20% or 70:30% of data is distributed across training and testing datasets. During evaluation stage, predefined metrics would be applied to the datasets to compute the accuracy and loss function of the model. In addition to this results obtained through model prediction would be compared to the ground truth provided by the experts in medical imaging for further validation of the model. Segmentation overlays also help in improving the auto evaluation methods by providing visual representation of the voxels in the neighborhood of the tumor and making it easy for the model to demarcate the patches within the images.

Followed by, in the remaining section of the paper, we have presented the performance of the cascaded image segmentation CNN model using two different challenging datasets taken from the Kidney Tumor Segmentation Challenge 2019 (KiTS'19) [45] and Brain Tumor Image Segmentation 2014 challenge (BRATS'14) [41] datasets using CIScnn. Through this study we have evaluated the CNN model and validated it for its accuracy to prediction of tumor using MRI images obtained from both the challenging datasets.

## 12.4 Findings from KiTS'2019 challenge MRI dataset

We have used the open-source data repository provided by the KiTS'19 [45] challenge portal to examine tumor from MRI images. We have considered kidney cancer as it happens to be most popular in the list of cancer types in men and women of varying age groups. Visualization of tumor will help nephrologists and radiologists to plan required surgical procedures to be followed or medication to be given. Considering the uneven kidney morphology, conventional image segmentation methods will not be sufficient to predict the tumors. We need a framework which can determine the tumor condition of the subject based on the kidney morphology. In this context semantic segmentation methods have shown better results as compared to others [35]. KiTS'19 [45] challenging datasets carry a collection of 300 kidney cancer patients CT scan images which will be used for prediction. However, for our study we have used only 80 CT scan images samples with the re-sampled image size, whereas the original datasets had 512 × 512 resolution with 210 slices. All the CT scans were provided with ground truth semantics by the experts which was labeled into three features capturing image intensity, target organ scanned and tumor to be predicted. A sample snapshot of MRI images showing all three features is given in Figure 12.3.

We have used 10 CT scan images provided with ground truth metadata by the KiTS'19 challenging portal and the rest 90 CT scan images were used to evaluate the prediction accuracy with the ground truth provided by the experts. We have processed the scan images to reposition the tumor details by applying normalization at the slice level to improve the training and prediction performance of model.

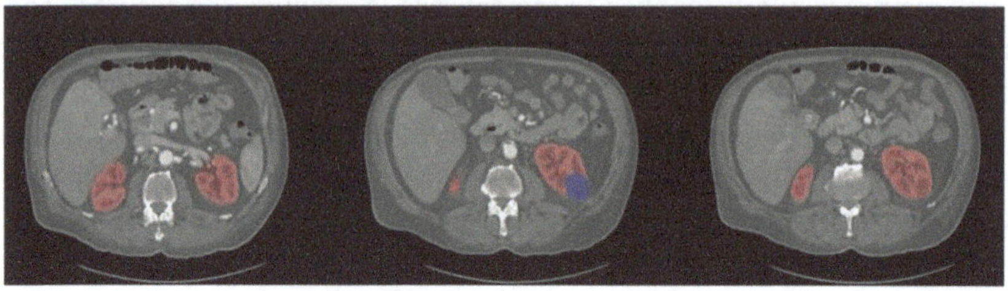

**Figure 12.3:** CT scan imagery of kidney tumor image segmentation dataset showing both kidneys with the ground truth (red) and tumor (blue).

Among the samples chosen for the study, we have selected 10 CT scan images with ground truth during training and 60 CT scan images without the semantic details was used for evaluation and rest of 10 CT scan images have to be discarded for non-uniformity of intensity level obtained in the image. All the scans were of NifTI formats with metadata on normalized slice thickness. We have generated MRI images and patch batches from the original format to train the CNN model.

## 12.4.1 Validation on the KiTS19 dataset

We have improved the CIScnn model by taking a subset of 120 CT scan images for normalization and run cross-validation to reduce model over-fit issues. This also helps to identify and fine-tune the model hyper-parameters. Validation of KiTS'19 datasets is done by applying three-fold cross-validation of the extracted datasets. A subset of 80 CT scan images was used to perform data augmentation in order to reduce model over-fitting issues. We did find statistically significant results from training and test datasets when subjected to three-fold cross-validation. And the same has been presented in the rest of the sections.

## 12.4.2 Improved CNN configurations

Cascaded image segmentation CNN model is configured to predict the multi-class patch analysis with a dimension of 80 × 160 × 160 with batch size 2. All the pixels of the image are subjected to Z-score normalization applied across all pixels of the image. During data augmentation, all the methods discussed in the earlier section were applied such as rotations, scaling, pixel intensity normalization along with Gaussian noise filter to control error and loss function metrics of the model. Overlapping patches of dimension 40 × 80 × 80 were used to predict the tumors. We also computed Tversky index with 800 epochs, optimized learning rate for comparative analysis of model performance. Based on the threefold cross validation applied to the KiTS'19 challenging dataset, the following results were obtained and these results are similar to the results obtained by other researchers who worked on the same dataset. Cross validation results fold-wise indicate the improvement performance of *f*-score among all the MRI datasets subjected to the deep neural network architecture as

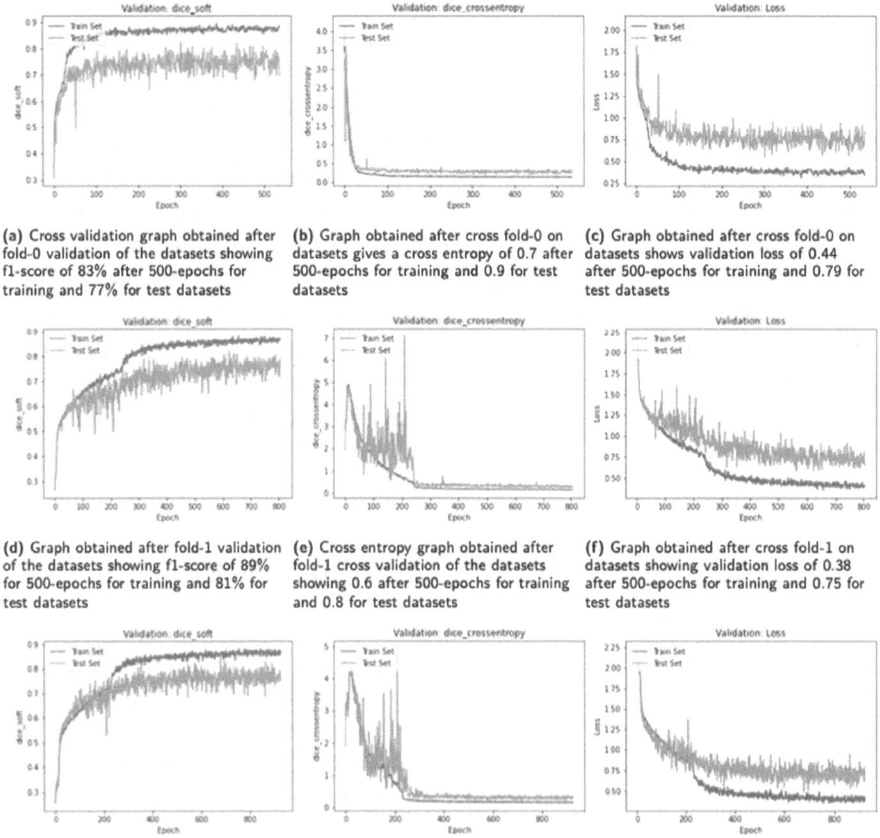


**(a)** Cross validation graph obtained after fold-0 validation of the datasets showing f1-score of 83% after 500-epochs for training and 77% for test datasets

**(b)** Graph obtained after cross fold-0 on datasets gives a cross entropy of 0.7 after 500-epochs for training and 0.9 for test datasets

**(c)** Graph obtained after cross fold-0 on datasets shows validation loss of 0.44 after 500-epochs for training and 0.79 for test datasets

**(d)** Graph obtained after fold-1 validation of the datasets showing f1-score of 89% for 500-epochs for training and 81% for test datasets

**(e)** Cross entropy graph obtained after fold-1 cross validation of the datasets showing 0.6 after 500-epochs for training and 0.8 for test datasets

**(f)** Graph obtained after cross fold-1 on datasets showing validation loss of 0.38 after 500-epochs for training and 0.75 for test datasets

**(g)** Graph obtained after cross fold-2 validation of the datasets showing f1-score of 91% after 500-epochs for training and 88% for test datasets

**(h)** Graph obtained after cross fold-2 cross entropy of the datasets showing 0.2 after 800-epochs for training and 0.3 for test datasets

**(i)** Graph obtained after cross fold-2 performed on datasets showing validation loss of 0.32 after 800-epochs for training and 0.71 for test datasets

**Figure 12.4:** Results showing the implication of threefold cross validation applied to the KiTS'19 datasets which shows improved *F*-scores as the cross-fold increases and gradual decrease in validation loss can be noticed in the above graphs.

shown in Figure 12.2. From the graphs we can conclude that kidney tumor image segmentation datasets showed better performance and accuracy through cross fold validation. The kidney dice coefficient obtained through cross validation was 95%, whereas the segmentation scores are higher as compared to dice index scores giving 79%. We have shown the performance metrics using dice index and other performance results using Figure 12.4(a–i).

We have used the standard U-net CNN architecture by retaining the default hyper-parameters for model training and testing datasets. We found the model perform reasonably well for non-uniform kidney morphology datasets. Although medical diagnosis with limited set of ground truth semantics provides weaker results. However, we notice the results obtained with KiTS'19 datasets seem to be better than existing research outputs which is promising for future studies. As it also helps other researchers to carry the results forward by improving the existing model. Considering only 30% of the original KiTS19 database, only 50 CT scans for training and rest 30 CT scans were used for testing. The remaining 10 CT scans were excluded due non-uniformity of intensity pixel values. However, our CIScnn data I/O pipelines to extract and perform data pre-processing on KitS'19 datasets was efficient to handle all the pre-processing tasks discussed in the earlier sections. And we also noticed the performance of the model to be in line with the performance of the existing CNN model used in Image segmentation tasks. Pytorch, Keras and Tensorflow frameworks work fine and deliver promising results as expected for all medical image segmentation tasks. As explained earlier in the section, open-source CNN frameworks allow for easy portability, extendable API feasability to scale up the model to handle more complex MRI images provided more GPU and higher memory requirements are satisfied.

## 12.5 Findings from BRATS'2014 challenge MRI dataset

Brain Tumor Segmentation (BRATS) Challenge 2014 MRI dataset was applied in the study to analyze the implications of Image segmentation technique to predict the tumor from MRI. Based on the cascaded CNN architecture models are proposed by [17]. In the current study we have implemented using BRATS'2014 dataset and compared it with KiTS'19 dataset on tumor prediction analysis using image segmentation. Cascaded CNN architecture is shown in Figure 12.5. We have used the same implementation to validate the tumor segmentation analysis with refinements to CNN architecture to enhance the tumor segmentation accuracy for the given challenging datasets. In the remaining sections of the paper we present the CNN architecture refined by concatenating the feature maps from different layers as another path to the CNN. By creating multi-channel computational paths, we were able to make CNNs perform different tasks through each path. We now present the two forms of CNN architecture adapted from existing literature to manifold the task of predicting brain tumors through image segmentation.

### 12.5.1 Receptive fields architecture

CNN architecture has two receptive fields comprising the smaller $7 \times 7$ and larger $13 \times 13$ form which is subjected to convolution layer through local-copy, input-concatenate and pre-output modes. The purpose of two different receptive fields is to improve the prediction of class labels from the image segments and patches within the batches to identify the tumor or voxel spaces with the batches. CNN architecture along with receptive fields has been shown in Figure 12.5. We allow the flow of two receptive fields image formats generated from top hidden layers is convoluted to $3 \times 3$ kernels of the second layer. This ensures that image formats obtained from all top hidden layers of two receptive fields are same across all layers of the CNN architecture. Feature maps obtained from both receptive fields are fed to the output layer giving a $5 \times 1 \times 1$ kernel. Using input concatenation to the image formats of receptive fields gives $224 \times 21 \times 21$ kernel convoluted to $21 \times 21$ kernel provided to fully connected layer run with softmax function gives the output $5 \times 1 \times 1$ kernel as shown in CNN architecture flow diagram in Figure 12.5.

Our approach also has a Cascaded Image Segmentation architecture which is done to overcome the problem of labeling the image segments separately. Instead all the segments within a batch that are spatially close to each other in the MRI image structure can be labeled together. By implementing a random selection field which can gracefully select all those labels located in close proximity would be considered for segmentation label. This procedure would also characterize the synthesized patch images and would also validate the classification accuracy of the model whether the patch has been classified to the correct class or not. Although, as the size of the segmentation increases it would also increase the computational time necessary to complete the task for the CNN. Besides, the storage requirements also have to be considered, as both memory and GPU have to be extended in order to consider for automated tumor segmentation CNN model.

### 12.5.2 Cascaded CNN architecture

We refer to this architecture as Cascaded Image Segmentation CNN Model considering the cascaded nature of the datasets flow pipeline. We have adapted this architecture based on the existing literature related to image segmentation [2, 9, 37]. We found that the CNN architecture proposed by other researchers had a few caveats related to segmentation labeling, prediction of tumor pixels and also showed low

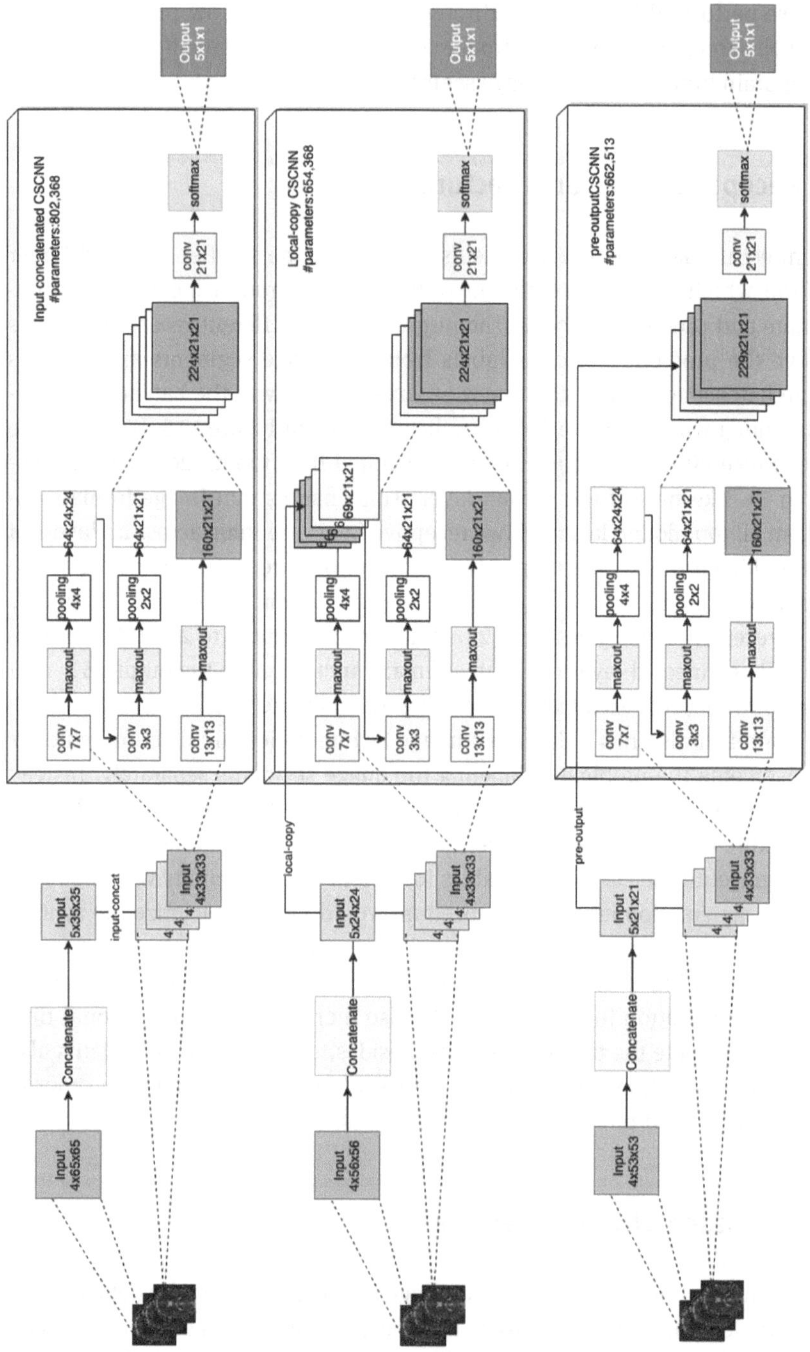


**Figure 12.5:** Cascaded CNN implementation adapted from Menze et al. [14] to predict brain tumor segmentation using deep neural networks architecture as shown in the flow diagram.

performance and accuracy scores [43, 44]. By improving segmentation labels and tumor voxel thickness, proximity of pixel locations within the extracted image patches within every batch. In order to implement such a flow our data I/O pipeline from input layer to hidden layer is convoluted using two streams of kernels as shown in CNN architecture diagram in Figure 12.5. And also from hidden layer to the output layer datasets will be passing through various stages of CNN model in cascaded form performing shuffling, batch normalization, max-pooling and finally reach the fully convoluted layer which runs the softmax function to further process the image segments using 5 × 1 × 1 kernel. In the remaining sections we have discussed the three forms of concatenation performed with data flow pipeline to improve the prediction of brain tumor image segmentation. A sample brain tumor segmentation pre-processed datasets are shown in Figure 12.6.

In the rest of the section, we presented the cascaded concatenation pipelines used in the CNN architecture.

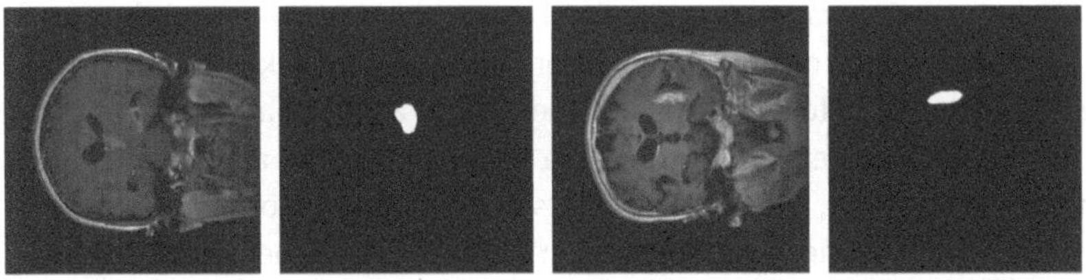

(a) Brain tumor segmentation image_766 snapshot has been shown. (b) Masked region shown in Brain tumor segmentation image_766 snapshot. (c) Brain tumor segmentation image_767 snapshot has been shown. (d) Masked region shown in Brain tumor segmentation image_767 snapshot.

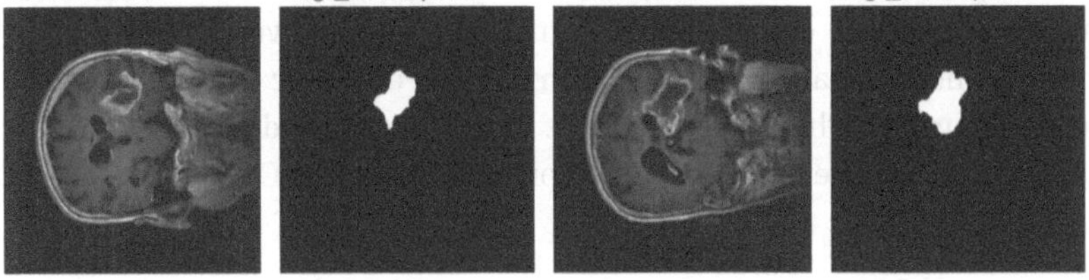

(e) Brain tumor segmentation image_768 snapshot has been shown. (f) Masked region shown in Brain tumor segmentation image_768 snapshot. (g) Brain tumor segmentation image_769 snapshot has been shown. (h) Masked region shown in Brain tumor segmentation image_769 snapshot.

**Figure 12.6:** BRAT's challenging dataset samples showing MRI brain tumor images along with masked region showing the tumor in the MRI.

#### 12.5.2.1 Input cascaded

Firstly, we present input-cascaded concatenation pipeline where the output of first layer is provided as direct input to the second layer of the CNN. In this approach,we will enable the additional image channel for the input patch images. And as discussed in earlier sections the kernels have been re-sampled to control the segmentation labeling issues. Details of cascaded pipeline are illustrated in Figure 12.5. We refer to this model as input cascade CNN.

#### 12.5.2.2 Receptive field cascaded

Second, we present the next layer which has two streams to process referred to as the receptive field pipelines. These streams will carry two formats of image kernels suitable for the segmentation labels. And perform concatenation of the first hidden layer with the second layer of the CNN model. Details of receptive field pipeline are illustrated in Figure 12.5. We refer to this model as receptive field cascade CNN.

#### 12.5.2.3 Pre-output cascaded

Lastly, we present the concatenation pipeline between hidden layers and the output layer. As we move from the central hidden layer to the outer layer kernels formed will be different from one layer to another layer. The weights assigned at each layer of the network will be varying depending upon the image patch size and segmentation label assigned. As a result, the kernel weights and dimensions of image segments will also vary as the pipeline moves from inner hidden layer to the next set of layers till it reaches the fully connected layer to produce the output kernel. However, the difference between the output at one layer will be influenced by the next set of layer to determine the possible convolutional kernel as output. Nevertheless, kernels formed in this architecture are different at each layer for a specific reason to improve the prediction accuracy of the image segments. Details of cascaded pipeline are illustrated in Figure 12.5. We refer to this model as pre-output cascade CNN.

## 12.6 Results and discussions

We found the cascaded CNN architecture pipeline to be better than the existing neural network models and provides promising results with the challenging datasets used for the study. We have found the implications of using cross-entropy, loss function to be helpful to understand the performance of segmentation model as the model was further refined based on the output. In addition to standard performance metrics we also have applied Tversky index, dice coefficient for both class-wise and soft categorical was used. We also computed the model fitting aspects to control under and over segmentation issues as discussed in the earlier sections. In addition to others, cross-entropy with three-fold cross validation was also carried and the results of cross-validation are shown in Figure 12.4(a–i). Cascaded CNN model showed better performance for kidney segmentation and our results are in line with the existing research. We found the kidney dice coefficient median to be 96% while the brain tumor segmentation accuracy was 94%. We also found varying performance metrics for low and high kidney morphology. Tumor segmentation performance of CNN model for a kidney with a low median structure gave 79%. Nevertheless, morphology of the organ also determines the prediction accuracy as noticed in our study. Besides the computed metrics, we have also evaluated our results with ground truth provided by experts. Considering the need for automated medical image segmentation models to put them in practice from research labs to real-world clinics accessible to physicians remedial measures. We believe that more researchers would further improve the CNN models and benefit the society to provide helpful hand to the needy cancer research providing the necessary recommendation to ease the implementation pipeline and wide accessibility of medical resources to harness better research impetus.

## 12.7 Future scope and limitations of current study

Current CNN architecture can be used for most common medical imaging formats such as DICOM, NifTI datasets and can be used to perform data augmentation, resampling, shuffling and batch management. However, we still need to extend the model to automate the prediction of noise or intensity-based distortion within the image, which is a more complex task to perform as it requires more computing time and storage resources as well which is one of the major limitations of our study and we plan to improve it by scaling the cascaded image segmentation architecture to a cloud computing environment such as Amazon S3 buckets, Google Cloud Computing and Microsoft Azure for model evaluation using large medical images. Another limitation would be to support extended API calls from other platforms as most of the existing models fail to integrate with most of the open-source technologies leaving a gap for the research community to address them as future scope of the work.

## References

[1] Moeskops, P., Veta, M., Lafarge, M. W., Eppenhof, K. A., & Pluim, J. P. (2017). Adversarial training and dilated convolutions for brain MRI segmentation. In Deep learning in medical image analysis and multimodal learning for clinical decision support (pp. 56–64). Springer, Cham.

[2] Hesamian, M.H., Jia, W., He, X., Kennedy, P. (2019). Deep learning techniques for medical image segmentation: Achievements and challenges. Journal of Digital Imaging, 32(4), 582–596.

[3] Shuvo, M.B., Ahommed, R., Reza, S., Hashem, M.M.A. (2021). Cnl-unet: A novel lightweight deep learning architecture for multimodal biomedical image segmentation with false output suppression. Biomedical Signal Processing and Control, 70, 102959.

[4] Eelbode, T., Bertels, J., Berman, M., Vandermeulen, D., Maes, F., Bisschops, R., Blaschko, M.B. (2020). Optimization for medical image segmentation: Theory and practice when evaluating with dice score or Jaccard index. IEEE Transactions on Medical Imaging, 39(11), 3679–3690.
[5] Kasuki, L., Raverot, G. (2020). Definition and diagnosis of aggressive pituitary tumors. Reviews in Endocrine & Metabolic Disorders, 21(2), 203–208.
[6] Goldbrunner, Roland, Pantelis Stavrinou, Michael D. Jenkinson, Felix Sahm, Christian Mawrin, Damien C. Weber, Matthias Preusser et al. "EANO guideline on the diagnosis and management of meningiomas." Neuro-oncology 23, no. 11 (2021): 1821–1834.
[7] Sekhar, A., Biswas, S., Hazra, R., Sunaniya, A.K., Mukherjee, A., & Yang, L. (2021). Brain tumor classification using fine-tuned Googlenet features and machine learning algorithms: IoMT enabled cad system. IEEE Journal of Biomedical and Health Informatics, 26(3), 983–991.
[8] Lin, F., Wu, Q., Liu, J., Wang, D., Kong, X. (2021). Path aggregation u-net model for brain tumor segmentation. Multimedia Tools and Applications, 80(15), 22951–22964.
[9] Khan, H., Shah, P.M., Shah, M.A., Islam, S.U., Rodrigues, J.J.P.C. (2020). Cascading handcrafted features and convolutional neural network for IoT-enabled brain tumor segmentation. Computer Communications, 153, 196–207.
[10] Khan, M.A., Ashraf, I., Alhaisoni, M., Damaševičius, R., Scherer, R., Rehman, A., Bukhari, S.A.C. (2020). Multimodal brain tumor classification using deep learning and robust feature selection: A machine learning application for radiologists. Diagnostics, 10(8), 565.
[11] Ramzan, F., Khan, M.U.G., Iqbal, S., Saba, T., Rehman, A. (2020). Volumetric segmentation of brain regions from mri scans using 3d convolutional neural networks. IEEE Access, 8, 103697–103709.
[12] Arabi, H., Bortolin, K., Ginovart, N., Garibotto, V., Zaidi, H. (2020). Deep learning-guided joint attenuation and scatter correction in multitracer neuroimaging studies. Human Brain Mapping, 41(13), 3667–3679.
[13] Henry, T., Carré, A., Lerousseau, M., Estienne, T., Robert, C., Paragios, N., & Deutsch, E. (2020, October). Brain tumor segmentation with self-ensembled, deeply-supervised 3D U-net neural networks: a BraTS 2020 challenge solution. In International MICCAI Brainlesion Workshop (pp. 327–339). Springer, Cham.
[14] Menze, B.H., Jakab, A., Bauer, S., Kalpathy-Cramer, J., Farahani, K., Kirby, J., Burren, Y., Porz, N., Slotboom, J., Wiest, R., et al. (2014). The multimodal brain tumor image segmentation benchmark (brats). IEEE Transactions on Medical Imaging, 34(10), 1993–2024.
[15] Heller, N., Isensee, F., Maier-Hein, K.H., Hou, X., Xie, C., Li, F., Nan, Y., Mu, G., Lin, Z., Han, M., et al. (2021). The state of the art in kidney and kidney tumor segmentation in contrast-enhanced ct imaging: Results of the kits19 challenge. Medical Image Analysis, 67, 101821.
[16] Coccia, M. (2020). Deep learning technology for improving cancer care in society: New directions in cancer imaging driven by artificial intelligence. Technology in Society, 60, 101198.
[17] Havaei, M., Davy, A., Warde-Farley, D., Biard, A., Courville, A., Bengio, Y., Pal, C., Jodoin, P.-M., Larochelle, H. (2017). Brain tumor segmentation with deep neural networks. Medical Image Analysis, 35, 18–31.
[18] Hatamizadeh, A., Tang, Y., Nath, V., Yang, D., Myronenko, A., Landman, B., … & Xu, D. (2022). Unetr: Transformers for 3d medical image segmentation. In Proceedings of the IEEE/CVF Winter Conference on Applications of Computer Vision (pp. 574–584).
[19] Sun, C., Dou, Z., Wu, J., Jiang, B., Iranmanesh, Y., Yu, X., Li, J., Zhou, H., Zhong, C., Peng, Y., et al. (2020). The preferred locations of meningioma according to different biological characteristics based on voxel-wise analysis. Frontiers in Oncology, 10, 1412.

[20] Shi, Z.-H., Xu, M., Wang, Y.-Z., Luo, X.-Y., Chen, G.-Q., Wang, X., Wang, T., Tang, M.-Z., Zhou, J.-X. (2017). Post-craniotomy intracranial infection in patients with brain tumors: A retrospective analysis of 5723 consecutive patients. British Journal of Neurosurgery, 31(1), 5–9.
[21] Hahn, A., Bode, J., Schuhegger, S., Krüwel, T., Sturm, V.J., Zhang, K., ... & Kurz, F. T. (2022). Brain tumor classification of virtual NMR voxels based on realistic blood vessel-induced spin dephasing using support vector machines. NMR in Biomedicine, 35(4), e4307.
[22] Chen, G., Li, Q., Shi, F., Rekik, I., Pan, Z. (2020). Rfdcr: Automated brain lesion segmentation using cascaded random forests with dense conditional random fields. NeuroImage, 211, 116620.
[23] Sharif, M.I., Li, J.P., Khan, M.A., Saleem, M.A. (2020). Active deep neural network features selection for segmentation and recognition of brain tumors using mri images. Pattern Recognition Letters, 129, 181–189.
[24] Yaqub, M., Feng, J., Sultan Zia, M., Arshid, K., Jia, K., Rehman, Z.U., Mehmood, A. (2020). State-of-the-art cnn optimizer for brain tumor segmentation in magnetic resonance images. Brain Sciences, 10(7), 427.
[25] Chen, H., Qin, Z., Ding, Y., Tian, L., Qin, Z. (2020). Brain tumor segmentation with deep convolutional symmetric neural network. Neurocomputing, 392, 305–313.
[26] Di Ieva, A., Russo, C., Liu, S., Jian, A., Bai, M. Y., Qian, Y., & Magnussen, J. S. (2021). Application of deep learning for automatic segmentation of brain tumors on magnetic resonance imaging: a heuristic approach in the clinical scenario. Neuroradiology, 63(8), 1253–1262.
[27] Di Ieva, A., Russo, C., Liu, S., Jian, A., Bai, M. Y., Qian, Y., & Magnussen, J. S. (2021). Application of deep learning for automatic segmentation of brain tumors on magnetic resonance imaging: a heuristic approach in the clinical scenario. Neuroradiology, 63(8), 1253–1262.
[28] Myronenko, A. (2018, September). 3D MRI brain tumor segmentation using autoencoder regularization. In International MICCAI Brainlesion Workshop (pp. 311–320). Springer, Cham.
[29] Zhuang, J., Tang, T., Ding, Y., Tatikonda, S. C., Dvornek, N., Papademetris, X., & Duncan, J. (2020). Adabelief optimizer: Adapting stepsizes by the belief in observed gradients. Advances in neural information processing systems, 33, 18795–18806.
[30] Khan, S.R., Sikandar, M., Almogren, A., Din, I.U., Guerrieri, A., Fortino, G. (2020). IoMT-based computational approach for detecting brain tumor. Future Generation Computer Systems, 109, 360–367.
[31] Nabizadeh, N., Kubat, M. (2015). Brain tumors detection and segmentation in mr images: Gabor wavelet vs. statistical features. Computers & Electrical Engineering, 45, 286–301.
[32] Qayyum, A., Lalande, A., Meriaudeau, F. (2020). Automatic segmentation of tumors and affected organs in the abdomen using a 3D hybrid model for computed tomography imaging. Computers in Biology and Medicine, 127, 104097.
[33] Dolz, J., Gopinath, K., Yuan, J., Lombaert, H., Desrosiers, C., Ayed, I.B. (2018). Hyper-densenet: A hyper-densely connected CNN for multi-modal image segmentation. IEEE Transactions on Medical Imaging, 38(5), 1116–1126.
[34] Hu, Y., Guo, Y., Wang, Y., Yu, J., Li, J., Zhou, S., Chang, C. (2019). Automatic tumor segmentation in breast ultrasound images using a dilated fully convolutional network combined with an active contour model. Medical Physics, 46(1), 215–228.
[35] Agarap, A.F. (2018). Deep learning using rectified linear units (relu). arXiv Preprint arXiv:1803.08375.

[36] Badrinarayanan, V., Kendall, A., Cipolla, R. (2017). Segnet: A deep convolutional encoder-decoder architecture for image segmentation. IEEE Transactions on Pattern Analysis and Machine Intelligence, 39(12), 2481–2495.
[37] Saito, K., Ushiku, Y., Harada, T., Saenko, K. (2017). Adversarial dropout regularization. arXiv Preprint arXiv:1711.01575.
[38] NifTI. (2019). Neuroimaging informatics technology initiative(NifTI). https://www.nitrc.org/projects/nifti/.
[39] Farahani, K., Menze, B., Brats, M.R. (2014). Challenge manuscripts, 2013. http://www.braintumorsegmentation.org/.
[40] Farahani, K., Menze, B., Reyes, M.. (2014). Multimodal brain tumor segmentation (brats 2013). http://martinos.org/qtim/ miccai2013/.
[41] Rodríguez, P., Bautista, M.A., Gonzalez, J., Escalera, S. (2018). Beyond one-hot encoding: Lower dimensional target embedding. Image and Vision Computing, 75, 21–31.
[42] Beers, A., Chang, K., Brown, J., Sartor, E., Mammen, C.P., Gerstner, E., Rosen, B., Kalpathy-Cramer, J. (2017). Sequential 3D U-nets for biologically-informed brain tumor segmentation. arXiv Preprint arXiv:1709.02967.
[43] Bertels, J., Eelbode, T., Berman, M., Vandermeulen, D., Maes, F., Bisschops, R., Blaschko, M.B. (2019). Optimizing the dice score and jaccard index for medical image segmentation: Theory and practice. In: International Conference on Medical Image Computing and Computer-Assisted Intervention. Springer, 92–100.
[44] KiTS'2019. Kidney tumor segmentation challenge(kits), 2019. https://kits19.grand-challenge.org/home/.

Abdulnasser Alwaheed*, Md Akmol Hussain, Akbar Sheikh-Akbari

# 13 Color constancy adjustment techniques

**Abstract:** This chapter presents an overview on color constancy adjustment techniques. The concept of color constancy within digital images is first introduced and then some of the recent color correction methods are discussed. Some publicly available benchmark standard image datasets, which are used by the researchers to assess the performance of the color correction methods are introduced. These datasets contain both real and syntactical images of scenes illuminated by a single or multiple light source/s. Color constancy quality assessment measures, which are widely used in the literature, are also detailed. Finally, the performance of different color correction methods on images of different benchmark image datasets is assessed and compared. The chapter demonstrates that the learning-based approaches outperform the statistical based algorithms at significantly higher computation costs. Moreover, their performances are very data-dependent, while recent statistical-based methods have slightly lower performance to those of the learning-based algorithms at significantly lower computation cost and data dependency.

**Keywords:** Color constancy, illuminant estimation, assessment criteria, image datasets

## 13.1 Introduction

Human vision can perceive and differentiate the color of the objects regardless of the intensity and the color of the scene illuminant. This ability of the human visual system is defined as the color constancy. The color of the objects in an image taken by a photo capturing device may not appear as it is perceived by the human visual system. To mimic the ability of the human vision to maintain the color of the objects regardless of their illuminant, digital cameras use a white balancing (WB) function to adjust the color of the image. WB function in a digital camera can be simply defined as the process of adjusting the color of the image to make the image color as if the scene is uniformly lit by a canonical illuminant. However, the presence of intense illumination can change the object color in the scene acquired by a capturing device (Qian et al., 2016). Figure 13.1 shows images taken under different illumination conditions, which exhibit color cast.

*Corresponding author: **Abdulnasser Alwaheed,** Leeds Beckett University, School of Built Environment, Engineering and Computing, Caedmon Hall, Leeds, LS6 3QS, United Kingdom, e-mail: a.alwaheed6446@student.leedsbeckett.ac.uk
**Md Akmol Hussain, Akbar Sheikh-Akbari,** Leeds Beckett University, School of Built Environment, Engineering and Computing, Caedmon Hall, Leeds, LS6 3QS, United Kingdom

https://doi.org/10.1515/9783110756722-013

(a) (b) (c) (d)

**Figure 13.1:** Digital images were taken under different light sources: (a) bluish -, (b) yellowish-, (c) reddish- and (d) neutral-light source [79].

The color balancing function of the cameras may not fully alleviate the impact of the extreme illuminant. In the worst case, it may even color cast the image toward the dominant scene light color [1]. The purpose of the color correction techniques is to balance the color of the digital images so that it seems as if they have been taken under a uniform white illuminant source [2]. Furthermore, color balancing methods can be used to reveal the true color of the objects, which are important in many applications such as image forensic and object recognition. The color of the scene can be adversely affected by the color of the source illuminant in many ways. For example, a white color object within a scene may appear bluish in a digital image if it is taken under bright lighting condition, while it may demonstrate a red, yellow or green color cast when illuminated under different lighting conditions [3]. Digital cameras were primarily engaged in estimating objects' color based on the comprehension of the human vision system. The increased use of digital cameras has derived the scholars to develop more effective color-balancing techniques and the challenge is evaluating the color of the scene illuminant using the digital image information, which are usually in red, green and blue (RGB) format [4].

## 13.2 Digital image formation

A digital image is created by a digital camera when the reflected light from the object reaches the optical sensor of the camera via the lens. Most digital cameras represent digital images in RGB color space; however, the real color of the object can be changed by the spectral color of the illuminant. This is due to several parameters, for example, the color of the illumination, reflection property of the object and sensitivity function of the capturing device sensor. Figure 13.2 shows a general digital image creation scenario. Digital cameras mostly create images in RGB format. On the Lambertian surface and reflection properties, $R$, $G$ and $B$ are pixel values at the location $(x,y)$, which can also be denoted as the function: $f(x,y)$, as shown in equation (13.1):

$$f_c(\lambda) = \int_{\lambda_1}^{\lambda_2} s(\lambda)e(\lambda)c(\lambda)\, d\lambda \tag{13.1}$$

where $[\lambda_1, \lambda_2]$ refers to the illuminant wavelength range, $s(\lambda)$ is the spectral sensitivity of the camera sensor, $e(\lambda)$ is the illuminant energy as a function of the electromagnetic wavelength and $c(\lambda)= R(\lambda),\ G(\lambda)\ B(\lambda)$ [80]. They also assume that the scene is uniformly lit, for example, $I(x,\lambda) = I(\lambda)$ for all $x$ locations in an image. Then, the perceived color of the illuminant source relies on the characteristics of the illuminant and the sensitivity function of the camera sensor.

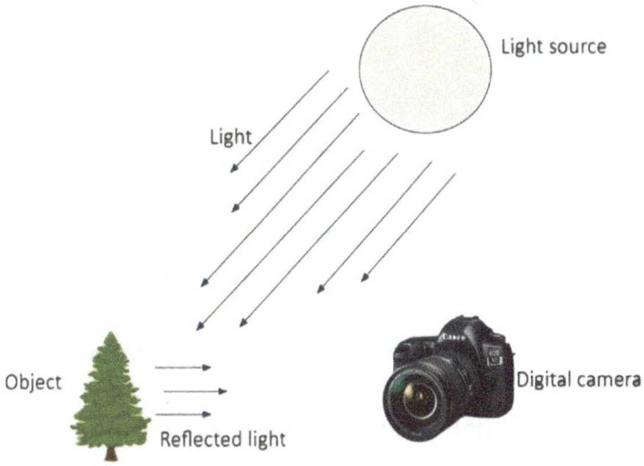


**Figure 13.2:** Creation of a digital image.

## 13.3 Illuminant

The color of scene illuminant has a significant effect on the fidelity of the objects' color within a digital image. The image scene can be lit by either one or more light source/s.

### 13.3.1 Single illuminant

Sample images of scene illuminated by a single light source from various datasets are shown in Figure 13.3, where they have been color corrected using the built-in camera color balancing function, which uses the image statistics to perform color correction. These color correction methods' main assumption is that the image scene is uniformly illuminated by only one illuminant. They perform quite well. However, their performance can be compromised, where there are wide uniform color regions within the image.

**Figure 13.3:** Example images of different scenes lit by one illuminant from different image datasets [3].

## 13.3.2 Multiple illuminant

In real world, an object in a scene can be illuminated by different lighting conditions, for example, brightness, indoor and outdoor lighting, daylight via glass and different color light bulbs. These circumstances are in contrast with the assumption that the scene is uniformly illuminated by a single light source. Hence, the color adjustment methods based on these assumptions may fail to deliver a high-quality color corrected image [5]. Therefore, it is important to consider and deal with the scenario that the scene is illuminated by multiple illuminant. Figure 13.4 shows example images from flying gray ball image dataset [6], where the scenes are illuminated by more than one light source.

**Figure 13.4:** Sample images of the scene lit by more than one illuminant from the flying gray ball image dataset (Aghaei and Funt, Color Constancy Adjustment Techniques).

In recent years, many color balancing approaches have been reported in the literature to solve the problem of color cast within digital images. These techniques can be classified as statistical-based, gamut-based and machine learning-based algorithms. In the following sections, these methods are discussed.

### 13.3.3 Statistical-based methods

Statistical-based color correction techniques employ the statistic of image pixels' data to determine the image color correction parameters. These techniques' assumptions are that the image scene is uniformly lit by single uniform illuminant. Gray World (GW) [7] and max-RGB or white patch methods [8] are the most common techniques of this category. The GW is among the most used simple methods for estimating the color of the scene illuminant. It assumes that a scene, on average, is a neutral gray, and its hypothesizes remain valid, where there is a sufficient spread of colors in the image scene. It then computes the mean of the RGB image color pixels and assumes that the resulting means represent the color of the scene color illumination. Hypothetically, the mean values of the RGB colors elements of the image scenes are roughly identical and the average values of every color component are adjusted by calculating three gain factors. The GW algorithm works well when the image scene consists of stuffiest color variation, otherwise, it biases the image color toward the dominant color patches within the image. For example, GW technique does a decent job in the images shown in Figure 13.5; however, the images have developed a blue tint color cast.

**Figure 13.5:** Color constancy adjustment using the Gray World (GW) technique: (a–b) original images and (c–d) their color corrected images using the GW method.

Kim & Lee [81] improved the GW algorithm's performance by using the R–B and G–B color coordinate system assuming that, when the mean values of image color components fall within the coordinates, the image contains adequate color variation for illuminant estimation. Their approach can determine true in-focus state even when images were taken from a cathode ray tube monitor or a light source, and it can compensate the luminance intensity of the image color channels without degrading the color of the image. The GW approach is a very basic method and easy to implement; however, it delivers a poor performance when dealing with images of scene illuminated by more than one illuminant or there are huge uniform color regions within the image. Beigpour and Riess [82] reported a color constancy method, which uses the GW method to estimate the colors of multiple illuminants. Their proposed method divides the input image into small regions and assumes that the GW assumption is fulfilled in each resulting region. Figure 13.6 shows the division of an image into small regions. If the resulting regions satisfy the GW assumptions, the proposed method estimates the color of the illuminant and its spatial distribution using an energy minimization-based framework. Their empirical result shows that their algorithm gives higher performance than assuming the scene is lit by one illuminant and the recent developed model reported in Gijsenij et al. [83], which deals with color correction of images of senses illuminated by more than one illuminant. Since the performance of color correction methods is dependent on the estimation of the color of the illuminant, Qian & Nikkanen [84] proposed a method that splits the input image into some areas and calculates the color of the illuminant for each resulting region. The color correction parameters of the image are then determined by combining the weighting factors of different patches regulated of their sizes.

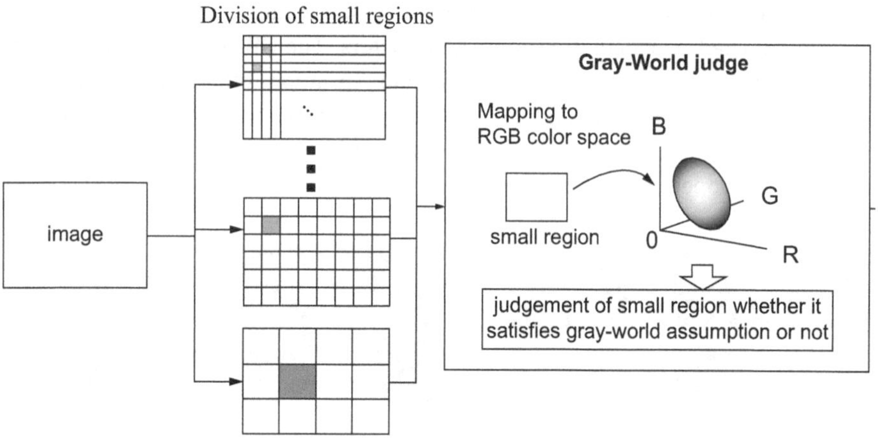


**Figure 13.6:** Division of small regions of the image [10].

Since the GW method simplifies the color constancy problem by estimating the color of the illuminant using the mean value of the image color components, it ignores some of the key parameters of the scene, for example, color variation of the objects, the effect of the dominant illuminant and huge uniform color region within the image, on estimating the color of the scene illumination. Therefore, the GW approach may sometimes fail to deliver the expected color corrected image. Hence, Jiang et al. [9] introduced an extended GW based on image saliency detection that split the input image into foreground and background, where the background consists of dominant color, such as grass, land, and sky, and they used a saliency detection algorithm to determine the foreground pixels. They experimentally showed that the resulting foreground areas of the input image have the color variation that fulfil the GW assumption. Hence, they used the foreground image pixels to determine the color correction parameters for the whole image using the GW method. The GW limitations were also considered by Kaur and Banga [11]. Kaur and Banga proposed an illuminate normalization method using Bayer filter. Zhang et al. [12] mentioned some challenges, where various existing algorithms failed to capture and attain accurate saliency detection. Some of these are occlusions, cluttered backgrounds and appearance changes. Their proposed algorithm utilizes light field to integrates several visual signals cues to determine the salient regions. Once saliency prior maps are generated from multiple illumination fields using super pixel-level intra-cue distinctiveness including color, depth and flow inherited from different focal planes as well as the viewpoint from multiple sources, pre-defined location is introduced to improve the saliency map. Jahanifar et al. [13] reported a higher performing form of supervision saliency detection technique based on Discriminative Regional Feature Integration. Their proposed method combines the multi-level segmentation, regional contrast property and background descriptor, and random forest regressor to identify a saliency score for different segments of the input images. The proposed algorithm overcomes the deficiencies and difficulties in the dermoscopy images including color inconsistency, hair occlusion and dark corner. It was stressed that the algorithm performance is the most powerful deep learning methods. A recent study by Kawamura [10] argued that the GW algorithm can be extended to achieve higher performance. The author introduced the layered GW algorithm to estimate the color of the illuminant. Their method divides the input image to several regions with different sizes. The standard deviation of different color component of each region was calculated as a criteria to define if that region carries adequate color variation to contribute to the color adjustment of the whole image. The author reported that their proposed layered GW algorithm produces higher color constancy than the traditional GW algorithm on images of the mandarin pattern of reddish dataset [6].

Yang et al. [14] reported a color correction technique using smaller number of gray pixels to estimate the illuminant. They argued that almost all-natural images consist of some intrinsic gray pixels if the scene is lit uniformly by a canonical illuminant. If it could retrieve the gray pixels, the resulting gray pixels can be employed for efficient scene illuminant estimation. To retrieve intrinsic gray pixels from a color-biased image, they defined an illuminant invariant measure (IIM) according to their spatial relationship to the local regions of the image. To identify the intrinsic gray pixels of the image, they computed the standard deviation of the local regions of the image in logarithmic space, which can be assumed to be an IIM. However, the exact gray pixels within the uniform region have lower chance of being detected because they do not have much local differences to be detected using the IIM. Hence, to accurately detect the gray pixels, they proposed using two constraints: (1) the gray pixels should be in small gray areas with a size of greater than $3 \times 3$ pixels and (2) these small gray areas representing non-uniform color areas within the image. These constrains enable determining the IIM and excluding areas that have uniform color. The proposed technique first transforms the input RGB image into a logarithmic RGB space. It then computes the standard deviation of each pixel using a $3 \times 3$ window, generating a relative standard deviation for pixels within its three color components. The IIM of the gray pixels within the image are very close to zero, where the IIM of the uniform areas are zero, which are excluded from further processing. Woo et al. [85] reported some concerns about the accuracy of the methods used to estimate the color adjustment factors in the presence of ambient light and claimed that in the presence of the ambient light the selected specular pixels using the existing statistical-based methods such as GW result in estimating dichromatic lines, which are thick or wavy. This may result in significant errors in the illumination estimation. Therefore, they proposed an adopting plane reflection model to derive the properties, which explain the relationship between peculiar pixels and formation of dichromatic lines under ambient light environment. Their experimental results show that the secularity of the image is sufficient even for real-world images. Laakom [15] reported an algorithm to lower this restraint by assuming that the color of the light within the convex hull of the pixel colors available in the image can be identified using the convex mixture of the image scene colors. Thus, the color correction techniques can be considered as a probability estimation problem. Their proposed method employs a graph-based approach to define the contribution of pixels of different image regions to estimate the color of the scene illumination. The color correction weighting factors for different pixels are then calculated using two main assumptions: pixels with similar colors offer the same contribution, while darker pixels contribute less. They represented the input image by a graph, $G=(V,E)$, where $V$ and $E$ represent the node and connection of the graph, respectively, using a graph-based learning method. The probability of each pixel, that is, each node of the graph is estimated. The higher the probability of a pixel the more information that pixel caries about the illuminant, and pixels with the same color carry the same information about the illuminant. Hence, they have the same probability value. Furthermore, the dark pixels encode less information about the illuminant. Hence, they have low probability values. Thus, the probability estimation is expressed as a two terms optimization problem, where the first term tells that the pixels with the same pixel values have the same probability and the second terms implies that there is a prior knowledge that darker area of the scenes is less affected by the illuminant. The GW method is a simple and popular algorithm. It assumes that the average reflectance of surfaces is achromatic. Basically, its popularity is due to its simplicity and its low computational cost compared to many other color constancy algorithms. However, like many other color constancy methods, the GW approach faces many challenges and shortcomings, which researchers have tried to resolve over the past years. One of its challenge is handling images of scenes lit by more than on light source. A Multi-Scale Adaptive Gray World (MAGW) algorithm was proposed by Li et al. [16]. This method uses a wavelet transform to generate a multi-scale images, and then the resulting image wavelet sub-bands are used to estimate the color of the scene illuminant. Finally, the original images are mapped under a uniform white light with supervision of an adaptive reliability using the estimated illuminant. Another wavelet multi-scale analysis for color constancy adjustment was proposed by Kyung, Kim and Ha [17]. Over the past years, researchers have proposed various methods to identify gray pixels within the images. Qian et al. [18] proposed the mean shifted gray pixel method. The Max-RGB algorithm, which is also known as the White Patch algorithm, which assumes that there is at least one pixel in the input image with the highest reflection that representing the scene light source. It uses their pixel values to determine the color of the illuminant [86]. Generally, the White Patch algorithm computes the image brightest point by integrating the highest values of the image three color components pixels to determine the color correction multiplier factors, as shown in equation (13.2):

$$\frac{\int f_{c_{\max}}(x)dx}{\int d_x} = ke_c = f_c \tag{13.2}$$

where $f_{c_{\max}}(x)$ is the maximum pixel value, $c \in$ RGB, $k$ is the multiplicative gain factor and $e$ is the input image illuminant and $f_c$ is the color corrected image. Although intuitively the idea of Max-RGB makes sense, when using the Max-RGB technique on images, it frequently generates inaccurate estimation of the light source due to the existence of even a single an-accurate pixel, excessive noise or the restricted illumination field of the capturing devices [19]. To resolve this problem, Banić and Lončarić [87] suggested to pre-process the input image by first removing overexposed pixels from the image using a median filter. They experimentally showed significant improvement in the performance of the anchor Max-RGB method when the image is pre-processed. Their approach was prompted by the Retinex-based method's performance in employing high subsampling rates without a significant accuracy reduction, and they used a subsampling method to reduce the redundancy as well as suppressing the noise-contaminated pixels. To maintain the simplicity of the hardware implementation of the algorithm, N random pixel samples were taken from the image and the Max-RGB color correction method was used to determine an estimation for whole image scene light source. To decrease the impact of noisy pixels on estimating the illuminant, the author used average of the M samples to determine the maximum pixel value for each image color band. They demonstrated that their proposed method is more robust to noise than its anchor method and the GW technique.

Because the pre-processing approaches do not guarantee the removal of all spurious pixels, Hussain and Sheikh-Akbari [3] proposed a color balancing method, which increases the robustness of the color correction algorithms by eliminating the uniform color patches within the image of being contributed to determine the color correction parameters. Images may not have that constant illumination, but a mixture of illuminant can be present. Hence, the application of general color constancy adjustment methods may also add color cast to the input image. The image may present a scene, which is not uniformly lit and may have large uniform color areas. Their proposed technique is aiming to lessen the impact of wide unvarying color regions of the image on the color constancy correction of the whole image by splitting the input image into non-overlapping blocks and excluding blocks with uniform color areas of contributing into the color constancy adjustment of the whole image. They used average absolute difference [AAD] value of each block's color component pixels to determine if the block possesses satisfactory color variation to be used for the correction for the entire image. The regions with an AAD value above a threshold value are assumed to have adequate color contrast, which can be employed for the illuminant estimation. Finally, the White Patch method is used to compute the color balancing parameters for the whole image using the pixels of the chosen. They experimentally demonstrated that their algorithm's images exhibit significantly higher color constancy to those of the anchor Max-RGB approach, when the images contain large uniform color areas.

Since the Max-RGB method uses the values of bright pixels to estimate the illuminant color, its effectiveness is limited when dealing with low intensity images. Hence, the Shades of Gray method was proposed by Finlayson & Trezzi [20], which allocate a generalized power mean identified as Minkowski Norm [p] to enhance the performance of the anchor approaches, when dealing with low intensity images. They showed that the Max-RGB and the GW algorithms are the special case of Minkowski $p$-norm for $p = 1$ and $p = \infty$, respectively. The Shades of Gray color constancy method generate its highest performance at $p = 6$. Similarly, Celik and Yetgin [21] unified the GW and the gray edge color constancy algorithms under discrete wavelet framework to examine the performance of color constancy estimation mainly using images containing colorful objects. The strength of their method is on its ability to exploit multi-scale structure of discrete wavelet to achieve color constancy. Experimentally they showed that their algorithm is simple and accurate in estimating the color of illuminant compared to the GW and gray edge algorithms when dealing with synthetic and real-world images.

Many of the proposed color constancy adjustment methods main assumption is that the scene lit uniformly by one illuminant. Since thus assumption is not hold in most of the images, for example, scenes are usually illuminated by more than one light source. Hence, researchers have introduced different methods to handle the color constancy adjustment problem under real illumination conditions, for example, the gray edge technique. The gray edge method assumes that the mean edge discrepancy in a digital image is achromatic, and this assumption enables computation of the light colors from the mean color derivative of the input image pixel values. Its color corrected images exhibit color constancies very similar to those of more advanced algorithms, but the gray edge algorithm is more computationally affordable, easier to implement and more time-effective in terms of computation performance compared to those sophisticated methods. Edge detection constitutes an essential pre-processing step in the context of the image processing. Much research has been conducting on developing effective accurate robust edge detection techniques with less sensitivity to noise. There are generally two types of edges derived from a 2D image of a 3D scene, namely: (a] edges that depend on the viewpoint, which may undergo alterations in keeping with viewpoint modifications and are generally indicative of scene geometry, for example, mutual occlusion of objects and (b) edges that do not depend on the camera viewpoint, which are usually indicative of 3D object intrinsic attributes, for example, surface markings and surface shape. The gray edge assumption is that the mean discrepancy of the edges in the image of a scene is achromatic. It seems that the Shades of Gray method's assumption that the mean of different image color bands' pixels are achromatic has been adopted by the gray edge algorithm and applied to image derivatives. The gray edge approach assumption is that the division of image pixels' color derivatives can be used to determine the maximal fluctuations in the pathway of the illuminant source. Hence, the direction of the illuminant is estimated with the help of the Minkowski norm of the derivatives. The technique is expanded through the introduction of higher-order derivatives or various types of derivations, for example, convolution with Gaussian filters. The color distribution within the image pixels facilitates the extraction of gradient information from the image. However, if a gradient operator is unable to discern the color gradient information, then the gray edge hypothesis does not ensure color constancy. One study applied this hypothesis to a wavelet domain, with multi-resolution analysis being conducted to achieve color constancy based solely on details (or edges). However, no automated technique for determining the Minkowski norm estimation was introduced, although an adaptive technique was suggested for the selection of the number of wavelet decomposition levels. Beigpour et al. [22] introduce an algorithm that generates a per-pixel illuminant source color estimation by utilizing the conditional random field technique, which takes into consideration both the spatial distribution of local illuminant estimation and their colors. The authors reported a framework that uses a combination of the statistics and physics-based color adjustment algorithms to estimate the light source color, which uses some sort of energy minimization technique. Compared to the study of Lee et al. [88] who proposed two-step multi-illuminant algorithms for color constancy of outdoor images, they claimed that the existing multi-illuminant algorithms still face the challenge of supporting diverse multi-illuminant scenes both indoors and outdoors. The proposed algorithm tries to combine sunlight and shade light. While the sunlight is already taken care of by the previous single-illuminant algorithm, the shade light still requires further correction. Using RGB color space, authors can determine the correct R and B ratio. Their results demonstrated that the two-step method yields less angular error compared to the gray edge. Likewise, the natural transition of the colors has been observed at the border of light and shade. The demarcation observed was mainly due to the color distinction that the algorithm produces. Generally, the proposed algorithm can be seen as the extension of the conventional single illuminant method to support multi-illuminants. Recently, Xinyu et al. [89] insisted that the problem of image edge information such as highlight edge, shadow edge and material edge and high frequency noise on light color estimation received has still not well identified by traditional gray edge hypothesis. They replaced bilateral filter with a Gaussian filter in estimating illumination chromatics in diagonal transformation to correct color deviation. The developed hybrid model shows exemplary performance over the traditional gray edge hypothesis on estimating the color constancy of image illuminant. It was further justified the power of traditional gray edge algorithm, which was improved by using a Gaussian filter to suppress the noise and preserve image high frequency information. This can be verified as the image pixel values near the edges were not significantly altered. In Hussain et al. [23] authors proposed a color adjustment method for uniformly/non-uniformly lit scenes using the image texture. Their proposed method uses the image texture to identify pixels with adequate color variation that can be employed in image color correction. Their algorithm employed a histogram-based approach to define required number of image segments to perform color correction. It then performs the clustering K-mean++ approach to divide the input image into pre-determined regions. They assumed that each resulting image segment represents a dominant scene lit source. They then use each resulting segment's pixel textures to determine if the segment has adequate color variation to be employed for image color correction. The proposed method then determines initial color correction factors for segments with sufficient color variations using a shelf algorithm. Finally, the color balancing factors for each pixel of the image are calculated by regulating the color constancy factors of the selected segments using the Euclidean distance of the pixel from the centers of the chosen segment. Their experimental results on images of the MIMO real, MIMO laboratory and multiple light source dataset and single illuminant dataset show significantly greater efficiency to those of the statistical methods and extremely competition outcomes to the learning-based methods at a portion of their computation cost. In [23], authors reported a color correction method for non-uniformly illuminated scenes' images. The proposed technique performs the K-mean++ classification approach on the input image, splitting the image to several regions, where the required number of segments is determined by employing a histogram-based technique. They used the Normalized Average Absolute Difference (NAAD) of segments pixels as an assessment criterion to determine if the segment has adequate amount of color variation. Segments with sufficient color variations are selected to be used for color correction. They then used the Grey World color correction method to determine initial colors correction factors for every selected region. The resulting initial color correction factors for different chosen segments are then consolidated to produce adjustment factors for every pixel of the image. They empirically demonstrated that their technique outperforms statistical algorithms and gives competitive results to those of the learning-based methods.

### 13.3.4 Gamut mapping-based methods

The gamut mapping methods assume that under a single light source, just few colors can be perceived, which is known as a canonical gamut. The color variations in an image are due to the source light deviation. The gamut mapping color correction approach attempts to map an image color into the gamut of colors that are observed under canonical light. The color correction is performed in two steps. In the first step, a 3D convex set of diagonal mappings is recovered, which show all the possible scene lights. A single mapping is then chosen from the resulting set of illuminant colors as an estimation of the unknown light source. Possible colors of the light source are shown as diagonal matrices, which map the gamut of an RGB image to a set of canonical-recorded RGB under a uniform light source. Although this algorithm's images exhibit high color constancy on images of art, matte, uniformly lit scenes, its color corrected images of more realistic scenes demonstrate much lower color constancy. It faces problem particularly with specularities and the method is computationally complex. To address this issue, Finlayson [24] modified the technique to operate in 2D chromaticity space. This simplifies the computation complexity of the method and features the confused 3D gamut mapping such as specularities, shape information and change of the light source do not affect and influence the algorithm's operation. Another modified version of gamut mapping was proposed by Finlayson et al. [25] that addresses the problem of null solution of typical gamut mapping methods caused due to the failure of the diagonal model of illumination change on which the methods are based. Their method utilizes a new diagonal offset that consists of six parameters rather than three parameters which provides opportunity to deal with a wider range of imaging conditions. The gamut-based approaches are one of the well-established solutions for color constancy adjustment, which has potential to deliver high performance. The fact that gamut mapping's model is based on a diagonal illumination change, which is not always satisfied in practice, the implementation of the algorithm is a non-trivial, and it is sometimes sensitive to real image data, limiting the application of the algorithm. Moreover, the gamut mapping-based approaches are sometimes failed to determine any estimation for the illuminant [26].

To improve the capability of the gamut mapping-based methods, Barnard et al. [90] introduced two extensions to the 3D gamut mapping technique. Their method consists of two separate phases. Possible solutions are constrained at the first stage. This phase is a function of the illumination change of the diagonal model, which is dependent of the camera sensor [91]. The other phase of the gamut mapping approach is to select a solution from the possible set of solutions. Therefore, the hybrid approach that combines the advantages of both techniques deliver superior results to both techniques. Otherwise, the proposed method can be re-formulated in a convex programing framework to address some of its implementation problems. However, the proposed model may not produce an estimate for the illuminant as it uses a diagonal model for the illuminant [27]. To address these challenges, Finlayson et al. [25] proposed a color constancy method known as cubical gamut mapping. This method is simple but outperforms most of the known available method in the literature in terms of the median illumination estimation error. Furthermore, it is possible to tune the method to minimize the maximum error. However, this could significantly increase the method's median error [28]. Alsam and Farup [29] presented a computational efficient, artefact-free, spatial color gamut mapping algorithm. The proposed algorithm offers a compromise between the color metrically optimal gamut clipping and an ideal spatial gamut mapping.

The study of Zamir [92] proposed spatial gamut mapping algorithm, which relied on a perceptual-based variation framework. This algorithm is an extension of gamut mapping, which aims to avoid the exploitation of full potential of the rendering device in term of color radiation. The standard gamut mapping algorithm is used when transforming the color of an input image to the color of the target device. The algorithm also aims to develop an extension that is capable of accurately expanding the colors from a smaller source gamut to a larger destination gamut. Furthermore, the model tried to mimic the basic properties of human visual system. Both subjective and objective quality assessments were conducted to evaluate the performance of the proposed algorithm. The results of the experiment showed its robustness compared with standard gamut mapping algorithms. It is a recommended algorithm for color constancy estimation for both film and broadcast industries. In another study Gatta and Farup [30] also raised concern about various technical issues to be considered when gamut mapping is used. This includes color space, direction and magnitude of the mapping and to what extent gamut colors should be altered. The spatial invariant algorithm was proposed that treats all the image pixel independently on their position. For that reason, it was capable to create virtual image with no artifacts and haloing. The power of the algorithm was associated by an iterative diffusion function, which it uses. They mentioned that the proposed algorithm can be incorporated into a multi-resolution scheme to speed it up consistently, making it suitable for very large images.

### 13.3.5 Machine learning-based methods

Machine learning-based color constancy adjustment techniques determine an estimation of the color of the scene light employing a model. These techniques first train their models using a sub-set of the input image dataset and then the trained model is used to color correct the query image. With this definition the Gamut-based color correction techniques can be considered as a sub-category of the learning-based methods. However, because of their vast use, they were introduced in a separate section. Neural networks and deep learning-based color correction methods are examples of the machine learning techniques. However, these methods, when correctly trained, deliver superior color correction to those of the statistical-based methods, even if there are very few distinct color surfaces in the image. To train the model these techniques require many images, and their performances are very data–dependent [31]. This applies to support vector regression [32] and linear regression. The linear regression method intends to directly learn a regression mapping from image representation to its corresponding vector of illumination. The machine learning-based methods consist of two main stages, namely training and testing. In the training phase, the methods learn the relationship between input and output data. The trained model is then used to color correct the unseen images in the testing phase. The performance of the learning-based color correction algorithm is dependent on the images used to train their models. Therefore, the pre-processing of their training images is vital to avoid their disagreeable predictions, which is due to outliers. Neural network-based color constancy adjustment methods were the first group of this kind, for example, Cardei et al. [32] introduced a color constancy adjustment technique using multi-layer perceptron, feedforward neural network approach. Their proposed neural network architecture has 400 input nodes and 40 neurons in each of its first and second hidden layers and the two output neurons. They used both synthetic and natural images to assess the performance of their technique. Similarly, Ebner [33] adopted neural network to perform a parallel color correction algorithm in RGB color space. Ebner proposed a parallel algorithm for color constancy adjustment, which uses a two-dimensional mesh processing element. The proposed algorithm is scalable to arbitrary sizes to generate a large natural image dataset from a small image dataset. Funt and Cardei [34] proposed a statistical color correction method called bootstrapping algorithm, which assumes that the image has large number of colors, and a set of images were captured from unknown illuminant with an uncalibrated camera. It takes sample of the captured images to generate a large training dataset to train the model. The author showed that the algorithm's color corrected images display higher color constancy than other methods' images. Moreover, Joze and Drew [35] reported an algorithm that employs both the RGB pixel values and image texture features to deduct the neighboring surface from the training data. They then merged this information to determine color correction factors for the entire image. Baron [36] recognized that weighting the image's color components could induce translation in the image log-chromaticity histogram. This facilitates the use of techniques like Convolutional Neural Network and structural prediction. Bianco et al. [5] developed a three-stage color correction algorithm to estimate the color of the illuminant. The CNN was employed to define the amount of illuminant in the scene. They then used a non-linear local aggregation method to refine the locally estimated illuminant. This results in a global illuminate estimation when the scene is lit by a single illumination. Koscevic et al. [37] reported an iterative CNN algorithm to estimate the color of the illuminant, which aims at suppressing the effect of the saturated light sources. Their proposed technique's training process was conducted in several iterations using the sigmoid function to compute an intermediate incremental illumination estimate of the scene. They reported higher illuminant estimation accuracy to other methods, where they used the statistical measures to evaluate the quality of their color corrected images. Their method is like that of the Hernandez-Juarez et al. [38] method, which uses a multi-hypothesis approach to estimate the color of the scene illuminant. Their proposed method uses a CNN, which was trained by a Bayesian algorithm, to determine a color estimation for scene light source. They showed that the use of the Bayesian algorithm for training the CNN is sufficient to validate the benefit of their model. Almost the same accuracy of color constancy can be attained by employing a deep neural network (DNN) or multi-layer neural network as reported in [38]. Their multiple hidden layers assist to significantly increase the accuracy of the estimated color of the illuminant. The authors used a training data from different devices to draw a generalizable conclusion. Moreover, they used rectified linear unit (ReLU) as the training function of their algorithm. Their experimental results show that their reported method outperforms the previous algorithms. Guo et al. [93] reported a color correction method for images of non-uniformly lit scenes illuminated by multiple light sources. Their technique uses the TK-means partition algorithm before performing source color estimation using a convolutional neural network to transform images of non-uniformly lit scenes by several light sources' images to the images of uniformly lit scenes by a single light source. Their experimental results demonstrate that their color corrected images demonstrate superior color constancy to those of the GW and Max-RGB in terms of angular error. In contrary to Laakom et al. [39] method, which uses substantial amount of data to train the CNN to generalize the capability of their method, they employed a fine-tuned encoder in an unsupervised pre-trained technique along with a composite loss function to improve the capability of their method. Their performance of their proposed method was assessed using the images of the tree dataset, which contains images of both natural and artificial scenes. They experimentally demonstrated that their model delivers a generalized color correction results across large number of images. The use of the pre-training to improve the accuracy of scene illumination estimation was also proposed by Qian et al. [40]. Their method uses a pre-trained squeezed-Net backbone with a 2D chroma histogram layers and Exit-Multilayer Perceptron to improve the effectiveness of the model. The Auto-White Balance and Cube + dataset were used to generate experimental results for their proposed method. Their results revealed that their model can boost the generalization of their color constancy method. The quality of their color corrected images was in contrast to Ratnasingam [41] argument that the use of traditional convolutional neural network involves various processing steps for image reconstruction, which could result in accumulation of residual error that could destroy the quality of the color corrected images. Therefore, they proposed a full convolutional neural network algorithm to reduce noise and number of the defected pixels. The uniqueness of their algorithm is on avoiding using a sequential signal processing, which is performed by traditional convolutional neural network algorithms. They used both stochastic and adaptive gradient algorithms during the training and testing of their algorithm. Their results show that their method outperforms other methods under complicated circumstances. Likewise, Choi and Yun [42] claimed that the shallow convolutional neural networks suffer from poor learning capability. To resolve this issue, they introduced a multi-stream deep neural network based on convoluted mixture of deep experts (CMoDE) fusion approach to estimate the color of the local illuminant. Technically, the CMoDE fusion technique was employed for extraction of the partial and spectral characteristics of the images. Their technique separates both series and parallel layers to identify a path that is difficult to determine under traditional convolutional neural network. They reported higher performance when dealing with this sort of images. Recent research on the use of neural network for color constancy adjustment has used common parameters, for example, using multiple sources of image data, validating the experiment results using benchmark image datasets. However, their differences are mainly on the training functions that they use.

## 13.4 Datasets

In this section 10 publicly available image datasets, which are widely used for assessing and comparing the performance of the color constancy adjustment techniques, are introduced.

### 13.4.1 SFU laboratory images

The Simon Fraser University (SFU) released an RGB image dataset of various objects [43]. They used 11 different light sources to capture the image of each object several times. They identified four subgroup such as objects with slight specularities (comprising 223 images of 22 scenes), objects with minimum one clear dielectric specularity (98 images of 9 scenes), objects with metallic specularities (149 images of 14 scenes) and objects with glowing surfaces (59 images of 6 scenes). However, the minimal specularities and the dielectric secularities images are normally used. Although images contain various scenes and illuminant, the images variety is reduced. Sample images of this dataset are illuminated in Figure 13.7.

**Figure 13.7:** Images samples from SFU Laboratory image dataset [43].

### 13.4.2 SFU Gray Ball image dataset

SFU Gray Ball image dataset was also released by SFU laboratory as the continuation of SFU Laboratory Images released in 2002 [44]. This dataset comprises 11,340 images of size 240 × 360 from a different scene. The images were captured from the scenes naturally illuminated or illuminated by more than one light source and a gray ball is in the scene. The advantage of this dataset is that the ground truth images were also captured using the gray ball mounted on top of the camera to increase the visibility inside the image. However, there is a possibility of image correlation when the dataset is used for learning algorithm; therefore some restrictions were imposed while using them. Example images of the Gray Ball dataset are illustrated in Figure 13.8.

**Figure 13.8:** Four sample images from the Gray Ball image dataset [44].

### 13.4.3 Color-checker original

This dataset was released by Vision Group at Microsoft Research Cambridge [45]. It was generated from a group of the real-world images collected by calibrated cameras. There is a color-checker in the scene of all images to help creating their ground truth image. This dataset comprises 568 indoor and outdoor images of people, sceneries and a range of objects. A Macbeth color-checker chart is in all the scenes used for generating this dataset. Figure 13.9 demonstrates four example images of the color-checker. Original image dataset.

**Figure 13.9:** Four sample images from the color-checker. Original image dataset [45].

### 13.4.4 Barcelona images

The Barcelona calibrated image dataset was generated by Parraga et al. [94] to study the color constancy algorithms. This dataset comprises many natural images, where a gray sphere is in the lowest left corner of all its images. This dataset consists of 83, 58 and 68 images of Urban Scenery, Forest & Motorways images, and Snow & Seaside scenes. Sample images of this dataset are shown in Figure 13.10.

**Figure 13.10:** Four sample images from the Barcelona image dataset (Parraga et al., 2009).

### 13.4.5 SFU HDR images

The SFU HDR images [19] represent the latest Computational Vision Laboratory image dataset. It contains 105 images of high dynamic range captured by calibrated cameras, where up to nine images per scene is used to create the high dynamic range images. Four color-checkers were placed at various angles in the scene. These color-checkers are used to determine the scene illuminant. Figure 13.11 shows four sample images of SFU HDR images dataset.

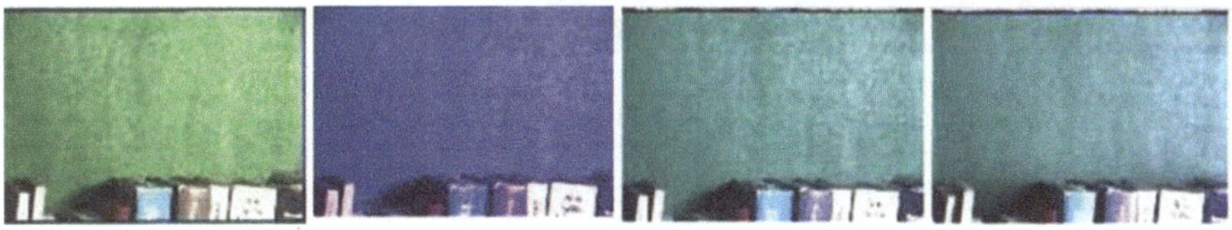

**Figure 13.11:** Four sample images of SFU HDR images dataset [19].

### 13.4.6 Color-checker reprocessed

The study of Shi and Funt [96] reprocessing the original color-checker dataset is introduced by Vision group [96]. Since the dataset of original color-checker was created from RAW data using automatic settings, Shi and Funt [96] were worried about artifacts that could have been introduced into the images. Therefore, they show the original RAW data in 12-bit lossless PNG image format. Figure 13.12 shows eight sample images of this dataset.

**Figure 13.12:** Eight sample images from color-checker image dataset (Shi and Funt, 2010).

## 13.4.7 Multiple light sources dataset

Multiple light sources image dataset was introduced by Gijsenij et al. [97]. This dataset consists of 58 images of various sizes recorded under laboratory setting and 9 images captured outdoors. Two indoor and two outdoor sample images of this dataset are illustrated in Figures 13.13 and 13.14. Different techniques including placing gray balls in different places of the scene and also manually color adjusting the images were used to generate the ground truth of the images.

**Figure 13.13:** Two indoor sample images from Multiple Light Source image dataset [97].

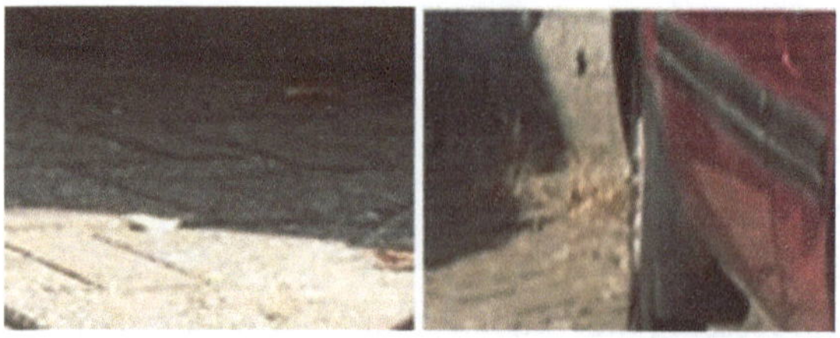

**Figure 13.14:** Two outdoor sample images from multiple light source image dataset [97].

### 13.4.8 Yet Another Colour Constancy Database Updated

The Yet Another Color Constancy Database Updated consists of images of low dynamic range (LDR) and high dynamic range (HDR) scenes [98]. Figure 13.15 shows tow example images of this dataset.

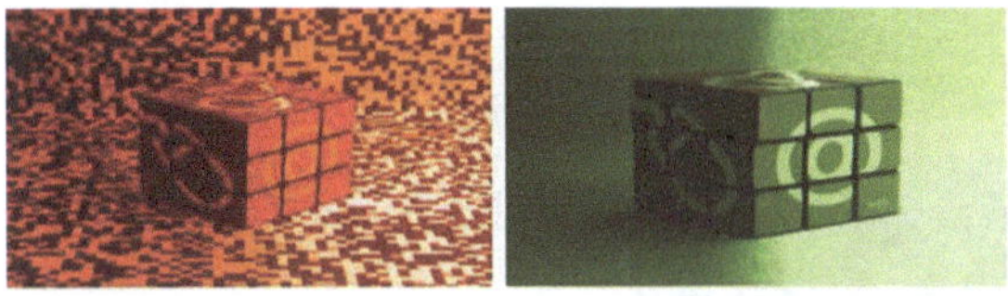

**Figure 13.15:** Yet Another Colour Constancy Database Updated (2014) LDR and HDR scene image [98].

### 13.4.9 Multi-illuminant multi-object

Multi-illuminant multi-object (MIMO) image dataset was reported by Beigpour et al. [99]. The dataset has 58 indoor laboratory images captured under controlled illumination and 20 real-world images captured under uncontrolled illumination. Figures 13.16 and 13.17 show tow example images of the MIMO laboratory and tow MIMO real-world dataset, respectively.

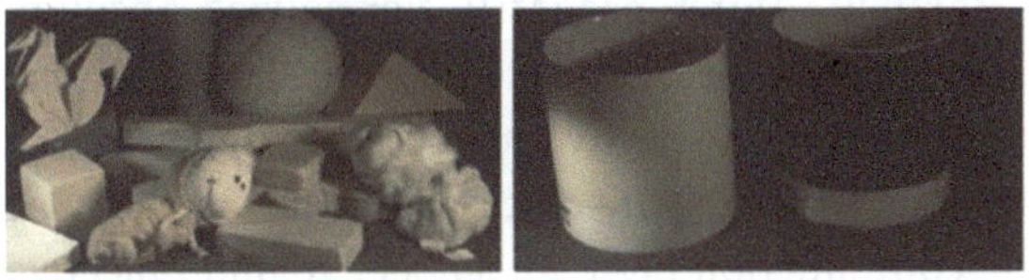

**Figure 13.16:** Two sample images form MIMO Laboratory image dataset [99].

**Figure 13.17:** Two sample images form MIMO real-world image dataset [99].

### 13.4.10 Cube dataset

Banić et al. [46] created the cube image dataset which contains 1365 outdoor images captured within the area of Croatia, Slovenia and Austria using a Canon EOS 550D camera in different seasons of the year. Figure 13.18 illustrates two example images of cube dataset.

**Figure 13.18:** Two sample images form the cube dataset [46].

## 13.5 Evaluation matrices for color constancy algorithms

The quality of the color corrected images can be evaluated by objective and subjective assessment. While the subjective assessment is based on human observation, the objective assessment is conducted based on various measures, for example, angular error (AE), peak signal-to-noise ratio (PSNR), Normalized Cross Correlation (NCC), Structural Content (SC), Maximum Difference (MD), Normalized Absolute Error (NAE), Mean Squared Error (MSE), Average Difference (AD) and Euclidian Distance (ED). The approaches of objective and subjective methods, as a measure of the performance of the color correction methods, are discussed in sub-sections 4.1 and 4.2, respectively. Finally, the contradiction between subjective and objective methods are explained in Section 4.3.

### 13.5.1 Objective methods

#### 13.5.1.1 Angular error

Angular Error (AE) is another objective assessment criterion that measures angular difference between the light source vector of the color adjusted image and the respective ground truth vector, and it can be calculate using equation (13.3):

$$\text{Angular error} = cos^{-1}\left(\frac{e_c \cdot e_g}{\|e_c\| \cdot \|e_g\|}\right) \tag{13.3}$$

where $e_c \cdot e_g$ represents the dot product between the true color of the illuminant vector, $e_g$, and the estimated color of the light source $e_c$ and $\|.\|$ is the Euclidian norm.

To assess the capability of the color correction methods, the mean and median of the angular error across numerous of their color corrected images are computed. The technique that its images have smaller mean or median angular error, delivers higher color balancing performance.

#### 13.5.1.2 Peak signal-to-noise ratio

The PSNR represents the alteration between the color adjusted image and its ground truth image in decibel, which can be calculated using equations (13.4–13.5):

$$\mathrm{PSNR} = 10 \times \log_{10}\left(\frac{\mathrm{MAX}^2}{\mathrm{MSE}}\right) \tag{13.4}$$

$$\mathrm{MSE} = \frac{1}{mn}\sum_{i=1}^{m}\sum_{j=1}^{n}\left(A_{ij} - B_{ij}\right)^2 \tag{13.5}$$

where MAX is the maximum pixel value, $m$ and $n$ are the number of row and column of the image and $A_{ij}$ and $B_{ij}$ are the color corrected and the ground truth image pixel values.

#### 13.5.1.3 Normalized cross correlation

Normalized cross correlation (NCC) is employed to assess the difference/similarity among the color corrected and the ground truth images. The NCC value is confined in the range between −1 and 1 [47] and can be computed using the equation (13.6):

$$\mathrm{NCC} = \sum_{i=1}^{m}\sum_{j=1}^{n}\left(A_{ij} - B_{ij}\right)^2 \tag{13.6}$$

where $m$ and $n$ are the number of row and column of the images, $A_{ij}$ and $B_{ij}$ are the colour corrected and the ground truth image pixel values.

#### 13.5.1.4 Structural content

The structural content (SE) is the estimation of similarity/difference among the color corrected and the ground truth images. It discloses resemblance, which exist between two images beyond the capability of human eye. The higher the value of the SC.

The lower the color constancy of the color corrected image. When the SE value is 1, which is its maximum possible value, it means that the two images are identical [48]. The equation (13.7) can be used to compute SC.

$$\mathrm{SC} = \frac{\sum_{i=1}^{m} \sum_{j=1}^{n} \left(A_{ij}\right)^2}{\sum_{i=1}^{m} \sum_{j=1}^{n} \left(B_{ij}\right)^2} \tag{13.7}$$

where $n$ and $m$ are the number of row and column of the images, $A_{ij}$ and $B_{ij}$ are the colour corrected and the ground truth image pixel values.

#### 13.5.1.5 Maximum difference

The maximum difference (MD) is interpreted as the maximum value of the difference between the color corrected and the ground truth image pixels. The greater the MD value the lower the color constancy of the color corrected image. The MD value can be calculated using equation (13.8):

$$\mathrm{MD} = \mathrm{Max}\left(\left|A_{ij} - B_{ij}\right|\right) \tag{13.8}$$

where $A_{ij}$ and $B_{ij}$ are the color corrected and the ground truth image pixel values and Max represent its maximum value.

#### 13.5.1.6 Normalized absolute error

The NAE is another objective assessment criterion to quantify the performance of the color adjusted image. It can be calculated using equation (13.9):

$$\mathrm{NAE} = \frac{\sum_{i=1}^{m} \sum_{j=1}^{n} \left(\left|A_{ij} - B_{ij}\right|\right)}{\sum_{i=1}^{m} \sum_{j=1}^{n} \left(A_{ij}\right)} \tag{13.9}$$

where $m$ and $n$ are the number of row and column of the images, $A_{ij}$ and $B_{ij}$ are the color corrected and the ground truth image pixel values.

#### 13.5.1.7 Mean squared error

Mean square error (MSE) is also employed to measure the quality of the color corrected images. It determines normalized energy difference between the color corrected image and its counterpart ground truth. MSE can be calculated using equation (13.10).

$$\text{MSE} = \frac{1}{mn}\sum_{i=1}^{m}\sum_{j=1}^{n}\left(A_{ij} - B_{ij}\right)^2 \tag{13.10}$$

where $m$ and $n$ are the number of row and column of the images, $A_{ij}$ and $B_{ij}$ are the color corrected and the ground truth image pixel values. The smaller the value of the MSE the higher the quality of the color corrected image, or the color corrected image represents smaller difference to its ground truth image.

#### 13.5.1.8 Average difference

The average difference (AD) determines the normalized average difference value among pixels of the color corrected image and its counterpart ground truth image. The smaller the value of the AD the closer the color corrected image to its ground truth image. The AD can be calculated using equation (13.11).

$$\text{AD} = \frac{1}{mn}\sum_{i=1}^{m}\sum_{j=1}^{n}\left(A_{ij} - B_{ij}\right) \tag{13.11}$$

where $m$ and $n$ are the number of row and column of the images, $A_{ij}$ and $B_{ij}$ are the color corrected and the ground truth image pixel values.

#### 13.5.1.9 Euclidian distance

Euclidian distance (ED) is an objective assessment criterion that represent the difference between the color corrected image and its ground truth image. The smaller the ED the closer the color adjusted image to its ground truth image. To determine the ED, the normalized $r$, $g$ and $b$ color components of both images are first computed employing equation (13.12):

$$r = \frac{R}{R+G+B}, g = \frac{G}{R+G+B}, b = \frac{B}{R+G+B} \tag{13.12}$$

The Euclidian distance between the normalized $r$, $g$ and $b$ values of the color corrected image and its ground truth is then calculated using equation (13.13):

$$\Delta E_{Q,I} = \sqrt{(r_Q - r_I)^2 + (g_Q - g_I)^2 + (b_Q - b_I)^2} \tag{13.13}$$

where $\Delta E_{Q,I}$ is the Euclidian distance between the color corrected image and its ground truth, $r_Q$, $g_Q$, $b_Q$ and $r_I$, $g_I$ and $b_I$ are the normalized color components of the color adjusted image and its ground truth image.

### 13.5.2 Subjective evaluation

The quality of an image is considered as final and reliable when it is assessed by human eyes due to the extraordinary capability of the human eyes. Hence, subjective assessment of the quality of the color corrected images have been wildly used by the researchers to measure the performance of the color correction methods. Mean opinion score (MOS) is one of the subjective assessment measures, which has been widely used in the literature. To generate a MOS for images of different techniques, their images are shown to some observers and the observers are asked to give a score in a predefined range, for example, ≤ 1, 5 ≥ or ≤ 1, 10 ≥, to color corrected images of different techniques based on their color correction quality. The average of each technique's images is calculated and used as a subjective comparison measure.

## 13.6 Performance analysis of existing color constancy techniques

This section represents objective performance analysis of various state of the art colour constancy algorithms employing widely used benchmark image datasets. To measure the performance of these approaches, mean, median and trimean angular error of the color corrected images of different methods are used as the assessment criterion. A lower angular error value means higher performance. The Grey World, Max-RGB, Shades of Gray, Gray Edge (1st-Order and 2nd-Order), Exemplar-Based Approaches [35], MIRF [99], CNN [36], ASM [100], CNN [5], CCATI [3], Combining Bottom-Up and Top-Down Visual Mechanism [101] and Multi-Stream Deep Neural Network [42] were performed on the image of the Gray Ball image dataset, and the average mean, median and trimean angular errors of their colour corrected images were computed and tabulated in Tables 13.1 and 13.2. From Table 13.1, the statistic-based methods including Grey World, White Patches, SOG and gray edge method are the least performer. Among the methods in Table 13.1, the CCATI method and the Combining Bottom-Up and Top-Down Visual Mechanism demonstrate noticeable improvement over other method. In overall comparison, the machine learning-based method such as the Combining Bottom-Up and Top-Down Visual Mechanism is the best performer among these color constancy methods in terms of mean, median and trimean angular error performance metrics.

Table 13.2 shows the performance of the state-of-the-art color constancy adjustment algorithms on the color-checker image dataset. From Table 13.2, the learning-based techniques, for example, CNN, angular GAN and multi-stream deep neural network, perform better compared to the statistics-based algorithms. Among the methods presented in Table 13.2, multi-stream deep neural network outperforms all other methods in all three mean, median and trimean angular error performance metrics.

**Table 13.1:** Average mean, median and trimean angular errors of the color corrected images of the Gray Ball image dataset for different color constancy adjustment methods.

| Method | Mean | Median | Trimean |
|---|---|---|---|
| Gray world | 7.8° | 7.1° | 7.14° |
| Max-RGB | 6.8° | 5.3° | 7.14° |
| Shades of Gray | 6.1° | 5.3° | 7.14° |
| 1st-Order Gray Edge | 5.9° | 4.7° | 5.1° |
| 2nd-Order Gray Edge | 6.1° | 4.9° | 5.3° |
| Exemplar-Based | 7.3° | 4.8° | 5.3° |
| ASM | 4.7° | 3.8° | – |
| CNN | 4.8° | 3.7° | – |
| CCATI | 3.8° | 3.8° | – |
| Combining Bottom-Up and Top-Down Visual Mechanism | 3.2° | 1.7° | 2.2° |

**Table 13.2:** Average mean, median and trimean angular errors of the color corrected images of the color-checker dataset for different color constancy adjustment methods.

| Method | Mean | Median | Trimean |
|---|---|---|---|
| Gray World | 6.3° | 5.6° | 6.2° |
| White Patch | 7.5° | 5.3° | 6.3° |
| Shades of Gray | 4.9° | 4.1° | 4.2° |
| 1st-Order Gray Edge | 5.3° | 4.5° | 4.7° |
| 2nd-Order Gray Edge | 5.1° | 4.4° | 4.6° |
| Exemplar | 8.2° | 5.4° | – |
| CNN | 2.7° | 1.9° | 2.1° |
| CCATI | 2.3° | 1.6° | 1.9° |
| Angular GAN | 6.0° | 2.5° | – |
| Multi-Stream Deep Neural Network | 2.0° | 1.0° | 1.4° |

Table 13.3 represents mean and median angular errors of the Gray World, White Patch, Shades of Gray, 1st-Order Gray Edge, 2nd-Order Gray Edge, the MIRF, the CCATI and Combining Bottom-Up and Top-Down Visual Mechanisms real and laboratory color corrected images from the MIMO dataset. Presented results also show that the Combining Bottom-Up and Top-Down Visual Mechanism method's color corrected "real-world" images contain the least average mean and median angular errors. Table 13.3 also demonstrates that the CCATI algorithm's color corrected laboratory images have the smallest mean angular error, while 2nd-Order gray edge's method's colour-corrected laboratory images represent the least median angular errors that amount all presented techniques.

**Table 13.3:** Average mean and median angular errors of the color corrected images of real and laboratory images of the MIMO image dataset for different colour constancy adjustment methods.

| Method | MIMO (real) | | MIMO (lab) | |
|---|---|---|---|---|
| | *Mean* | *Median* | Mean | Median |
| Gray World | *4.2°* | *5.2°* | 3.2° | 2.9° |
| White Patch | *5.6°* | *6.8°* | 7.8° | 7.6° |
| Shades of Gray | | | | |
| 1st-Order Gray Edge | *3.1°* | *5.3°* | 3.1° | 2.8° |
| 2nd-Order Gray Edge | *4.7°* | *6.0°* | 3.2° | 2.1° |
| MIRF | *4.1°* | *3.3°* | 2.6° | 2.6° |
| CCATI | *3.8°* | *3.8°* | 2.6° | 2.6° |
| Combining Bottom-Up and Top-Down Visual Mechanism | *2.5°* | *2.8°* | 2.9° | 3.8° |

## 13.7 Conclusion

In this chapter, the concept of color constancy in digital images was discussed. Various color constancy adjustment methods were studied. Both subjective and objective measures that are widely used in the literature to evaluate the color constancy of the color corrected images were detailed. Moreover, publicly available standard benchmark image datasets, which are used to generate experimental results for performance comparison across different methods, were introduced. The performance of different methods on the images of benchmark datasets was assessed and compared. Experimental results showed that the learning-based approaches images exhibit higher color constancy than those of the statistical methods. However, the learning-based method's performance were somehow data-dependent and were computationally complex. Statistical-based methods provide acceptable color constancy at significantly lower computation cost without needing to be trained and their data dependency is significantly low. Hence, they could be used in mobile devices.

## References

[1] Gijsenij, A., Gevers, T., Weijer, J.V.D. (2011). Computational color constancy: Survey and experiments.. IEEE Transactions on Image Processing, 20(9), 2475–2489. IEEE Transactions on, vol. 20, no. 13, pp. 2475–2481, 3 September 2011.

[2] Banic, N., Loncaric, S. (2015 March). Color dog-guiding the global illumination estimation to better accuracy. Visapp, (1), 129–135.

[3] Hussain, M.A., Akbari, A.S. (2018). Color constancy adjustment using sub-blocks of the image. IEEE Access, 6, 46617–46629.

[4] Cheng, M.M., Mitra, N.J., Huang, X., Torr, P.H., Hu, S.M. (2014). Global contrast based salient region detection. IEEE Transactions on Pattern Analysis and Machine Intelligence, 37(3), 569–582.

[5] Bianco, S., Cusano, C., Schettini, R. (2017). Single and multiple illuminant estimation using convolutional neural networks. IEEE Transactions on Image Processing, 26(9), 4347–4362.

[6] Aghaei, H., Funt, B. (2020). A flying gray ball multi-illuminant image dataset for color research. Color and Imaging Conference, 2020(28). Society for Imaging Science and Technology.

[7] Buchsbaum, G. (1980). A spatial processor model for object colour perception.. Journal of the Franklin Institute, 310(1), 1–26.

[8] Land, E.H. (1977). The retinex theory of color constancy.. Scientific American, 237(6), 108–129.

[9] Jiang, H., et al. (2013). Salient object detection: A discriminative regional feature integration approach. Proceedings of the IEEE conference on computer vision and pattern recognition.

[10] Kawamura, H. (2019). Multiple illuminants' color estimation using layered gray-world assumption. Electronic Imaging, 2019(14), 94–1.

[11] Kaur, S., Banga, V.K. (2016 February). Enriched image demosaicing using illuminate normalization and content based color filter array. 2016 International Conference on Information Communication and Embedded Systems (ICICES), 1–10. IEEE.

[12] Zhang, R., Zhu, J.Y., Isola, P., Geng, X., Lin, A.S., Yu, T., Efros, A.A. (2017). Real-time user-guided image colorization with learned deep priors. arXiv Preprint arXiv:1705.02999.

[13] Jahanifar, M., Tajeddin, N.Z., Asl, B.M., Gooya, A. (2018). Supervised saliency map driven segmentation of lesions in dermoscopic images. IEEE Journal of Biomedical and Health Informatics, 23(2), 509–518.

[14] Yang, K.-F., Gao, S.-B., Yong-Jie, L. (2015). Efficient illuminant estimation for color constancy using grey pixels. Proceedings of the IEEE Conference on Computer Vision and Pattern Recognition.

[15] Laakom, F., et al. (2020). Probabilistic color constancy. 2020 IEEE International Conference on Image Processing (ICIP). IEEE.

[16] Li, B., Xu, D., Lee, M.H., Feng, S.H. (2007). A multi-scale adaptive grey world algorithm. IEICE Transactions on Information and Systems, 90(7), 1121–1124.

[17] Kyung, W.J., Kim, D.C., Ha, Y.H. (2014). Enhancement of faded images using integrated compensation coefficients based on multi-scale gray world algorithm. The Journal of Korean Institute of Communications and Information Sciences, 39(8), 459–466.

[18] Qian, Y., Pertuz, S., Nikkanen, J., Kämäräinen, J.K., Matas, J. (2018). Revisiting gray pixel for statistical illumination estimation. arXiv Preprint arXiv:1803.08326.

[19] Funt, B., Shi, L. (2010). The rehabilitation of MaxRGB. Color and imaging conference, 2010(1). Society for Imaging Science and Technology.

[20] Finlayson, G.D., Trezzi, E. (2004). Shades of gray and colour constancy. Color and Imaging Conference, 2004(1). Society for Imaging Science and Technology.

[21] Celik, T., Yetgin, Z. (2015). Grey-wavelet: Unifying grey-world and grey-edge colour constancy algorithms. Signal, Image and Video Processing, 9(8), 1889–1896.

[22] Beigpour, S., Riess, C., Van De Weijer, J., Angelopoulou, E. (2013). Multi-illuminant estimation with conditional random fields. IEEE Transactions on Image Processing, 23(1), 83–96.

[23] Hussain, M.A., Sheikh-Akbari, A., Halpin, E.A. (2019). Color constancy for uniform and non-uniform illuminant using image texture. IEEE Access, 7, 72964–72978.

[24] Finlayson, G.D. (1996). Color in perspective. IEEE Transactions on Pattern Analysis and Machine Intelligence, 18(10), 1034–1038.

[25] Finlayson, G.D., Hordley, S.D., Xu., R. (2005). Convex programming colour constancy with a diagonal-offset model. IEEE International Conference on Image Processing (Vol. 3, pp. III–948). IEEE.

[26] Hordley, S.D. (2006). Scene illuminant estimation: Past, present, and future.. Color Research & Application: Endorsed by Inter-Society Color Council, the Colour Group (Great Britain), Canadian Society for Color, Color Science Association of Japan, Dutch Society for the Study of Color, the Swedish Colour Centre Foundation, Colour Society of Australia, Centre Français de la Couleur, 31(4), 303–314.

[27] Finlayson, G., Xu, R. (2003). Illuminant and gamma comprehensive normalisation in logrgb space. Pattern Recognition Letters, 24(11), 1679–1690.

[28] Gijsenij, A., Gevers, T., Van De Weijer, J. (2010). Generalized gamut mapping using image derivative structures for color constancy. International Journal of Computer Vision, 86(2–3), 127–139. 10.1007/s11263-008-0171-3.

[29] Alsam, A., Farup, I. (2011). Spatial colour gamut mapping by means of anisotropic diffusion. In: International Workshop on Computational Color Imaging, Berlin, Heidelberg, Springer.

[30] Gatta, C., Farup, I. (2017). Gamut mapping in RGB colour spaces with the iterative ratios diffusion algorithm. Electronic Imaging, 2017(18), 12–20.

[31] Uma Maheswari, V., Aluvalu, R., Chennam, K.K. (2021). Application of machine learning algorithms for facial expression analysis. Machine Learning for Sustainable Development, 9, 77.

[32] Cardei, V.C., Funt, B., Barnard, K. (2002). Estimating the scene illumination chromaticity by using a neural network. JOSA a, 19(12), 2374–2386.

[33] Ebner, M. (2004). A parallel algorithm for color constancy.. Journal of Parallel and Distributed Computing, 64(1), 79–88.

[34] Funt, B.V., Cardei, V.C. (1999). Bootstrapping color constancy. In: Human Vision and Electronic Imaging IV, vol. 3644. International Society for Optics and Photonics.

[35] Joze, H.R.V., Drew, M.S. (2014 May). Exemplar-Based Colour Constancy and Multiple Illumination. IEEE Transactions on Pattern Analysis and Machine Intelligence, 36(5), 860–873.

[36] Barron, J.T., 2015. Convolutional colour constancy. Proceedings of the IEEE International Conference on Computer Vision, 379–387.

[37] Koščević, K., Subašić, M., Lončarić, S. (2021). Iterative convolutional neural network-based illumination estimation. IEEE Access, 9, 26755–26765.

[38] Hernandez-Juarez, D., et al. (2020). A multi-hypothesis approach to color constancy. Proceedings of the IEEE/CVF conference on computer vision and pattern recognition.

[39] Laakom, F., et al. (2019). Color constancy convolutional autoencoder. 2019 IEEE Symposium Series on Computational Intelligence (SSCI). IEEE.

[40] Qian, Y., et al. (2021). SDE-AWB: A generic solution for 2nd International Illumination Estimation Challenge. Thirteenth International Conference on Machine Vision, vol. 11605. International Society for Optics and Photonics.

[41] Ratnasingam, S. (2019). Deep camera: A fully convolutional neural network for image signal processing. Proceedings of the IEEE/CVF International Conference on Computer Vision Workshops.

[42] Choi, -H.-H., Yun, B.-J. (2020). Deep learning-based computational color constancy with convoluted mixture of deep experts (CMoDE) fusion technique. IEEE Access, 8, 188309–188320.

[43] Barnard, K., et al. (2002). A comparison of computational color constancy algorithms. ii. experiments with image data. IEEE Transactions on Image Processing, 11(9), 985–996.

[44] Ciurea, F., Funt, B. (2003 January). A large image database for color constancy research. Color and Imaging Conference, 2003(1), 160–164. Society for Imaging Science and Technology.

[45] Gehler, P.V., et al. (2008). Bayesian color constancy revisited. 2008 IEEE Conference on Computer Vision and Pattern Recognition. IEEE.

[46] Banić, N., Koščević, K., Lončarić, S.. (2017). Unsupervised learning for color constancy. arXiv preprint arXiv:1712.00436.

[47] Rao, Y.R., Prathapani, N., Nagabhooshanam, E. (2014). Application of normalized cross correlation to image registration. International Journal of Research in Engineering and Technology, 3(5), 12–16.

[48] Rajkumar, S., Malathi, G. (2016). A comparative analysis on image quality assessment for real time satellite images. Indian Journal of Science and Technology, 9, 34.

[49] Vazquez-Corral, J., Párraga, C., Baldrich, R., Vanrell, M. (2009). Color constancy algorithms: Psychophysical evaluation on a new dataset. Journal of Imaging Science and Technology, 53(3), 31105–1.

[50] Moore, A., Allman, J., Goodman, R.M. (1991). A real-time neural system for color constancy.. IEEE Transactions on Neural Networks, 2(2), 237–247.

[51] Akbarinia, A., Parraga, C.A. (2017). Colour constancy beyond the classical receptive field. IEEE Transactions on Pattern Analysis and Machine Intelligence, 40(8), 1–14.

[52] Wilcoxon, F. (1992). Individual comparisons by ranking methods. In: Breakthroughs in Statistics. New York, NY, Springer, 196–202.

[53] Kim, Y., et al. (2002). A video camera system with enhanced zoom tracking and auto white balance.. IEEE Transactions on Consumer Electronics, 48(3), 428–434.

[54] Barnard, K., et al. (2002). A data set for color research. Color Research & Application: Endorsed by Inter-Society Color Council, the Colour Group (Great Britain), Canadian Society for Color, Color Science Association of Japan, Dutch Society for the Study of Color, the Swedish Colour Centre Foundation, Colour Society of Australia, Centre Français de la Couleur, 27(3), 147–151.

[55] Hussain, M.A., Akbari, A.S. (2018). Color constancy algorithm for mixed-illuminant scene images. IEEE Access, 6, 8964–8976.

[56] Shi, L., Xiong, W., Funt, B. (2011). Illumination estimation via thin-plate spline interpolation.. JOSA A, 28(5), 940–948.

[57] Barnard, K. (2000). Improvements to gamut mapping colour constancy algorithms. In: European Conference on Computer Vision. Berlin, Heidelberg, Springer.

[58] Gijsenij, A., Gevers, T., Lucassen, M.P. (2009) Perceptual analysis of distance measures for color constancy algorithms. Journal of the Optical Society of America A, 26(10), 2243–2256.

[59] Guanghua, C., Xiaolong, Z. (2015). A method to improve robustness of the gray world algorithm. 4th International Conference on Computer, Mechatronics, Control and Electronic Engineering.

[60] Qian, Y., Chen, K., Nikkanen, J., Kamarainen, J. K., Matas, J. (2017). Recurrent color constancy. In Proceedings of the IEEE international conference on computer vision (pp. 5458–5466).

[61] Hussain, M.A., Akbari, A.S., Ghaffari, A. (2016 August). Colour constancy using K-means clustering algorithm. 2016 9th International Conference on Developments in eSystems Engineering (DeSE), 283–288. IEEE.

[62] Fan, Y., Carré, P., Fernandez-Maloigne, C. (2015 September). Image splicing detection with local illumination estimation. 2015 IEEE International Conference on Image Processing (ICIP), 2940–2944. IEEE.

[63] Banić, N., Lončarić, S.. (2014). Improving the white patch method by subsampling. 2014 IEEE International Conference on Image Processing (ICIP). IEEE.

[64] Qian, Y., Kämäräinen, J.K., Nikkanen, J., Matas, J. On Finding Gray Pixels-Supplement.

[65] Gijsenij, A., Lu, R., Gevers, T. (2011). Color constancy for multiple light sources. IEEE Transactions on Image Processing, 21(2), 697–707.

[66] Gijsenij, A., Lu, R., Gevers, T. (2011). Color constancy for multiple light sources. IEEE Transactions on Image Processing, 21(2), 697–707.
[67] Woo, S.M., Lee, S.H., Yoo, J.S., Kim, J.O. (2017). Improving color constancy in an ambient light environment using the phong reflection model. IEEE Transactions on Image Processing, 27(4), 1862–1877.
[68] Barnard, K. (2001). A comparison of computational color constancy algorithms-part II: Experiments with image data. IEEE Transactions on Pattern Analysis and Machine Intelligence, 23(11), 1209–1221.
[69] Sun, X., Xie, K., Lin, S., Yang, A., Zhao, H. (2020 April). Research on calculation method based on gray edge color constant. 2020 3rd International Conference on Advanced Electronic Materials, Computers and Software Engineering (AEMCSE), 390–393. IEEE.
[70] Zamir, S.W., Vazquez-Corral, J., Bertalmı, M. (2014). Gamut mapping in cinematography through perceptually based contrast modification. IEEE Journal of Selected Topics in Signal Processing, 8(3), 490–503.
[71] Li, X., Wu, J. (2013). Improved gray world algorithm based on salient detection. In: Chinese Conference on Image and Graphics Technologies (pp. 315–321). Berlin, Heidelberg, Springer.
[72] Finlayson, G.D., Zakizadeh, R., Gijsenij, A. (2016). The reproduction angular error for evaluating the performance of illuminant estimation algorithms.. IEEE Transactions on Pattern Analysis and Machine Intelligence, 39(7), 1482–1488.
[73] Fan, Y., Carré, P., Fernandez-Maloigne, C. (2015 September). Image splicing detection with local illumination estimation. 2015 IEEE International Conference on Image Processing (ICIP), 2940–2944. IEEE.
[74] Liu, X., He, P., Chen, W., Gao, J. (2019). Improving multi-task deep neural networks via knowledge distillation for natural language understanding. arXiv Preprint arXiv:1904.09482.
[75] Fischer, G., Sajjaa, M. (2008). whitebalPR – A new method for automatic White Balance. Conference on Colour in Graphics, Imaging, and Vision, 2008(1). Society for Imaging Science and Technology.
[76] Fischer, G., Kolbe, K., Sajjaa, M. (2008, March). WhitebalPR: automatic white balance by polarized reflections. In Digital Photography IV (Vol. 6817, pp. 115–125). SPIE.
[77] Shi, L. (2000). Re-processed version of the gehler color constancy dataset of 568 images. http://www.cs.sfu.ca/~color/data/.
[78] Rizzi, A., Gatta, C., Marini, D. (2002). Color correction between gray world and white patch. In: Human Vision and Electronic Imaging VII, vol. 4662. International Society for Optics and Photonics.
[79] Joze, H., Vaezi, R., Drew, M.S. (2013). Exemplar-based color constancy and multiple illumination. IEEE Transactions on Pattern Analysis and Machine Intelligence, 36(5), 860–873.
[80] Gijsenij, A., Lu, R., & Gevers, T. (2011). Color constancy for multiple light sources. IEEE Transactions on image processing, 21(2), 697–707.
[81] Ikeuchi, K. (Ed.). (2014). Computer Vision || Image Enhancement and Restoration. Cham: Springer International Publishing.
[82] Kwag, H. K., Kim, S. H., Jeong, S. H., & Lee, G. S. (2002). Efficient skew estimation and correction algorithm for document images. Image and vision computing, 20(1), 25–35.
[83] Beigpour, S., Riess, C., van de Weijer, J., Angelopoulou, E. (2014). Muti-illuminant estimation with conditional random fields. Image Processing, IEEE Transactions on, 23, 83–96.
[84] Gijsenij, A., Gevers, T., Van De Weijer, J. (2012). Improving color constancy by photometric edge weighting. IEEE Transactions on Pattern Analysis and Machine Intelligence, 34(5), 918–929.
[85] Qian, Y., Chen, K., Nikkanen, J., Kamarainen, J. K., Matas, J. (2015). Recurrent color constancy. In Proceedings of the IEEE international conference on computer vision (pp. 5458–5466).

[86] Woo, S. M., Lee, S. H., Yoo, J. S., Kim, J. O. (2017). Improving color constancy in an ambient light environment using the phong reflection model. IEEE Transactions on Image Processing, 27(4), 1862–1877.
[87] Banić, N., & Lončarić, S. (2014, October). Improving the white patch method by subsampling. In 2014 IEEE International Conference on Image Processing (ICIP) (pp. 605–609). IEEE.
[88] Banić, N., & Lončarić, S. (2010). Green stability assumption: Unsupervised learning for statistics-based illumination estimation. Journal of Imaging, 4(11), 127.
[89] Lee, S. H., Woo, S. M., Choi, J. H., & Kim, J. O. (2017, September). Two-step multi-illuminant color constancy for outdoor scenes. In 2017 IEEE International Conference on Image Processing (ICIP) (pp. 710–714). IEEE.
[90] Sun, X., Xie, K., Lin, S., Yang, A., & Zhao, H. (2020, April). Research on Calculation Method Based on Gray Edge Color Constant. In 2020 3rd International Conference on Advanced Electronic Materials, Computers and Software Engineering (AEMCSE) (pp. 390–393). IEEE.
[91] Barnard, K., Martin, L., Coath, A., Funt, B. (2001). A comparison of computational color constancy algorithms. ii. experiments with image data. IEEE transactions on Image Processing, 11(9), 985–996.
[92] Alessandro, E., Pouli, T., Artusi, A., Akyüz, A. O., Banterle, F., Mazlumoglu, Z. M., Reinhard, E. (2002). A gamut-mapping framework for color-accurate reproduction of HDR images. IEEE computer graphics and applications, 36(4), 78–90.
[93] Zamir, S. W., Vazquez-Corral, J., Bertalmı, M. (2014). Gamut mapping in cinematography through perceptually-based contrast modification. IEEE Journal of Selected Topics in Signal Processing, 8(3), 490–503.
[94] Guo, C., Li, C., Guo, J., Loy, C. C., Hou, J., Kwong, S., Cong, R. (2020). Zero-reference deep curve estimation for low-light image enhancement. In Proceedings of the IEEE/CVF Conference on Computer Vision and Pattern Recognition (pp. 1780–1789).
[95] Parraga, C. A., Vazquez-Corral, J., Vanrell, M. (2009). A new cone activation-based natural images dataset. Perception, 36(Suppl), 180–180.
[96] Funt, B., Shi, L. (2010, February). The effect of exposure on MaxRGB color constancy. In Human Vision and Electronic Imaging XV (Vol. 7527, pp. 282–288). SPIE.
[97] Gehler, P. V., Rother, C., Blake, A., Minka, T., Sharp, T. (2008). Bayesian color constancy revisited. In Computer Vision and Pattern Recognition, 2008. CVPR 2008. IEEE Conference on. IEEE, pp. 1–8.
[98] Gijsenij, A., Lu, R., Gevers, T. (2012). Color constancy for multiple light sources. IEEE Transactions on image processing, 21(2), 697–707.
[99] Rizzi, A., Bonanomi, C., Gadia, D., Riopi, G. (2013, February). YACCD2: yet another color constancy database updated. In Color Imaging XVIII: Displaying, Processing, Hardcopy, and Applications (Vol. 8652, pp. 77–86). SPIE.
[100] Beigpour, S., Riess, C., Van De Weijer, J., Angelopoulou, E. (2014). Multi-illuminant estimation with conditional random fields. IEEE Transactions on Image Processing, 23(1), 83–96.
[101] Akbarinia, A., Parraga, C. A. (2017). Colour constancy beyond the classical receptive field. IEEE transactions on pattern analysis and machine intelligence, 40(9), 2081–2094.
[102] Gao, S. B., Ren, Y. Z., Zhang, M., Li, Y. J. (2019). Combining bottom-up and top-down visual mechanisms for color constancy under varying illumination. IEEE Transactions on Image Processing, 28(9), 4387–4400.
[103] Qian, X., Nguyen, H. N., Song, M. M., Hadiono, C., Ogden, S. C., Hammack, C., ... & Ming, G. L. (2016). Brain-region-specific organoids using mini-bioreactors for modeling ZIKV exposure. Cell, 165(5), 1238–1254.

Henry Ndu, Akbar Sheikh-Akbari, Iosif Mporas

# 14 Hyperspectral imaging and its applications for vein detection: a review

**Abstract:** With advances in technology, hyperspectral imaging has become an emerging area of research due to its numerous advantages over conventional imaging techniques. HyperSpectral (HS) cameras generate images of high spectral as well as spatial resolution. Hence, HS images carry much more information from the scene than the conventional red, green and blue (RGB) images. This inspired researchers to use HS technologies for many different applications ranging from crime investigations to crop monitoring.

It is important to accurately detect veins during surgical treatments, but this often turns out to be difficult. Wrongly locating veins or anatomical variations could result in accidental injury to blood vessels. This could lead to a longer operation time or even create serious complications. Furthermore, for majority of medical procedures, it is necessary to accurately define the location of veins. Over the past years, various methods including near infrared (NIR) and multi-spectral image processing-based methods have been proposed to help with detecting and accurately locating the veins. However, the performance of these methods is limited and demand for more accurate and convenient methods are increasing. HS images are two-dimensional (2D) representation of the scene at many light spectral. This brings the challenge of processing high dimensional data, which require significant processing power to deal with them. Various methods such as principal component analysis (PCA), Moving Window-PCA and Folded-PCA, which are widely used to reduce the dimensionality of HS image data, are reviewed in this book chapter. Conventional RGB, HS, NIR and multispectral images are studied and then HS imaging systems are introduced. Different applications of HS imaging are reviewed and their potential for vein detection is highlighted. Different techniques for reducing high dimensional data are discussed, and finally, different vein detection methods and some of the existing vein benchmark datasets are also introduced.

**Keywords:** hyperspectral image, vein detection, hyperspectral cameras

**Henry Ndu, Akbar Sheikh-Akbari,** School of Built Environment, Engineering and Computing, Leeds Beckett University
**Iosif Mporas,** School of Engineering and Computer Science, University of Hertfordshire

https://doi.org/10.1515/9783110756722-014

## 14.1 Introduction

Figure 14.1 shows a general colour digital image, which was captured by a mobile camera. The wall in the image is colourful and the mobile camera captures a very similar replica of the scene with all vivid colours, which can be seen by the human eyes. However, there is a limit to what the human eyes can perceive and there may be aspects of the same image that can neither be captured by our naked eye nor by a camera.

**Figure 14.1:** A sample photo image [1].

There are different types of digital images, for example binary, greyscale and colour images. Each type of image, carry different amount of information, as shown in Figures 14.2 and 14.3.

**Figure 14.2:** Sample: (a) binary image; (b) grayscale image [2].

When capturing an image using a conventional digital camera, it generates image with three colour components named RGB from the intensity of the reflected visible electromagnetic light spectrum (approximately 564–580 nm (red), 534–545 nm (green) and 420–440 nm (blue)) [1], as shown in Figure 14.3.

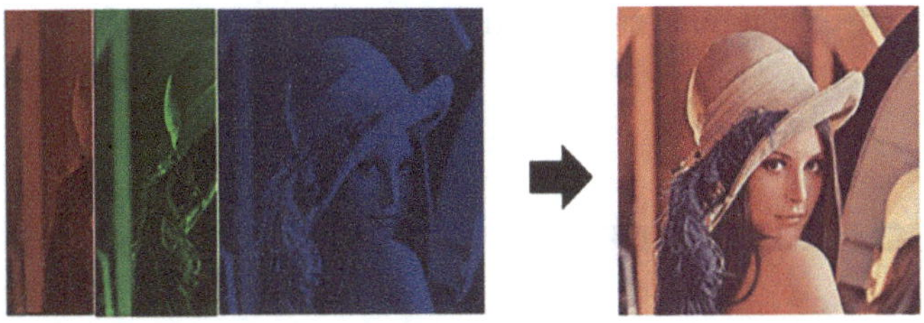

**Figure 14.3:** Sample red, green, blue (RGB) image (Original Lena image from [3]).

HyperSpectral (HS) image cameras generate high spatial and spectral resolution images. An HS image could have tens or even hundreds of bands across the electromagnetic spectrum, as shown in Figure 14.4. Therefore, an HS image carries significantly more information than its counterpart RGB image from the scene/objects. Hence, HS images can be used to extract features of the objects that are not possible to extract from their counterpart RGB images.

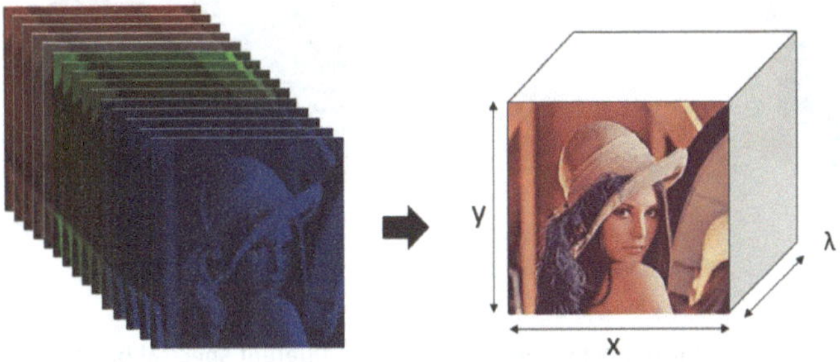


**Figure 14.4:** Hyperspectral image (Original Lena image from [3]).

Recently, there has been great research interest in the study of vein detection with considerations in relation to the electromagnetic spectrum. Researchers explore the NIR region (NIR) of the electromagnetic spectrum to acquire vein related data, which turns out to be efficient, and cost-effective [4]. This could help to reduce the challenges medical practitioners are facing in detecting vein, when giving treatments to patients. Nurses sometimes find it difficult to locate veins during venipuncture, intravenous drug delivery or cannulation treatments. Some of the reasons for this are the patients age (elderly or pediatrics), skin color or the patient's obesity level [4], which often result in multiple attempts in needle insertion. Where this can be very discomforting and painful to patients. This could even lead to bigger complications like bleeding, delayed treatments and infections. It is therefore of great importance having good mediums set in place for easy detection of human vein regions. With rich information that HS imaging technology could capture from the human body, application of the HS imaging in vein detection could significantly improve the accuracy of detecting and locating veins.

## 14.2 Hyperspectral imaging

HS imaging is a spectral sensing technique that can take hundreds of contiguous narrow images at several wavelengths within the visible and infrared regions of the electromagnetic spectrum [5]. The image pixels generate spectral vectors that portray the spectral components of the materials in the scene. HS imaging has consequently been applied for identification of materials and discrimination related purposes [5]. Information collected through HS imaging processes are in the form of data cubes, which are often referred to as *hypercubes*, as shown in Figure 14.5c [6].

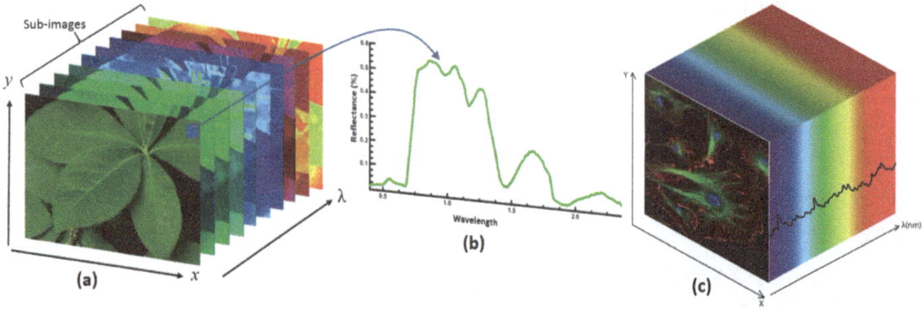


**Figure 14.5:** Hyperspectral image: (a) stacked hyperspectral images at different spectral bands, (b) image wavelength bands representation and (c) hypercube image [6, 7].

### 14.2.1 Hyperspectral image at different wavelengths

HS images are generally classified into ultraviolet (UV), visible (VIS), NIR, short wave infrared (SWIR), mid wave infrared (MWIR) and long wave infrared (LWIR) regions of the electromagnetic spectrum, as shown in Figure 14.6.

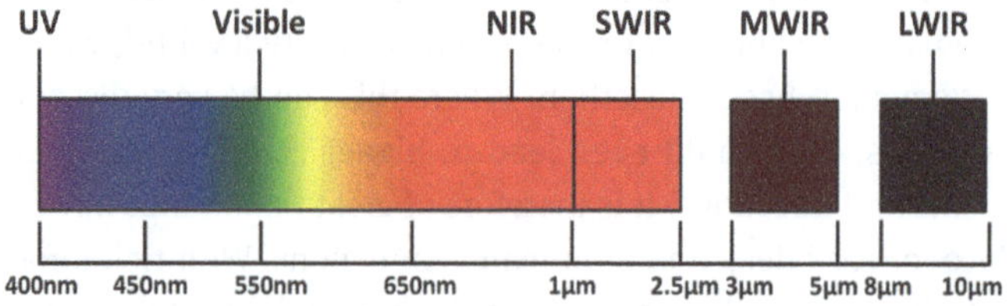


**Figure 14.6:** Hyperspectral image classification versus their electromagnetic spectrum [8].

#### 14.2.1.1 Near infrared

The NIR region of the electromagnetic spectrum commences from the wavelength just above the visible electromagnetic wavelength, which can be seen by human eyes. Available NIR systems usually operate between 750 and 1400 nm of this region. Electromagnetic radiation within this range can penetrate the human anatomy deeper than visible light (400–750 nm). NIR radiation can penetrate the human biological tissue up to a depth of 3 mm, in which veins are located [9]. Hence, most vein detection investigations have been conducted within this region. Hemoglobin and other fluids located in the human body acquire absorptive characteristics, the absorption difference between vascular structures and surrounding tissues can be detected using electro-optical technology. Scientists take advantage of this technology to solve several intravenous problems including vein detection. Recent studies show a 97% increase in detection of difficult intravenous sites with implementation of NIR technology [10]. Figure 14.7 shows mean number of visible sites for different age groups using NIR and conventional vein detection methods.

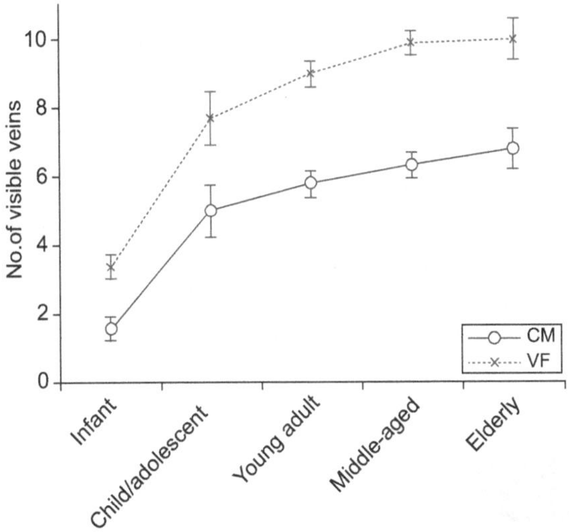


**Figure 14.7:** Infrared vein detection (marked as VF in the graph) vs. conventional vein detection (CM) visible sites across age groups [10].

Vision systems capable of providing spectral images within the NIR and visible spectrum ranges, together with three-dimensional information, yield reliable results for vein detection [10]. The resonon HS camera that will be used in the proposed research can capture and generate several HS images within the NIR and visible ranges, hence ensuring reliable results.

## 14.2.2 Multispectral and RGB versus hyperspectral imagery

RGB imaging system converts the visible electromagnetic wavelength into three colour bands, namely RGB, as shown in Figure 14.8, while multispectral images consist of a limited number of discrete spectral wavelengths images (ideally not more than 20). HS images may capture hundreds of wavelength bands in a continuous spectral range. Hence, multispectral images include images of several distinct spectral data, whereas HS images contain images of continuous spectrum data over a specified electromagnetic wavelength range, as shown in Figure 14.5b.

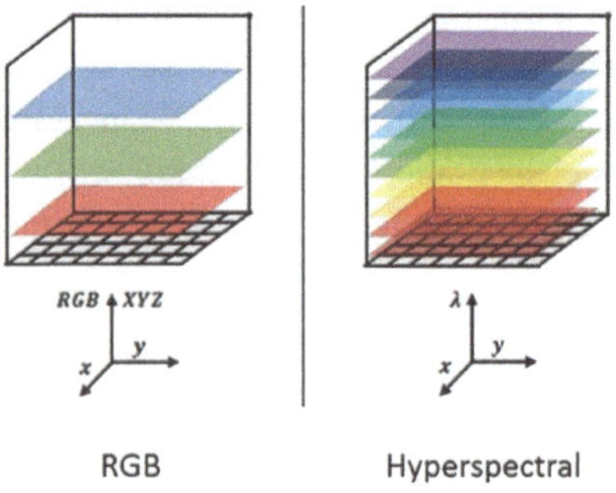


**Figure 14.8:** RGB image versus hyperspectral image [1].

For an image to be considered HS, it will generally have to be made up of more than 20 bands. If it consists of any fewer than that, it is considered as a multispectral image. There are not any strict definitions attached to these terms; however, image containing fewer than 20 bands is sometimes referred to as HS too. When an image consists of more than several thousands of bands it is often referred to as “ultraspectral” image [11]. Figure 14.9 shows the bands for multispectral and HS images.

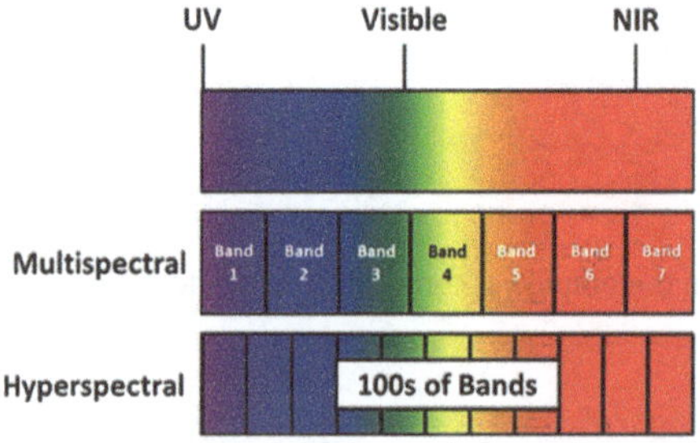


**Figure 14.9:** Multispectral versus hyperspectral image bands for visible electromagnetic wavelengths [8].

### 14.2.3 Hyperspectral cameras and image acquisition systems

Some of the techniques for HS image acquisition are area scan, whiskbroom, pushbroom and snapshot. Figure 14.10 shows an illustration of the operational modes for each of the HS image acquisition techniques.

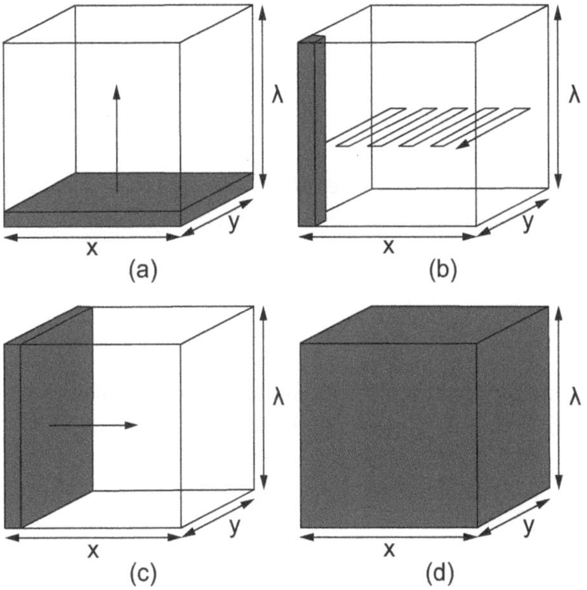


**Figure 14.10:** Hyperspectral image acquisition techniques: (a) area scan, (b) whiskbroom, (c) pushbroom and (d) snapshot [11].

#### 14.2.3.1 The pushbroom technique

The pushbroom technique is mostly used in HS imaging acquisition. Rather than capturing a single pixel at all wavelength band like some of the other techniques available, the pushbroom technique simultaneously captures an entire line of pixels at all wavelengths. The object or detector is then moved along the other spatial dimension [11]. This can be seen in Figures 14.10(c) and 14.11. Data are produced in the band interleaved by line (BIL) format. The hypercube is built with each spatial line being a 2D slice.

Hyper spectral Imaging (HSI) Technology fundamentally begins with the hardware for acquiring the images and software for processing the images [13]. The actual configuration is determined by the approach used for the HSI system. The essential components of an HS imaging system include a detector, a light source, a wavelength dispersion device and a computer with software capable of performing

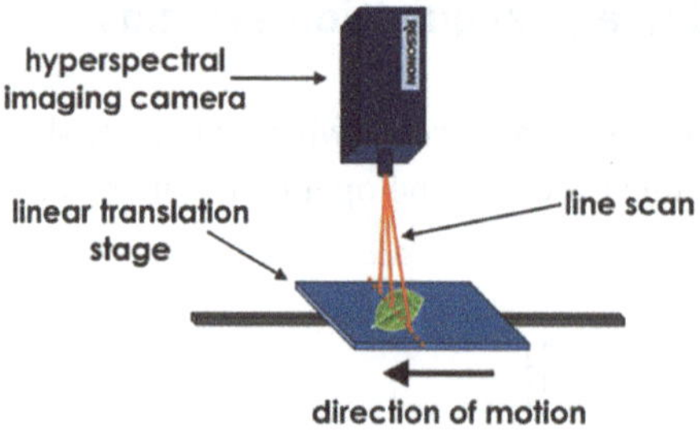


**Figure 14.11:** The pushbroom technique [12].

images acquisition task [13]. It is important to pay attention to the selection of each of these components as they determine the quality of the captured HS images.

#### 14.2.3.2 Hyperspectral cameras

An HS camera is made of an imaging spectrograph and a monochrome matrix array sensor. Figure 14.12 shows the basic building block of a pushbroom HS camera.

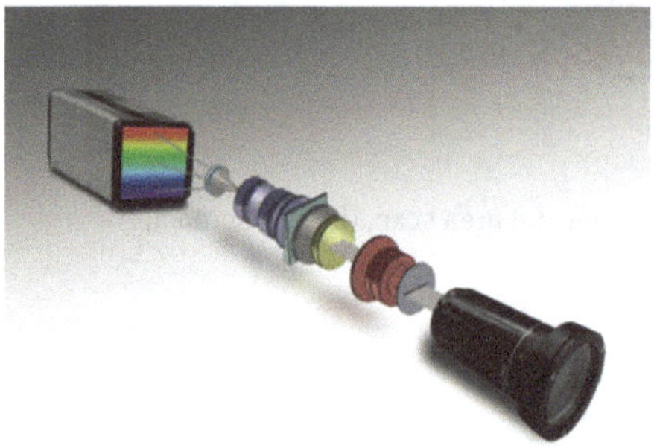

**Figure 14.12:** A schematic of pushbroom hyperspectral camera [14].

With the aid of a special lens, the sample is imaged on a slit of a transmission spectrograph. A spectrum image is generated on a focal plane array detector by the spectrograph, and this is done while preserving the location of specified points on the slit; thus, the points of the line on the sample stay the same [14]. A complete HS dataset is generated with successive line scans on the sample measured over the time. The hypercube is formed through this procedure and it contains information of the scene on spatial and spectral dimensions. The hypercube is usually processed using similar hypercube measurements of a white reference material that is highly reflective and the residual background signal; these measurements are later done again with no light falling on the focal plane array. The result of this is a corrected spectra produced in transmittance, reflectance or absorbance that is like the traditional spectroscopic measurements [14]. These results can be further processed, scaled, smoothed and finally compressed to obtain information that can be used for measurements, such as composition, colour coordinates or thickness. HSI has advantage over conventional single-point spectroscopy with its capability of generating very large amount of data. It is necessary to keep up with the data stream produced by an HS camera to be able to provide compressed and processed data like pinpointing contaminants, identifying defects in a process and producing composition maps and other relevant technological information. To do this, special processing software and hardware are required. There are currently a range of computers and software options available for handling these data sets [14].

### 14.2.4 Applications of hyperspectral imaging

HS imaging is being implemented in several fields. It is mostly used in areas requiring enhanced spatial and spectral resolution and provision of accurate colour-coding for cases of materials having similar colours [15]. Some of the areas that HS imaging has been used are:
- Agriculture: for identification of defects, introduction of intelligent classification of food and determination of product quality [16].
- Remote sensing: for environmental monitoring and mapping land resources [17].
- Military and defence: for criminals and terrorist investigations [18].
- Medical sciences: for disease diagnosis and image-guided surgeries [19].
- Mining exploration and mineral processing: to conduct in-line inspection of raw materials and minerals [20].
- Astronomy: to enable distant inspection of planets or for conducting enhanced analysis of spatially resolved spectral images of stars [21].

## 14.3 Dimensionality reduction

### 14.3.1 Introduction to high dimensional data

Advances in technology have made acquisition of higher spectral resolution HS images possible. However, as spectral resolution increases, the dimensionality of HS data increases too. This significantly adds to the challenge of classifying large-size high-dimensional data. Classification algorithms, which are used to process this data, suffer from the Hughes effect [22]. This is mainly due to the limited amount of available training samples, which are used to estimate the classification algorithms' parameters.

Dimensionality reduction of HS data, whilst retaining their vital information, is an important pre-processing step to improve the performance of the classifiers.

There are four factors that generally determine classification performance, they include: the class separability, the training sample size, the dimensionality and the classifier type [23]. Part of the pre-processing stage of the HS image data discussed in this paper involves dimension reduction. The main idea of dimensionality reduction is to reduce the complex nature of an input data by prioritizing and preserving the essential information of the data. There are two approaches used to perform dimensionality reduction, namely feature selection and features extraction. The first involves selecting a small subset of features from the original feature space directly according to their level of class separability or with respect to classification criteria. This process is called feature selection or band selection. The second method, feature extraction, involves using a transform to map the data to a lower dimensional data space. This enables the substantial removal of redundant information within the input data without significantly losing important information. Figure 14.13 shows a three-band false-color image of an agricultural field taken with an HS camera, its ground truth classification image and its color code.

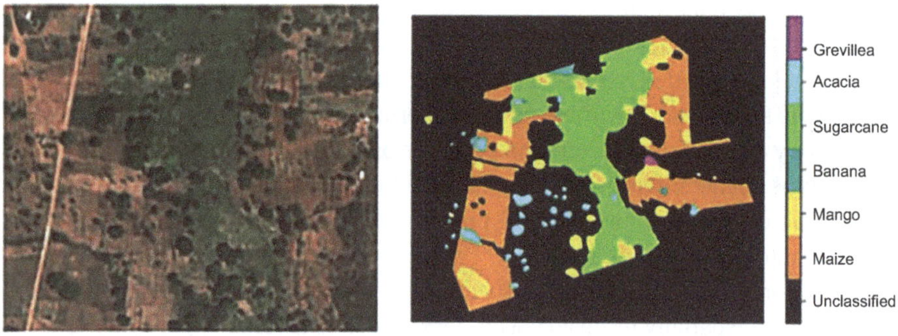


(a) (b) (c)

**Figure 14.13:** A hyperspectral data: (a) three band false-color image of an agricultural field in the Taita Hills; (b) its related ground truth (reference data); (c) color code [24].

The next step in HS data processing is classification, which entails partitioning of the whole spectral domain into $K$ exhaustive, separate regions, so that each point in the domain has a unique association to a $K$ class, as shown in Figure 14.13, and pixels are allocated to their classes according to their feature set. A one-dimension image is formed as output of this step. The importance of considering the classification steps in HS data processing will be discussed in Section 14.5.

#### 14.3.1.1 Overview of dimensionality reduction

The goal of dimensionality reduction is to reduce the size of the input data substantially without having to sacrifice significant information. It aims at projecting a high dimensional space to a low dimensional sub-space generated with most effective features [23]. The presence of increased spectral channels in HS images leads to greater chances of encountering more theoretical and practical problems when processing images of high dimensional data using conventional techniques. Dimensionality reduction could solve problems related to the Hughes phenomenon [22] to some extent. Hughes phenomenon indicates that the classification accuracy decreases as the number of spectral bands increases when using the same amount of training samples. The steps for HS image classification are shown in Figure 14.14.

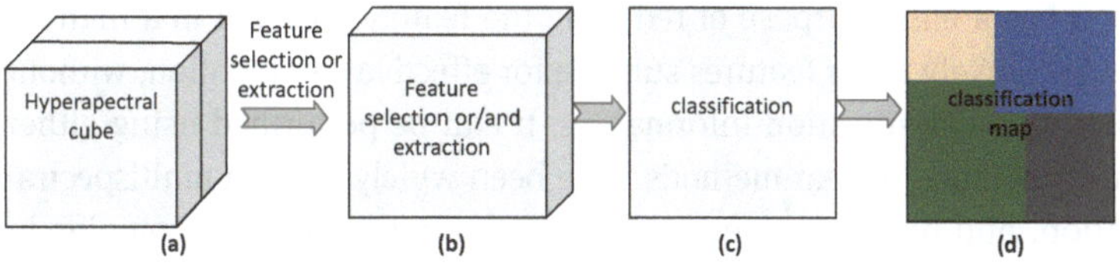


**Figure 14.14:** Hyperspectral image classification steps.

In feature selection-based methods, a sub-set of the input data is used to select some features based on pre-defined criterium. The extracted features are then used for classifying the input data. In feature extraction, the data within the original feature space is projected into a lower-dimensional subspace using a linear or nonlinear operator, $T$. Both approaches are briefly reviewed in the following sub-sections. Various dimensionality reduction methods were reviewed in [25].

### 14.3.2 Feature selection

In feature selection, the most effective subset of features, which carry the discrimination characteristics of different classes, is selected [26]. Thus, features that have no significant contribution to the classification are discarded from the original data. Various criteria are used in the selection of features that increase discrimination across classes. Some of the commonly used measurements of separability during feature selection process include: Divergence, Transformed Divergence, Mahalanobis-Bhattacharyya and Jeffries-Matusita [23]. In feature selection approach for dimension reduction, possible errors during classification are estimated for different feature combinations by performing separability analysis on the training data.

Feature selection with the use of statistical distance measures has been vastly studied and explored in different applications successfully. However, when data dimensionality increases, the amount of feature combinations to be inspected exponentially increases, which results in increased required computation power [23]. Assuming $n$ is the number of spectral bands in an input data, feature selection aims at deducing an optimal subset of $m$ with $m<n$. The number of feature combinations to be considered for selection equals $n!/((n-m)!m!)$. For a typical HS data, this number is usually very large and often have low computational efficiency. Several algorithms have been proposed to introduce feasibility during computation. These algorithms, which can identify an optimal or sub-optimal feature set include Sequential Forward, Branch-and-Bound, Sequential Backward and Max–Min Feature algorithms [23].

### 14.3.3 Feature extraction

Feature extraction has a main purpose of reducing the features present in a high dimensional data to relatively fewer features suitable for effective computation, without significant loss of vital classification information. It can be performed using either linear or nonlinear methods. Linear methods have been widely used in multispectral image classification, and nonlinear methods often turn out to be more complex in practical applications. However, in recent years, researchers have proposed different successful techniques for investigating complex nonlinear data [27]. The most common method used for feature extraction is the PCA [23]. In PCA, an orthogonal subspace projection is performed on the input HS image data, and a new sequence of uncorrelated images represented by their eigenvectors is generated. The first few eigenvectors with higher energy contain the most effective features, and the eigenvectors with smaller energy are ignored. This results in dimensionality reduction and improved classification speed. Although this technique effectively produces accurate classification results, it is prone to noise and needs to be implemented with the whole data set [23]. The basic idea of PCA is discussed further in Section 14.4.1.

Feature extraction can also be performed using the Discriminant Analysis Feature Extraction approach, which operates in contrast to the PCA by generating a transformed set of feature axes with optimum class separation. This approach uses a criterion function, which considers the ratio of “a between-class covariance matrix” and “within-class covariance matrix” [23]. Also, a transformation matrix applied to maximize this ratio, hence maximizing the class separability. Although the implementation of discriminant analysis works well for most cases, there are certain challenges associated with its use. First, in this method, features are delivered only up to the number of classes minus one. Second, the extracted feature vectors tend to be unreliable when the mean values are very close. Also, if the mean vector for a class is very different from the other classes, the between-class matrix generates biased information about this class reducing the effectiveness of its features [28]. According to [29], decision boundaries can be used to separate the useful features from the redundant features. The method is referred to as Decision Boundary Feature Extraction. They demonstrated that all reliable features are useful for accurate classification, where and effective decision boundary is defined. Basic predictions regarding intrinsic discriminant dimension and extraction of discriminative information from decision boundary was performed using a defined decision boundary matrix. The effective decision boundary is determined by first selecting majority of the training samples. For high dimensional data, the number of training samples could be large. When dealing with HS images, the available training samples are usually not sufficient to avoid the chance of singularity or obtain a satisfactory covariance estimation [23].

In line with signal processing, results of a Fourier transform (FT) are significantly useful because the frequency content of signal is vital. A spectrum of the frequency power of FT can efficiently localize information regarding global patterns of the spectral curve. The low-frequency content that gives the identity of the signal is equally localized [23]. Hsu and Tseng [30] proposed a method for obtaining spectral features called the Fourier feature extraction (FFE). In FFE, the spectral data for each pixel undergoes a discrete Fourier transform (DFT) and series of Fourier power spectrum are generated. Like the PCA, the first few components produced by the DFT may contain the most important information. Hence, during classification they can be considered as the most important feature [23].

# 14.4 Background of PCA

## 14.4.1 Principal component analysis

The PCA, also referred to as Hotelling transform [31] and principal components transformation [32, 33], is an optimal transform that is used to represent data according to its variance [34]. It can be described as a version of the Karhunen–Loeve transform (KLT), which is widely used in signal processing and communications [35]. PCA is a transform that uses eigenfunctions as basis function for the representation and decorrelation of data with respect to mean-squared error. In general, it is referred to as the Karhunen–Loeve expansion for representing a function as a series of eigenfunctions, and these eigenfunctions are continuous-time functions. Eigenfunctions, when sampled at discrete time instants, become eigenvectors in a discrete case, where KLT is reduced to PCA. The basic idea of the PCA can be described as follows:

Assuming $S = \{\mathbf{r}_i\}_{i=1}^{N}$ is a set of $L$-dimensional image pixel vector and $\boldsymbol{\mu}$ represents the mean of the sample pool, $S$, derived by $\boldsymbol{\mu} = (1/N)\sum_{i=1}^{N}\mathbf{r}_i$. Let $\mathbf{X}$ represents the sample data matrix that is formed by $\mathbf{X} = [\mathbf{r}_1, \mathbf{r}_2, \cdots, \mathbf{r}_N]$, and the sample covariance matrix of $S$ is derived by $\mathbf{K} = (1/N)[\mathbf{X}\mathbf{X}^T] = (1/N)\left[\sum_{i=1}^{N}(\mathbf{r}_i - \boldsymbol{\mu})(\mathbf{r}_i - \boldsymbol{\mu})^T\right]$. Assuming $\{\lambda_l\}_{l=1}^{L}$ indicates the set of eigenvalues derived from the covariance matrix $\mathbf{K}$ and $\{\mathbf{v}_l\}_{l=1}^{L}$ represents their respective unit eigenvectors, that is, $||\mathbf{v}_l|| = 1$, the diagonal matrix $\mathbf{D}_\sigma$ can be defined, with variances $\{\sigma_l^2\}_{l=1}^{L}$, as

$$\mathbf{D}_\sigma = \begin{bmatrix} \sigma_1^2 & 0 & \mathbf{0} \\ 0 & \ddots & 0 \\ \mathbf{0} & 0 & \sigma_L^2 \end{bmatrix} \tag{14.1}$$

and a matrix for eigenvector $\boldsymbol{\Lambda}$ specified by $\{\mathbf{v}_l\}_{l=1}^{L}$ as

$$\boldsymbol{\Lambda} = [\mathbf{v}_1, \mathbf{v}_2, \cdots, \mathbf{v}_L] \tag{14.2}$$

Therefore,

$$\mathbf{D}_\sigma = \boldsymbol{\Lambda}^T \mathbf{K} \boldsymbol{\Lambda} \tag{14.3}$$

A linear transform $\xi_\Lambda$ is defined by

$$\xi_\Lambda(\mathbf{r}) = \boldsymbol{\Lambda}^T \mathbf{r} \tag{14.4}$$

uses the eigenvector matrix $\boldsymbol{\Lambda}$ to transform all data sample $\mathbf{r}_i$ to a new data sample $\widetilde{\mathbf{r}}_i$ by

$$\widetilde{\mathbf{r}}_i \equiv \boldsymbol{\Lambda}^T \mathbf{r}_i \tag{14.5}$$

Then the mean of the new $\xi_\Lambda$-transferred data samples $\{\widetilde{\mathbf{r}}_i\}_{i=1}^{N}$ becomes $\widetilde{\boldsymbol{\mu}} = (1/N) \sum_{i=1}^{N} \widetilde{\mathbf{r}}_i$. As a result, its covariance matrix is reduced to a diagonal matrix as follows:

$$\widetilde{\mathbf{K}} = (1/N) \sum_{i=1}^{N} (\widetilde{\mathbf{r}}_i - \widetilde{\boldsymbol{\mu}})(\widetilde{\mathbf{r}}_i - \widetilde{\boldsymbol{\mu}})^T = \boldsymbol{\Lambda}^T \mathbf{K} \Lambda = \mathbf{D}_\sigma \tag{14.6}$$

Researchers have applied PCA technique in many high dimensional investigations to execute tasks effectively and increase classification performance. This works well for HS data with PCA enabling more effective dimensionality reductions.

PCA is an effective data reduction technique commonly used to extract data, discard excessive information, discover hidden features and interpret existing characteristics between features. It has been extensively applied in the study of multivariate statistical process monitoring due to its efficiency in extracting information in multi-variate environments.

### 14.4.2 Dynamic principal component analysis

In the conventional principal component analysis, the value at a certain time observed is assumed to be independent of the observation value at the past time [36]. However, monitored processes usually have a dynamic response and observations on the variable are of time-series correlative format [37]. Essentially, the current values of the monitored process variable depend on the past values. Ku et al. [38] proposed an extended version of the PCA, which accounts for the dynamic process behavior and improves monitoring performance of the anchor PCA, called the dynamic PCA (DPCA). The DPCA operates quite similarly to the PCA, the main contrast is in the presence of a "time lag shift" introduced in the DPCA allowing for the transformation of the original **X** matrix into a new extended matrix, $\widetilde{\mathbf{X}}$, with $l$ time-shifted duplicate vectors of every variable in **X**. PCA is then applied to $\widetilde{\mathbf{X}}$ matrix.

Ku et al. [38] study indicate that with a correct selection of $l$ and retained PCs, linear constraints should be revealed in the noise subspace (or approximate space for void data) with small eigenvalues. Therefore, Ku et al.'s [38] research insinuates that the $Q$ statistic (sum of squared errors, SPE, from the DPCA model) will be generated independently and will provide a means to monitor the dynamic system [37]. Ku et al. [38] indicated that the occurrence of $l$ is 1 or 2 normally but can be higher when dealing with nonlinear systems.

The DPCA essentially implies that the resulting PCs will generate an implicit vector autoregressive model of the data.

### 14.4.3 Moving window principal component analysis

The moving window PCA (MWPCA) is a technique developed to deal with dynamic problems or time varying systems. Some processes exhibit a non-stationary behavior, and this is due to the presence of time changing parameters such as the mean or covariance [39]. The MWPCA operates based on a defined data window of fixed length, which moves and changes in accordance with real time to update the PCA model on the detection of a new normal sample. Hence, for window-size $z$, the data matrix at time $t$ and $t+1$ can be defined as:

$X_t = [x(t-z+1), x(t-z+2), \ldots, x(t)]$ and $X_{t+1} = [x(t-z+2), x(t-z+3), \ldots, x(t+1)]$ respectively. The MWPCA algorithm updates for different parameters including parameters for the PCA model, the normalization parameters and monitoring the selected threshold for statistics control (Harkat et al., no date). The basic idea of the MWPCA technique for time variation monitoring involves first selecting training data sets and normalizing the training data using its mean and standard deviation. The covariance matrix is then computed, and eigen-decomposition is performed on the normalized data, generating its PCA model for the process. The number of components is then defined and the control limits for the monitoring statistic are specified. Second, a new testing sample is obtained and normalized using the scaling parameters of the moving window data. Then the PCA model is determined as it was done in the previous step, and the monitoring statistic for the normalized testing sample is evaluated and compared with the previously specified control limit. If it does not exceed the control limit, then the measurement is considered normal. Otherwise, the moving window is updated by shifting by one time sample (a new measurement is added and the oldest deleted). The procedure is considered faulty if up to three consecutive measurements turns out abnormal in which case measuring of data should be stopped and examined.

### 14.4.4 Folded principal component analysis

It can be challenging using PCA to perform covariance matrix calculations when dealing with data of very large spatial size. Also, PCA can sometimes be inaccurate, when obtaining details of the individual contribution of F bands, and during covariance matrix calculation, it considers all bands of the HS image equally [40]. Thus, folded principal component analysis (FPCA) offers a technique to reduce the size of the covariance matrix and effectively extract all the local structure within the hypercube spectral domain. The approach for folded-PCA is shown in Figure 14.15.

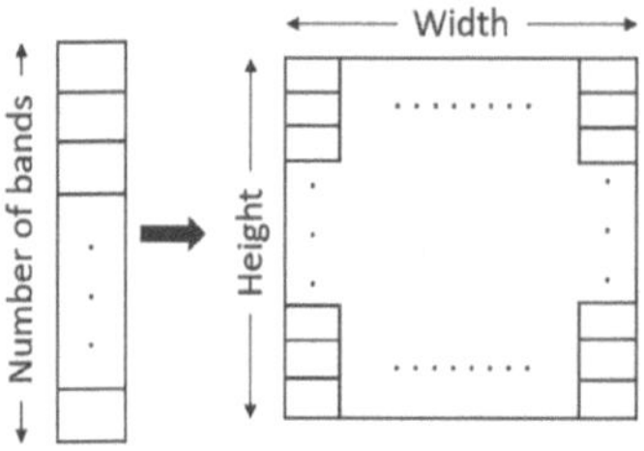


**Figure 14.15:** Folded-PCA technique [41].

The FPCA technique involves the transformation of each mean-adjusted spectral vector $M_n$ to a $H \times W$ matrix $D_n$ that is in the form of the following expression:

$$D_n = [d_{n1}, d_{n2} \ldots, d_{nh}]^{\mathrm{T}} \tag{14.7}$$

Finally, by accumulating these partial matrices, an overall covariance matrix is generated for the whole dataset. This is done using the following expression:

$$D_F = 1/S \sum_{n=1}^{S} C_n \tag{14.8}$$

The projection matrix generated is then computed after eigen decomposition on $D_F$ has been performed.

In traditional PCA, all the bands in the HS data are treated equally when performing calculations for the covariance matrix. Calculating in this way may fail to completely exploit all the individual contributions of the featured $F$ spectral bands (Zabalza et al., 2014). This method is sometimes unable to distinctly collect each spectral signature contribution which may lead to the overlook of relevant local characteristics of the HS image. Another setback of using the PCA to generate the covariance matrix is that when dealing with a very large number of pixels, $S$, usually over 100 k, when using software like MATLAB this increases the tendency of experiencing a software crash due to heavy computational load (Zabalza et al., 2014). Hence, the folded-PCA is introduced to address each of these limitations of using the PCA. The FPCA provides a unique way of effectively extracting information on the correlation between bands and their groups, generating a reduced covariance matrix size. However, early inspiration for the FPCA was drawn from investigations related to a 2D-PCA [42], which was used for analyzing facial recognition images.

In the study performed by Zabalza et al. in (Zabalza et al., 2014a), every mean-adjusted spectral vector $\mathbf{I}_n$ is transformed into a 2D matrix with each row of the matrix containing a group of $W$ bands, where the total bands $F$ are folded into $H$ groups or segments [43]. Assume that $\mathbf{A}_n$ is a transformed data matrix with a size $H \times W (F = H \times W)$ for each spectral signature, then $\mathbf{A}_n = [\boldsymbol{a}_{n1}\boldsymbol{a}_{n2} \ldots \boldsymbol{a}_{nH}]^{\mathrm{T}}$, where $\boldsymbol{a}_{nh} = [I_{n(1+W(h-1))} I_{n(2+W(h-1))} \cdots I_{n(W+W(h-1))}]$ and $h \in [1, H]$ [43]. The selected values for $H$ and $W$ used may determine the sort results produced. The covariance matrix, $\mathbf{C}_n$, for each $\mathbf{A}_n$ is processed computationally as $\mathbf{C}_n = \mathbf{A}_n{}^{\mathrm{T}}\mathbf{A}_n$, where the size of $\mathbf{A}_n$ is represented by $H \times W$ and for $\mathbf{C}_n$ is $W \times W$. The overall covariance matrix for the entire dataset, $\mathbf{C}_{\mathrm{FPCA}}$, is then generated using the following expression:

$$\mathbf{C}_{\mathrm{FPCA}} = \frac{1}{S}\sum_{n=1}^{S} \mathbf{C}_{\mathrm{n}} = \frac{1}{S}\sum_{n=1}^{S} \mathbf{A}_{\mathrm{n}}{}^{\mathrm{T}}\mathbf{A}_{\mathrm{n}} \tag{14.9}$$

## 14.4.5 Mutual information

The idea of mutual information (MI) is to provide a measure for calculating the similarity between two random variables referred to as the mutual dependence [40]. The amount of information comprised in a random variable is called the entropy, and this is considered when determining mutual information. By considering the Shannons information theory, the entropy for a discrete random variable A is expressed as follows:

$$\mathrm{H(A)} = \sum_{\mathrm{a}\in\mathrm{A}} p(a) \log p(a) \tag{14.10}$$

where $p(a)$ represents the probability density function of A. Hence, the mutual information $\mathrm{I(A, B)}$ of the random variables A and B is expressed as

$$\mathrm{I(A, B)} = \mathrm{H(A)} + \mathrm{H(B)} - \mathrm{H(A, B)} \tag{14.11}$$

where $\mathrm{H(A, B)}$ represents joint entropy of A and B. Finally, the mutual information for both random variables is defined by the following expression:

$$\mathrm{I(A, B)} = \sum_{\mathrm{b}\in\mathrm{B}} \sum_{\mathrm{a}\in\mathrm{A}} p(a, b) \log \frac{p(a, b)}{p(a)p(b)} \tag{14.12}$$

where $p(a)$ and $p(b)$ represent the function for marginal probability distribution of the random variables A and B, respectively. If both random variables have no similar information, the resulting MI value will be zero and if there are similar information in A and B, the MI value will be maximum [40].

From the equation, $I_{CS}(A, B)$ value is zero if A and B are independent of each other, and it is positive if otherwise.

### 14.4.6 Support vector machine

Support vector machine (SVM) is one of the most popular methods for effective classification of images. This technique is applied when a set of training images are to be grouped into two classes [41]. The optimal separating hyperplanes are built using kernel functions (K), where each side of the hyperplane identifies one of the output classes. Some of the applications that SVM have been successfully applied are three-dimensional object recognition, biomedical investigations and image compression [41]. Deepa and Thilagavathi [41] used SVM classifier together with the PCA and FPCA to perform HS image classification. Figure 14.16 shows a block diagram of their proposed algorithm.

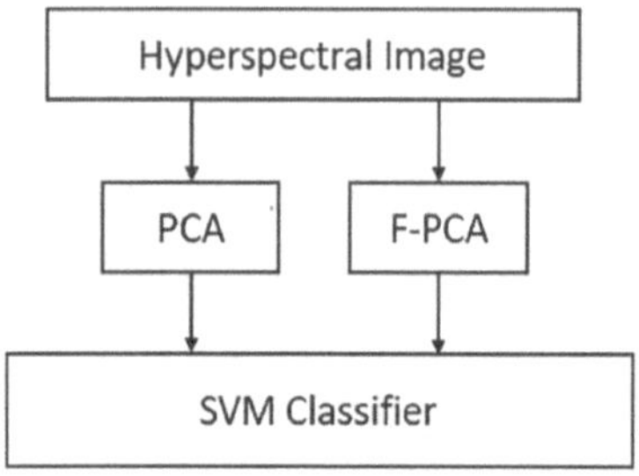


**Figure 14.16:** Experimental setup for Deepa and Thilagavathi dataset [41].

It can be seen from Figure 14.1 that they used PCA and FPCA feature extraction algorithm to reduce the dimensionality of the input data and then they applied SVM classifier on the resulting data to perform classification.

## 14.5 Hyperspectral image classification

HS image classification is the process of assigning each hyper spectral image pixel to a class. Considering the redundant nature of features is present in the HS images, pre-processing step is needed to extract/select effective features from the input image, so that the separability between the classes is maximized. An HS image does not only have dimensionality issues but also has data redundancy problem mainly due to its high inter-band correlation. Mapping the high dimensional data into a lower dimensional subspace, whilst ensuring the features of the HS data is not altered, is necessary. Dimensionality reduction can be considered as a linear or a non-linear transformation of the data. PCA is one of the most used techniques for linear unsupervised dimensionality reduction. The dimensionality reduction procedure is achieved by selecting the first few PCs. A segmented PCA-based algorithm was proposed in [44]. The proposed technique produces satisfactory results at reduced computational cost. Similar alternatives include the folded PCA [45] and the structured covariance PCA, both were proposed to deliver fast computation. Also, there has been considerable interest in the independent component analysis (ICA), which has widely been used for HS image dimensionality reduction. The aim of these methods is to identify a linear transformation that separates the input image into a set of statistically independent images. The minimum noise fraction (MNF) [46] and non-negative matrix factorization (NMF) [47] are similar approaches. Two cascaded PCA transformations are used by the MNF to project high dimensional input data into a new space grouped in two parts: noise and the signal data. Dimensionality reduction can be achieved with MNF through segregation of the noise-dominated part in the data. Unlike the PCA, ICA can extract only the holistic features of the input image, and NMF is a learning-based algorithm. It applies a non-negative matrix, which is decomposed into two simpler non-negative matrices that are easier to analyze. A set of basis vectors is represented by one of the decomposed matric, which can be employed in projecting the original data into a subspace with lower dimension. The linear approaches introduced earlier can be extended to become non-linear through the implementation of kernel methods at higher computation costs, for example, PCA [48], kernel MNF [49], kernel NMF [50] and kernel ICA [51].

For supervised methods of dimensionality reduction, prior information about the labeled training samples is required to perform discriminative analysis, unlike the unsupervised methods. A standard technique for supervised dimensionality reduction for HS data investigations is linear discriminant analysis (LDA), where the Raleigh quotient is maximized to attain highest degree of class separability, that is, a ratio of the scatter matrix for the between-class and the within-class. There are tendencies of conventional LDA failing if the training samples for all classes in the HS image are not available. As they require the within-class scatter matrix to be non-singular, this problem can be solved by introducing a preliminary step-like PCA to generate a modified version of the original image in a lower dimensional space before the LDA is applied so that the within-class scatter matrix is in a non-singular form. However, by using this procedure, PCA achieves the dimensionality reduction, and LDA only helps to increase the separability of the classes in the PCA modified space. Another issue of LDA is the unreliable nature of its selected features, when the difference within the class-mean vectors is small. If there is a large distance between a mean vector and the others, its class obscures the others when the scatter matrix for in-between classes is generated. Hence, a technique called non-parametric weighted feature extraction was proposed in [52] to make sample weighting possible to obtain a weighted mean and defining of new non-parametric in-between and within the class scatter matrices so that more features can be obtained. The supervised dimensionality reduction methods, like the kernel-based methods, have problems of slow speed because of high computational cost.

In addition, other than performing dimensionality reduction procedures, signal decomposition, in both spectral and spatial domains, can also be a reliable means for feature extraction. A technique called singular spectrum analysis (SSA) was proposed for signal decomposition within spectral domain in (Zabalza et al., 2014a; Zabalza et al., 2015b). The proposed SSA method found to be effective when it was used to extract spectral feature from HS data. Moreover, the SSA method significantly outperformed an anchor method named empirical mode decomposition (EMD) [53, 54]. The EMD is a non-linear and non-stationary signal decomposition method, which is used for implementation of the Hilbert–Huang transform. The EMD method decomposes the input signals into a finite number of intrinsic mode functions [55]. Although reasonable efficacy has been realized through the implementation of the EMD in time series decomposition cases (e.g., speech recognition), results from [54] indicate the possibility of EMD potentially reducing the classification accuracy when used for extraction of spectral features [55]. To address this issue, an improved version of the method called the ensemble EMD (EEMD) was reported in [56, 57]. Although there has been an uprise in spectral feature extraction, there are limited contributions that do not consider the spatial characteristics present in the HS image. The extended 2D EMD [54, 58] and SSA (Zabalza et al., 2015a) have been used to analyze individual band of HS images, experimental results show extracted spatial information yields great improvements. Various machine learning techniques have been used for HS image classification, for example, artificial neural networks, maximum likelihood (also known as Bayesian estimation) and decision trees [55].

## 14.6 Vein detection

Veins are blood vessels that are responsible for returning blood from all parts of the body to the heart [59]. Theoretically it is possible to use any part of the body for visual extraction of vein pattern, but hand is mostly used because it is usually an uncovered body part with reduced pilosity, and skin veins are at shallow depth [60]. A basic prototype for detecting vein is usually composed of a high power NIR light emission diode (NIR-LED) as the light source and a compact infrared-sensitive camera, to capture the image [4]. Some other research prototypes contain incorporation of additional parts for image enhancement purposes, like the use of a digital light processing that permits visualization of vein image on skin surface. Figure 14.17 shows the basic components for vein detection.

The vein detection principles rely on two illumination types: reflected and trans-illumination light. The reflected light method operates like that of the commercial hand-vein scanner, where the light from the source is reflected at the site, showing

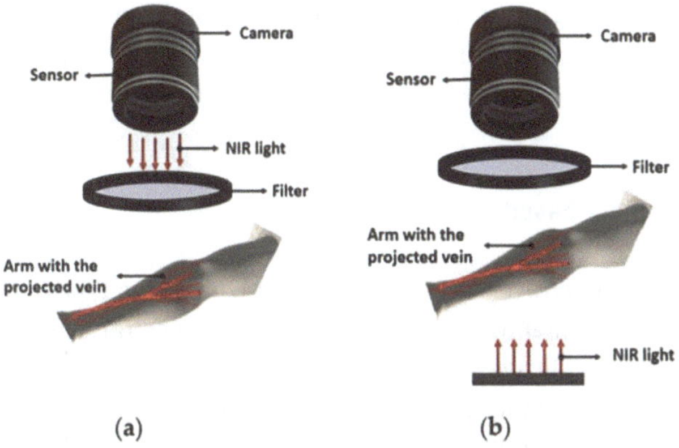


**Figure 14.17:** Basic components for vein detection: (a) reflected light type (b) transillumination type [4].

an example of the hand surface and a capture is then taken by the camera. For the transillumination method, the light penetrates the skin and tissue of the site and then the camera captures the image. The image of veins in transillumination type appears as dark lines on the skin surface that can be easily located by medical practitioners, and this sort of illumination is due to the presence of deoxyhemoglobin in venous blood that absorbs the red light [4]. This process occurs in two stages: the penetration of the IR light in the human tissue and the absorption of the incident light by the deoxyhemoglobin in venous blood more than the surrounding tissue. Vein detection using the reflected light method is shown in Figure 14.17a. Transillumination vein detection method requires higher light intensity using more power and is non-compact with the scanning process needing arm/hand to be placed in between the light source and the camera as shown in Figure 14.17b.

## 14.6.1 Vein detection methods

Growth in vein detection studies has available on a range of methods for vein assessment to reduce difficulties in intravenous procedures [10]. Some of the new techniques and devices commonly used for locating veins include: colour vision, ultrasound (US), NIR spectroscopy, visible light transilluminator, pressure sensor, multispectral camera and robotics systems [10]. The degree to which vein detection challenges are experienced today is reflected in the wide variety of devices designed for carrying out this procedure. Despite the availability of these techniques and devices, valuable time and money still get wasted on different surgical cases and people continue to experience trauma from multiple needles sticking [10]. Researchers take advantage of the technology of NIR spectroscopy to provide cheaper and more reliable means of detecting veins. In this section some of the state-of-art vein detection methods, both on basis of intrinsic hardware artefacts and software related features, are discussed.

#### 14.6.1.1 Intrinsic hardware artifact-based methods

To further extend knowledge in the intravenous field [61] proposed a method for vein detection based on multispectral imaging with the use of a multispectral camera (Spectrocam) for taking multispectral images at different wavelengths to acquire spatial resolution data. Wang et al. [61] technique is relied on a background removal and normalization/subtraction scheme algorithm that can be applied to multispectral vein images to overcome the limitations of existing methods and generate high-contrast subcutaneous vein detection. Despite the reliable nature of digital cameras, there are still tendencies of encountering internal imperfections due to manufacturing process limitations and the materials used in production. In [59] the author used median filter together with a Gaussian filter and Wiener filter to suppress the effect of the noise in the image. Ayoub et al. [59] proposed a vein detecting method based on a Nikon D810 camera (resolution 36.66 MP) with a Zomei 720 infrared filter for capturing images of the subjects hand after illuminating two infrared light sources.

#### 14.6.1.2 Vein detection algorithms

Some of the existing algorithms for vein detections are discussed in the following:

a) *Wang et al. [61] normalization/subtraction algorithm*
   In [61] the algorithm takes a five-band multispectral image of an arm from 475, 546, 570, 615 and 850 nm spectral. It improves their vein-to-skin contrast ratio, $M$, using equation (14.5):

$$M = \frac{|U_{\text{vein}} - U_{\text{skin}}|}{(U_{\text{vein}} + U_{\text{skin}})} \tag{14.13}$$

   where $U_{\text{vein}}$ represents average pixel intensity of the vein area and $U_{\text{skin}}$ represents average pixel intensity of intact skin around the veins. The authors generated a fused image with higher visual vein-to-background contrast, which works well on moderately hairy skins.

b) *Ayoub et al.* [59] *vein scanner algorithm*
   In Ayoub et al., [59] proposed algorithm and images of the subject's arm were captured from 940 and 850 nm.

## 14.6.2 Image datasets

To assess the performance of different vein detection algorithms, several publicly available image datasets have been generated and used by the researchers. In the following, some of these datasets are introduced.

### 14.6.2.1 Multispectral image dataset

[61] introduced a multi-spectral vein image dataset. This dataset contains five multispectral images of naked arms of human subjects with different skin tones and textures. The images were taken under normal indoor ambient light and camera exposure time of 10 ms. The five images of each subject were then registered to enable image fusion. Figure 14.18 shows an example of the multispectral image dataset.

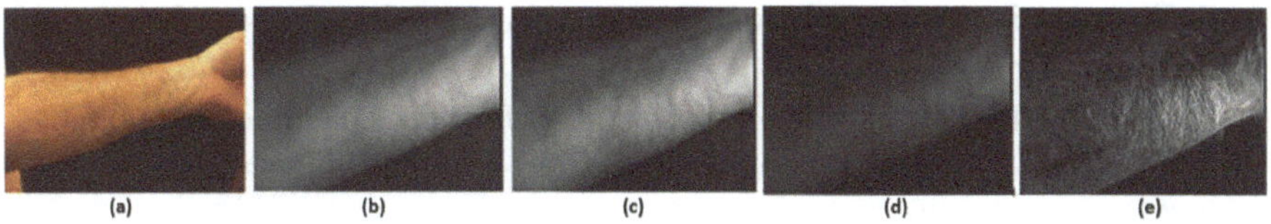


**Figure 14.18:** Example image of the multispectral image database: (a) an RGB image of the subject and (b–e) the four multi-spectral image band of the same subject [61].

### 14.6.2.2 Superficial vein scanner dataset

Superficial vein scanner dataset was introduced by Ayoub et al. in [59]. They used a Nikon D810 camera with a Zomei 720 infrared filter for capturing images of superficial veins inside the subject's hand, where the subject was lit by two infrared light sources (940 and 850 nm). The test was carried out on 10 different subjects. Figure 14.19 shows an example of the superficial vein scanner dataset.

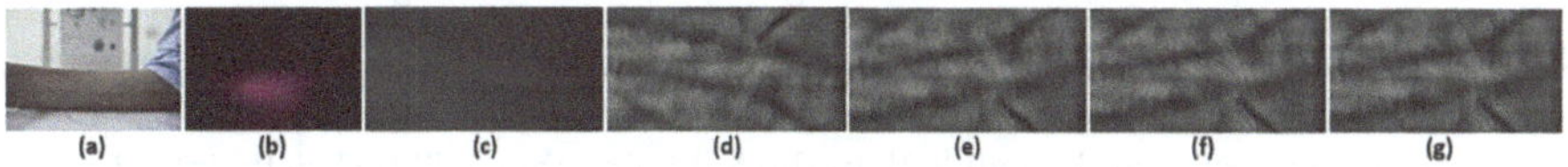


**Figure 14.19:** Example images from superficial vein scanner dataset [59].

## 14.7 Assessment methods

To evaluate the performance of different classification algorithms on determining vein regions, a range of statistical measures have been used by the researchers. This takes into consideration the classification metrics commonly used to evaluate models.

### Classification metrics

During classification predictions, the possible outcomes are of four types.

- **True positives** are when an observation is predicted to belong to a class, and it occurs to that class.
- **True negatives** are when an observation is predicted to not belong to a class, and it occurs not to belong to that class.
- **False positives** occur when an observation is predicted to belong to a class when it does not.
- **False negatives** occur when an observation is predicted to not belong to a class and it in fact does.
- After all predictions have been made and all four possible outcomes obtained, a plot of these outcomes is usually generated on a confusion matrix.

The following expressions are used to calculate the occurrence rate of each of the classification outcomes:

$$\text{True positive rate (TPR)} = \frac{\text{TP}}{\text{TP} + \text{FN}} \tag{14.14}$$

$$\text{True negative rate (TNR)} = \frac{\text{TN}}{\text{TN} + \text{FP}} \tag{14.15}$$

$$\text{False positive rate (FPR)} = \frac{\text{FP}}{\text{FP} + \text{TN}} \tag{14.16}$$

$$\text{False negative rate (FNR)} = \frac{\text{FN}}{\text{FN} + \text{TP}} \tag{14.17}$$

where TP is true positive, FN is false negative, TN is true negative and FP is false positive [62].

There are three main metrics used to evaluate a classification model, and these are accuracy, precision and recall [63]. In the following they will be briefly introduced.

#### Accuracy

Accuracy is an estimate of the percent of correct predictions for the test data [63]. It can be calculated simply by dividing the number of correct predictions by the total number of predictions as shown in the expression below:

$$\text{Accuracy} = \frac{\text{number of correct predictions}}{\text{total number of predictions}} = \frac{\text{TP} + \text{TN}}{\text{TP} + \text{TN} + \text{FP} + \text{FN}} \tag{14.18}$$

#### Precision

Precision is a fraction of the relevant examples of a class (true positive) by all examples predicted to belong to that class [63]. This is shown in the following expression:

$$\text{Precision} = \frac{\text{true positives}}{\text{true positives} + \text{false positives}} \tag{14.19}$$

#### Recall

Recall is a fraction of the examples predicted to belong to a class by all examples that truly belong in the class [63]. This is shown in the following expression:

$$\text{Recall} = \frac{\text{true positives}}{\text{true positives} + \text{false negatives}} \tag{14.20}$$

#### Receiver operating characteristics curve

There are various performance measurement criteria to assess the performance of the classification algorithms. Receiver operating characteristics (ROC) curve and its area under the curve (AUC) are the most common ones [64]. The AUC–ROC curve measures performance of classification problems at different stages. ROC is a probability curve and AUC show a measure of the separability, which demonstrates the capability of a classification method in distinguishing between the classes. The higher the AUC, the more reliable the model is [64]. The ROC curve is plotted with true positive rate on the *y*-axis and false positive rate on the *x*-axis as shown in Figure 14.20.

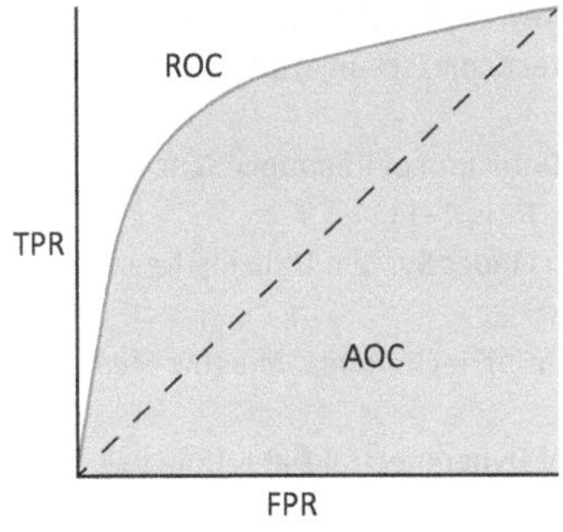


**Figure 14.20:** Receiver operating curve (ROC) [64].

## 14.8 Conclusions and future work

In this chapter, different types of images, which are used for vein detection including conventional RGB, HS, NIR and multispectral images were introduced. HS imaging systems and their applications were briefly explained. Existing image processing-based vein detection methods and vein image datasets were explored and the potential that HS imaging has for developing a robust and accurate vein detection were highlighted.

Some of the recent dimensionality reduction methods were explained and their application in HS image dimensionality reduction, feature extraction and image classification was reviewed. Measures for assessing the performance of the classification methods were detailed. It was found that the dimensionality reduction, effective feature extraction and HS images classification are ongoing challenges in this field and much research needed to be conducted to process HS images.

## References

[1] Panigrahi, S. (2019), What are hyperspectral images?, Medium, 15th November [online], available at: https://towardsdatascience.com/what-are-hyper-spectral-images-a5de5d9fa91 (Accessed on 5th June 2021).

[2] DISQUS, (2021), Black and white image using C, [online] Available at: https://abhijitnathwani.github.io/blog/2018/01/07/Black-and-White-Image-using-C (Accessed on 24th August 2021).

[3] Wired, (2021), Finding Lena, the Patron Saint of JPEGs, [online] Available at: https://www.wired.com/story/finding-lena-the-patron-saint-of-jpegs/ (Accessed on 23rd August 2021).

[4] Pan, C., Francisco, M., Yen, C., Wang, S., Shiue, Y. (2019). Vein Pattern Locating Technology for Cannulation A Review of the Low-Cost Vein Finder Prototypes Utilizing near Infrared (NIR) Light to Improve Peripheral Subcutaneous Vein Selection for Phlebotomy, Sensors, 16th August, 1–17.

[5] Peter, Y., Richardson, M. (2015). An introduction to hyperspectral imaging and its application for security, surveillance and target acquisition. ResearchGate, 19th April, Vol 58, 241–243.

[6] Hoasys (2019), Modular hyperspectral camera, [online] Available at: https://hoasys.fr/en/products/modular-hyper-spectral-camera (Accessed on 24th August 2021).

[7] Mishra, P., Shahrimie, M., Herrero-Langreo, A., Lohumi, S., Diezma, B., Scheunders, P. (2017), Close range hyperspectral imaging of plants: A review, ELSEVIER, [online], Volume 164, 49–67, Available at: https://www.sciencedirect.com/science/article/pii/S1537511017302635#sec1 (Accessed on 15th June 2021).

[8] Thomas, A. (2017). Atmospheric Correction and Anomaly Detection of Remotely Sensed Hyperspectral Imagery. University of Strathclyde, October 16th, 1–13.

[9] Tran, L., Pham, H. (2019). Designing and Building the Vein Finder System Utilizing Near-Infrared Technique. ResearchGate, 17th June, 383–387.

[10] Gali, C., Moshe, H., Zehava, O. (2016), Peripheral vein locating techniques, Imaging Med., ISSN 1755-5191, 84–87.

[11] Timothy, G. (2015). Techniques for Capture and Analysis of Hyperspectral Data. University of Strathclyde, December, 13–16.

[12] Resonon (2021), Benchtop system – reflectance, Resonon Inc. available at: https://resonon.com/hyperspectral-benchtop-reflectance-system (Accessed 23rd August 2021)
[13] Mohammed, K., Da-Wen, S. (2017). Introduction to hyperspectral imaging technology. ResearchGate, 18th November, Chapter, 5, 111–113.
[14] Middleton Spectral Vision, (2021), What is hyperspectral imaging?, [online] Available at: https://www.middletonspectral.com/resources/what-is-hyperspectral-imaging/ (Accessed on 23rd August 2021).
[15] Juan, L. (2020). Hyperspectral Imaging. Research Gate, University of Granada, July, 1–7.
[16] FDPP (2020). How can we improve food quality and reduce waste? [online] Available at: https://fdco.uk/2020/07/23/how-can-we-improve-food-quality-and-reduce-waste/ (Accessed on 21st August 2021).
[17] Harvieu, B. (2021). Proxies and remote-sensing: the pillars of digital agriculture. [online]. INRAE. Available at: https://www.inrae.fr/en/news/remote-sensing-dossier (Accessed on 24th August 2021).
[18] STAM (2021), Hyperspectral imaging, [online]. Available at: https://www.stamtech.com/services/airborne-hyperspectral-imaging/ [Accessed on 15th August 2021].
[19] Lu, G., Fei, B. (2014). Medical hyperspectral imaging: A review. Journal of Biomedical Optics, [Online], 19(1), 20th January. Available at: https://file:///C:/Users/c7215732/Downloads/010901_1.pdf Accessed on 15th August 2021.
[20] Headwall Photonics (2014), Mining and mineral exploration hyperspectral imaging, [online] Available at: http://www.analytik.co.uk/wp-content/uploads/2016/03/hyperspectral-imaging-mining-and-mineral-exploration.pdf (Accessed on 15th August 2021).
[21] Guilloteau, C., Oberlin, T., Berne, O., Dobigeon, N. (2019). Hyperspectral and multispectral image fusion under spectrally varying spatial blurs – Application to high dimensional infrared astronomical imaging. arXiv.org. [online], Volume 1, 26th December, 1–2. Available at https://arxiv.org/abs/1912.11868 Accessed on 23rd August 2021.
[22] Hughes, G.F. (1968). On the mean accuracy of statistical pattern recognizers. IEEE Transactions on Information Theory, IT(14), 55–64.
[23] Pai-Hui, H., Yi-Hsing, T., Peng, G. (2002). Dimension reduction of hyperspectral images for classification applications. Geographic Information Sciences, Vol. 8(1), June, 1–8.
[24] Su, P., Tarkoma, S., Pellika, K.E.P. (2020). Band ranking via extended coefficient of variation for hyperspectral band selection. Remote Sensing, 12, 3319, 12th October, 1–3.
[25] Maheswari, U., Aluvalu, R., Keerthi Chennam, K. (2021). Chapter 5 Application of machine learning algorithms for facial expression analysis. In: Kant Hiran, K., Khazanchi, D., Kumar Vyas, A., Padmanaban, S., eds. Machine Learning for Sustainable Development. Berlin, Boston, De Gruyter, 77–96. https://doi.org/10.1515/9783110702514-005.
[26] Ghamisi, P. (2015), Spectral and Spatial Classification of Hyperspectral Data, PhD thesis, University of Iceland, Haskoli Islands.
[27] Khodr, J., Younes, R. (2011), Dimensionality reduction on hyperspectral images: a comparative review based on artificial datas, IEEE 4th International Congress on Image and Signal Processing, December, 1875–1883.
[28] Tadjudin, S., Landgrebe, D. (1998), Classification of High Dimensional Data with Limited Training Samples, PhD thesis, Purdue University, West Lafayette, Indiana.
[29] Lee, C., Landgrebe, D. (1993), Feature Extraction and Classification Algorithms for High Dimensional Data, PhD thesis, Purdue University, West Lafayette, Indiana.
[30] Hsu, P.H., Tseng, Y.H. (1999), Feature extraction for hyperspectral image, Proc. 20th Asia Conference on Remote Sensing, Hong Kong, November, Volume 1, 405–410.
[31] Gonzalez, R.C., Woods, R.E. (2002). Digital Image Processing, 2nd edn. Eaglewood Cliffs, NJ, Prentice-Hall.

[32] Schowengerdt, R.A. (1997). Remote Sensing: Models and Methods for Image Processing, 2nd edn. New York, Academic Press.
[33] Richards, J.A., Jia, X. (1999). Remote Sensing Digital Image Analysis, 2nd edn. Berlin, Springer-Verlag.
[34] Chang, C. (2013). Hyperspectral Data Processing. Maryland, USA, University of Maryland, Baltimore County (UMBC), 170–177.
[35] Poor, H.V. (1994)). An Introduction to Detection and Estimation Theory, 2nd edn. New York, Springer-Verlag.
[36] Li, Z., Liang, L., Han, C., Guo, F., Sun, D. (2016). Fault detection for multi-rate sampling systems based on dynamic principal component analysis. IEEE World Congress on Intelligent Control and Automation (WCICA), June, 12–15, 1307–1311.
[37] Vanhatalo, E., Kulahci, M., Bergquist, B. (2017). On the structure of dynamic principal component analysis used in statistical process monitoring. ELSEVIER Chemometrics and Intelligent Laboratory Systems, Volume 167(15), August, 1–11.
[38] Ku, W., Storer, H.R., Georgakis, C. (1995). Disturbance detection and isolation by dynamic principal component analysis. Chemometrics and Intelligent Laboratory Systems, 30, 179–196.
[39] Hajarian, N., Movahedi, S.F., Sadjadi, S.J. (2020). An improved approach for fault detection by simultaneous overcoming of high-dimensionality, autocorrelation, and time-variability. PLoS ONE, 15(12), e0243146, 1–24. https://doi.org/10.1371/journal.pone.0243146.
[40] Mishu, Z.S., Ahmed, B. (2019). Hybrid subspace detection based on spectral and spatial information for effective hyperspectral image classification. International Journal of Computer Applications, 178(41), August, 0975–8887.
[41] Deepa, P., Thilagavathi, K. (2015), Feature extraction of hyperspectral image using principal component analysis and folded-principal component analysis, IEEE Sponsored 2nd International Conference on Electronics and Communication System, 656–660.
[42] Yang, J., Zhang, A., Frangi, F., Yang, J. (2004). Two-dimensional PCA: A new approach to appearance-based face representation and recognition. IEEE Transactions on Pattern Analysis and Machine Intelligence, 26(1), 131–137.
[43] Uddin, P., Mamun, A., Hossain, A. (2019). Effective feature extraction through segmentation-based folded-PCA for hyperspectral image classification. International Journal of Remote Sensing, 40(18), 7190–7220. 10.1080/01431161.2019.1601284.
[44] Jia, X., Richards, J.A. (1999). Segmented principal components transformation for efficient hyperspectral remote-sensing image display and classification. IEEE Transactions on Geoscience and Remote Sensing, Volume 37(Issue 1), January, 538–543.
[45] Jaime, Z., Jinchang, R., Mingqiang, Y., Yi, Z., Jun, W., Stephen, M., Junwei, H. (2014). Novel folded-PCA for improved feature extraction and data reduction with hyperspectral imaging and SAR in remote sensing. ISPRS Journal of Photogrammetry and Remote Sensing, 93, 112–122.
[46] Amato, U., Cavalli, M.R., Palombo, A., Pignatti, S., Santini, F. (2009). Experimental approach to the selection of the components in the minimum noise fraction. IEEE Transactions on Geoscience and Remote Sensing, 47(1), January, 153–159.
[47] Huang, X., Zhang, L. (2008). An adaptive mean-shift analysis approach for object extraction and classification from urban hyperspectral imagery. IEEE Transactions on Geoscience and Remote Sensing, 46(12), December, 4173–4183.
[48] Fauvel, M., Chanussot, J., Benediktsson, A.J. (2006), Kernel principal component analysis for feature reduction in hyperspectral images analysis, IEEE, 8TH January, Proceedings of the 7th Nordic Signal Processing Symposium. Reykjavik, Iceland, 238–240.
[49] Nielsen, A.A. (2011). Kernel maximum autocorrelation factor and minimum noise fraction transformations. IEEE Transactions on Image Processing, 20(3), March, 612–624.

[50] Gu, Y., Wang, Q., Wang, H., You, D., Zhang, Y. (2015). Multiple Kernel learning via low-rank nonnegative matrix factorization for classification of hyperspectral imagery. IEEE Journal of Selected Topics in Applied Earth Observations and Remote Sensing, 8(6), June, 2739–2750.
[51] Bach, F.R., Jordan, I.M. (2003). Kernel independent component analysis. The Journal of Machine Learning Research, 3, 1–48.
[52] Kuo, B., Landgrebe, D.A. (2004). Nonparametric weighted feature extraction for classification. IEEE Transactions on Geoscience and Remote Sensing, 42(5), May, 1096–1104.
[53] Wang, Y., Yeh, C., Young, V.H., Hu, K., Lo, M. (2014). On the computational complexity of the empirical mode decomposition algorithm. Physica A: Statistical Mechanics and Its Applications, Volume 400, 15th April, 159–167.
[54] Demir, B., Erturk, S. (2010). Empirical mode decomposition of hyperspectral images for support vector machine classification. IEEE Transactions on Geoscience and Remote Sensing, Volume 48(11), November, 4071–4083.
[55] Qiao, T. (2016). Advanced Spectral-Spatial Processing Techniques for Hyperspectral Image Analysis. University of Strathclyde, 7–23.
[56] Wu, Z., Huang, E.N. (2009). Ensemble empirical mode decomposition: a noise-assisted data analysis method. Advances in Adaptive Data Analysis, Volume 01(01), 1–41.
[57] He, Z., Shen, Y., Wang, Q., Wang, Y. (2014). Optimized ensemble EMD-based spectral features for hyperspectral image classification. IEEE Transactions on Instrumentation and Measurement, Volume 63(5), May, 1041–1054.
[58] Demir, B., Erturk, S., Gullu, K. (2011), Hyperspectral Image Classification Using Denoising of Intrinsic Mode Functions, IEEE Geoscience and Remote Sensing Letters, Volume 8, No. 2, March, 220–224.
[59] Ayoub, Y., Serhal, S., Farhat, B., Ali, A., Amatoury, J., Nasser, H., Ali, M. (2018), Diagnostic superficial vein scanner, International Conference on Computer and Applications (ICCA), IEEE Xplore, 321–324.
[60] Crisan, S., Tebrean, B. (2017). Low cost, high quality vein pattern recognition device with livenessdetection. Workflow and implementations. Elsevier, [Online], 108, 207–216, 19th May, Available at https://www.sciencedirect.com/science/article/pii/S026322411730341X Accessed on 24th August 2021.
[61] Wang, F., Behrooz, A., Morris, M., Adibi, A. (2013). High-contrast subcutaneous vein detection and localization using multispectral imaging. Journal of Biomedical Optics, May, 18, 1–3.
[62] Split (2021), False Positive Rate, [online], Available at: https://www.split.io/glossary/false-positive-rate/#:~:text=The%20true%20positive%20rate%20(TPR,as%20TN%2FTN%2BFP (Assessed on 24th August 2021).
[63] Jordan, J. (2017). Evaluating a machine learning model. [online] Available at: https://www.jeremyjordan.me/evaluating-a-machine-learning-model/ (Accessed on 124th August 2021).
[64] Narkhede, S. (2018). Understanding AUC-ROC Curve. [online] Medium, Available at: https://towardsdatascience.com/understanding-auc-roc-curve-68b2303cc9c5 (Accessed on 15th June 2021).

Sadu Chiranjeevi, Pradip K. Das

# 15 Detection of Augmented Facial Landmarks-based Face Swapping

**Abstract:** Facial images play a large part in visual communication. Such images furnish nonverbal messages such as intent, emotion and identity of a human being and have numerous applications in the fields of computer vision and graphics. Due to various image processing tools and the utilization of Deep Learning (DL) methods, it is easy even for non-experts to swap the faces in face images, aiming to construct face swap attacks in face images or videos. We focus on face swapping and its detection. Face swapping replaces a face in the destination image with a face in the source image, while maintaining image attributes. It has numerous applications, including privacy protection, computer games and entertainment. However, it could also be used for fraudulent or illicit purposes. A new method for swapping the facial images and its detection based on augmented facial landmarks and Weighted Local Magnitude Patterns (WLMP) is proposed. The 81-augmented facial landmarks are extracted and performed face swapping. The extracted WLMP features from the facial images are provided to the Support Vector Machines (SVM) classifier to detect the presence of face swap attacks. The effectiveness of the proposed approach is estimated on a real-world face image dataset. The performance of the proposed approach is evaluated with various types of SVM classifiers. The overall results show that the proposed approach effectively swaps the two facial images and detects the swapped face images from the original with accuracy of 95%.

**Keywords:** face landmark points, face swap, face swap attack detection

## 15.1 Introduction

Nowadays, facial images play an important role in visual communication both in personal and social media platforms.

These pictures naturally offer non-verbal messages that include the identity, emotion and purpose of a human being. To extract facts of the facial images, localization of facial landmarks is a key process and has numerous applications in the fields of computer vision and graphics such as face swapping, face recognition, face animation

**Sadu Chiranjeevi,** Department of Computer Science and Engineering, Rajiv Gandhi University of Knowledge Technologies Nuzvid, India
**Pradip K. Das,** Department of Computer Science and Engineering, Indian Institute of Technology Guwahati, India

https://doi.org/10.1515/9783110756722-015

and face image synthesis [1, 2]. The performance of such applications depends on the localization of precise facial landmark points. For example, the alignment of faces depends on the spatial locations of facial landmark points in many face recognition systems. In such cases, the localization of imprecise facial landmark points may result in inaccurate face alignment and performance degradation. Further, 2D facial landmarks can be used for deforming the 3D meshes for realistic face performances [3].

Facial landmarks detection (FLD) algorithms localize facial landmarks points on face images. The landmarks are either dominant points denoting different places on facial components (such as the nose) or an interpolated point linking those dominant points around the facial contour and facial components. FLD algorithms locate such locations on important facial features as the nose, eyes, lips and jawline. Given a face image, say I, FLD algorithm extracts the locations of M facial landmarks $X = \{(x_1, y_1), (x_2, y_2) \ldots (x_m, y_m)$ where $(x_i, y_i)$ represents the image coordinates of the facial landmark point. Although many efforts have been made to automatically locate facial landmark locations, it remains a difficult task for a variety of reasons: (1) significant differences in facial appearance across subjects under various head poses and facial expressions, (2) natural conditions such as illumination would affect the facial appearances in facial images and (3) variation in facial occlusion would result in incomplete information of facial appearance. The primary facial image-based contributions for FLD are presented in [4]. As a result, the detection of accurate facial landmark points is still an active research problem.

Face swap attempts to make realistic swapped face images by replacing the faces of two face images. Its applications include digital forensics [5], face emotion recognition [6], preserving privacy [7] and entertainment [8]. For example, face swap results based on 68-landmarks and the augmented 81-landmarks are shown in Figure 15.1. Face swapping has been linked to popular Internet memes and has been used for a variety of fraudulent or immoral purposes [5]. Furthermore, it also provides a way for intruders or cyber attackers to gain unauthorized access to identification or authentication systems. DeepFakes [9], for example, employs a deep learning model to create fake videos of people performing actions that never happened.

To appropriately replace faces in images, the authors of [10] suggested an approach to analyze the 3D-textured model of a face and its key scene features such as lighting and location. A fully automatic method for face replacement with 3D face model is proposed in [11]. In [12], a deep learning method based on Convolutional Neural Network (CNN) is proposed for obtaining the facial appearance of the destination face image from an unstructured collection of images. It is not an appropriate alternative for a range of applications because the network must be trained for each destination facial image. Despite the fact that there are numerous image processing tools and deep learning models for face swapping, the research community remains interested due to the following challenges: (1) An accurate framework for face swapping has to be adjusted to arbitrary faces; (2) The swapped face image should preserve the identity of the source face; (3) The facial attributes such as lighting, posture and expression of the swapped face image has to be consistent with the destination face attributes.

It is challenging when swapping faces in unconstrained and arbitrarily selected images. In such cases, there is no guarantee on the similar appearance of expressions, viewpoints, genders or any attribute when faces are swapped. Thus, robust and effective approaches are required for face swapping.

In this chapter, an extension to the work [13] is proposed. Given a face image, the first 68-facial landmarks are localized using [14] and the additional 13-facial landmarks are localized on the forehead by using the Surrey Face Model (SFM) [15]. The augmented 81-facial landmarks are extracted as shown in Figure 15.2. The 68-facial landmarks and SFM are shown in Figure 15.3. Next, face swapping is performed based on the extracted augmented 81-facial landmark points. Finally, we propose Face Swap Attack Detection (FSAD) based on WLMP features [16] from face images and SVM [17] for the detection of the swapped face images from genuine images. The proposed system is trained and tested on a real-world face image dataset and uses various types of SVMs for detection. The experimental results for both swapping face images and detection are presented. Figure 15.1 will show the source,destination images and swapped images.

**Figure 15.1:** Source and destination images in the first row and the swapped face images based on 68-landmarks and the augmented 81-landmarks in the second row.

The summary of our work is as follows:
1. An approach is proposed for the detection of the augmented 81-facial landmarks of a face image by utilizing the method in [14] and SFM [15].
2. Face swap is performed based on the detected augmented facial landmarks of the source and target face images.

3. FSAD is incorporated based on WLMP and SVM to detect the face images with face swap attacks from the original. Finally, we demonstrate the performance of our system on a real-world dataset and used different types of SVMs for evaluation of the detection. The experimental results are presented.

The rest of the chapter is structured as follows: Related works are discussed in Section 15.2. The proposed approach is elaborated in Section 15.3. Experimental results are presented in Section 15.4. Finally, Section 15.5 concludes the chapter.

**Figure 15.2:** The augmented 81-facial landmark points.

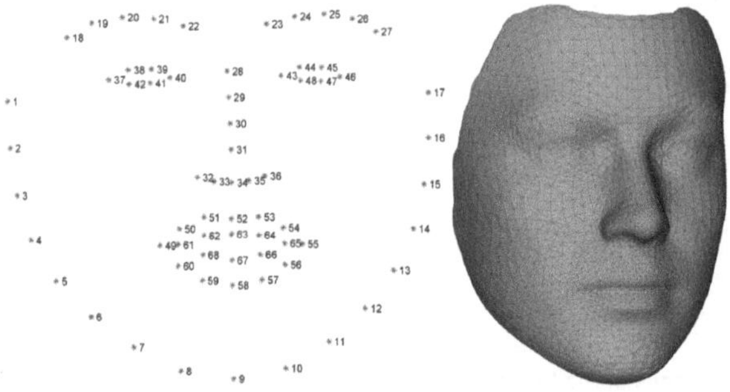

**Figure 15.3:** The benchmark 68-facial landmarks and Surrey Face Model to extract additional landmarks.

## 15.2 Related works

Over the past years, many methods have been proposed for targeting image manipulation and its detection. However, all approaches that address image manipulation may not address the swapping of the face images and its detection. Therefore, we briefly present state-of-the-art methods for the localization of facial landmark points, face swapping and its detection.

### 15.2.1 Facial landmarks detection

Due to its extensive applications, there is a demand for totally autonomous, highly efficient and robust algorithms for locating facial landmarks. Despite the fact that the task of detecting facial landmarks remains difficult due to large differences in facial appearances, partial occlusions and illumination, numerous efforts have been made by various researchers toward automatic FLD based on various testing approaches, performance measures and datasets. The first FLD algorithms concentrated on detecting the form of large face components such as the chin, mouth, eyes and nose and returning a bounding box or contour around them [18]. These techniques are limited to face images with minor lighting and posture modifications. Over time, FLD methods aim to localize the facial landmarks of even face images with changes in facial appearance including expression, occlusion, pose and lighting. Based on how they simulate shape patterns and facial appearance, these approaches can be split into three categories.

#### 15.2.1.1 Holistic methods

For the recognition of face landmarks, these methods explicitly use facial appearance information as well as global facial shape patterns. The facial appearance represents the intensity patterns of the unique pixel around the facial landmark points or in the entire face region, whilst the face shape patterns indicate the shape patterns determined by the locations of the facial landmarks and their spatial interactions. The authors introduced Active Appearance Model (AAM), a new method for matching statistical appearance models to images, in [19]. A collection of model parameters learned from a training set controls the shape and gray-level variation modes. The limitation is that an AAM model presupposes linearity in face appearance and shape fluctuation. The authors of [20] used ensemble models to improve the performance. For sequential model fitting, a series of AAMs are trained in a cascaded fashion. Some significant advances in the AAMs are detailed in [21].

#### 15.2.1.2 Constrained local model methods

For the recognition of face landmarks, Constrained Local Model (CLM) algorithms use global facial shape patterns and independent local information from each facial landmark [22]. In comparison to the holistic model, it is easier to extract face landmarks and more resistant to occlusion and light. A novel regression-based CLM technique is proposed in [23].

#### 15.2.1.3 Regression-based methods

These methods directly map the appearance of a facial image to the positions of landmarks. These strategies differ from Holistic and CLM in that they do not explicitly develop a global model for the faces, but the face shape criteria are certainly embedded. In order to recognize facial points, a combination of regression-based technique and a probabilistic model based on face shape patterns is proposed [24]. The authors of [25] proposed an approach based on privileged information that is only available during the training phase to train the random forests for facial features. It predicts face attributes first and then estimates the location of attribute-dependent landmarks.

Other contributions to FLD have been proposed in recent years. In [26], an interactive approach for face landmarks localization from a single image is proposed. On high-resolution face images with abnormalities in expression, position and illumination, it accurately extracts facial landmarks. Belhumeur et al. [27] describe a unique strategy for detecting face component locations that combines the output of SIFT local detectors with a non-parametric collection of global models. The authors of [14] suggested an approach based on an ensemble of regression trees that locates facial landmarks faster and more precisely. Other advancements in FLD algorithms are discussed in [4].

While these methods provide impressive results, still there is a need for an algorithm to precisely detect facial landmarks that can be suited for an effective face swapping under different poses, color and illuminations.

### 15.2.2 Face swap

The most difficult part of swapping face images is detecting facial traits or facial features and associating them with target face image feature features while maintaining photo authenticity. The state-of-the-art approaches for face swapping are discussed in this section.

The authors introduced an approach in [10] that calculates the textured 3D structure and its scene properties such as position and lighting to appropriately transfer faces in face images. In [11], an idea for replacing a face without requiring a 3D face model is provided. It's a fully automated method for swapping face photos.

It used face detection software to extract all of the faces in the face image library and then aligned them to a similar coordinate system. The destination face image is chosen from a large image library that is identical to the source face image in appearance and position. The algorithm then adapts the target images' qualities such as lighting, colour and poses to match the source face images' attributes. The main limitation is that it can't swap two faces at random. The authors of [28] proposed an algorithm for automatically selecting face photos and replacing them with stock faces. The method [29] uses face region of interests (ROIs) for face swapping in referenced images with similar shapes and features as the source face.

Deep learning models have yielded significant results in image processing and related applications, resulting in a variety of new face swapping approaches. [12] suggested a CNN-based approach for extracting the destination face image appearance from a collection of unstructured face images. Because the CNN must be trained for each target face image, it is not a feasible option for many face swapping or replacement applications. The authors of [37] suggested a new algorithm that uses Deep Neural Network (DNN) instead of facial textures to allocate the region of the face to a latent variable. The authors used DNN to learn the face and hair regions in large-scale image datasets in [38]. In [30], a new face re-enactment method based on recurrent neural networks (RNN) that manages both expression and position variations in face photos or videos is presented.

### 15.2.3 Face swap detection

Instead of using raw pixels, the approach [31] used Bag of Words (BoW) and Speeded Up Robust Features (SURF) to extract features. To illustrate the detection's performance, the authors tested a variety of classifiers, including Neural Networks, Random Forests and SVMs. It correctly identified the swapped face images with an accuracy of 92%. The authors did not compare the quality of swapped face images to that of other datasets, and they only used a relatively small dataset of 10,000 images. The authors of [32] produced a 53,000-image face image dataset from 150 videos, demonstrating the generalization capabilities of state-of-the-art CNN-based and texture-based fake face image detection approaches. A new approach based on WLMP features, which are the same as Local Binary Patterns (LBP) features and SVM is reported in [16]. The performance was demonstrated on a video dataset.

The approach [31] used BoW and SURF for the extraction of features rather than raw pixels. They used various types of classifiers like Neural Networks, Random Forests and SVMs to demonstrate the performance of the detection. It detected the swapped images with 92% accuracy. The authors have not compared the quality of swapped face images with other datasets and also done on a relatively small dataset, consisting only of 10,000 images. [32] creates a face image dataset from 150 videos, consists of 53,000 images and demonstrates the generalization capability of the state-of-the-art CNN-based and texture-based detection techniques of the fake face images. In [16], a new algorithm based on WLMP features, which are the same as LBP features and SVM is presented. The authors demonstrated the performance on the video dataset. In [33], a deep learning-based method for detecting face swap assaults in video datasets is proposed. The authors of [34] recently presented a DNN-based solution for detecting the presence of the face swap attack in face images via transfer learning.

The proposed method differs from the current state-of-the-art. It extracts new augmented 81-facial landmarks from the face images including forehead landmarks. Then, rather than randomly searching for the target face image in the dataset, face swap is performed by arbitrarily selecting the source and target face images from the image dataset. The swapping results produced with the extracted augmented 81-facial landmarks are compared to the 68-facial landmarks swapping results. The extracted feature descriptors are provided to various types of SVMs to distinguish the swapped face images from the original.

## 15.3 The proposed method

The steps involved in the proposed system are outlined in this section: (1) The localization of the augmented 81-facial landmarks that cover landmarks on the forehead, (2) Face swapping is done using extracted facial landmark points, and (3) FSAD technique based on WLMP and SVM is proposed for detecting swapped face images.

### 15.3.1 Augmented facial landmarks detection

The faces of the source and target face images are detected by using Histogram of Oriented Gradients and linear SVM [35]. Then, the augmented 81-facial landmarks of the source and target faces are localized. That is, all retrieved facial landmarks' coordinates $(x, y)$ are detected. As illustrated in Figure 15.4, the first 68 face landmarks are identified using a regression-based approach provided in [14], while the last 13 facial landmarks on the forehead are recognized using SFM [15]. SFM is created from 3D face scans. Each model is represented in the form of a graph, where spatial locations on the surface of face are vertices and the edges connect the vertices to form a triangular mess. Preferred additional 13 facial landmarks are the coordinates of some corners of each triangle in the mess. Considering the facial landmark points on the forehead help to achieve full face swapping and better results where the swapped image looks more realistic.

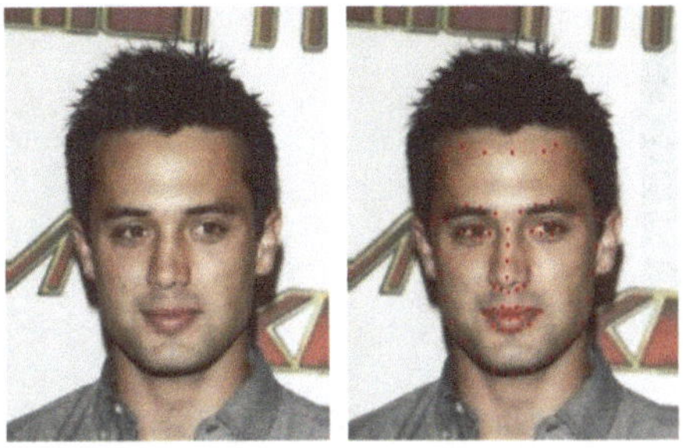

**Figure 15.4:** A source image and the augmented 81 facial landmark points.

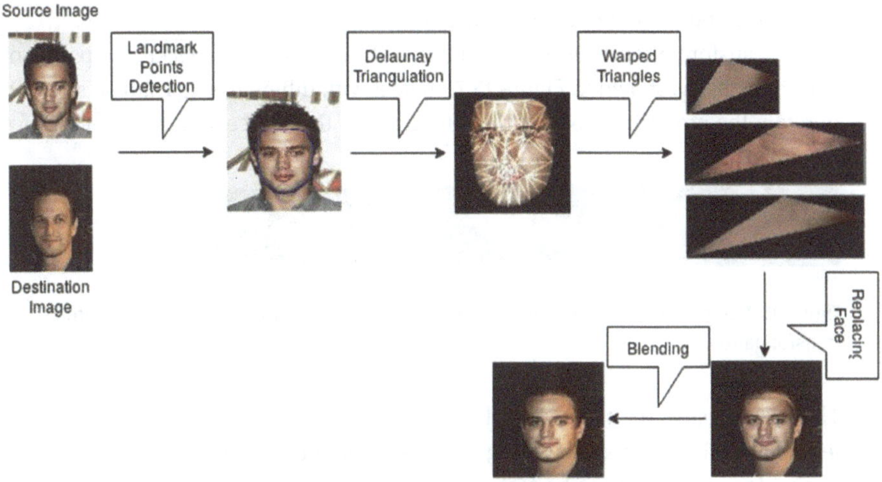


**Figure 15.5:** The flowchart of the proposed face swapping method.

## 15.3.2 Face swapping

The proposed approach for face swapping is depicted in Figure 15.5 as a flowchart. The following is a full description of the face swapping procedure:

1. The 2D planes of the facial landmarks can be utilized to estimate the 3D faces after the facial landmarks have been extracted. That is, by changing a small section of a 2D plane into another 2D plane, the 3D information of images can be approximated. As seen in Figure 15.6, the source face is split into multiple triangles using Delaunay triangulation.
2. For the target face, repeat Step 1. The benefit of constructing triangles is that we can keep the proportions of the faces while transforming each triangle of the source face into a triangle of the target face.

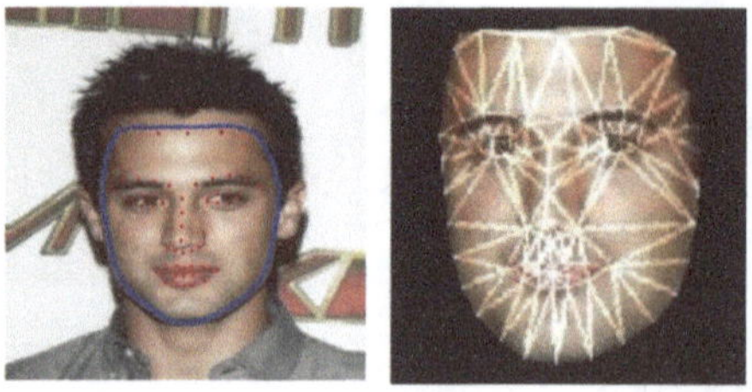

**Figure 15.6:** Convex hull of facial landmark points and Triangles using Delaunay Triangulation.

The matching of triangles is done using the triangle indices. Directly matching the triangles of two faces, on the other hand, is not appropriate if the size of the source and target faces differs or vice versa.

3. Affine transformation is used to produce warped triangles between the points of two triangles, as shown in Figure 15.7, to deal with differences in triangle size and to retain the proportions between the triangles of the source and target faces.

**Figure 15.7:** A triangle of the source face, destination face and Warped triangle after affine transformation.

4. The target face is then exchanged with the source face image's face. Due to differences in skin tone and lighting conditions of both face images, the final face image does not appear to be realistic after swapping, as seen in Figure 15.8(b).

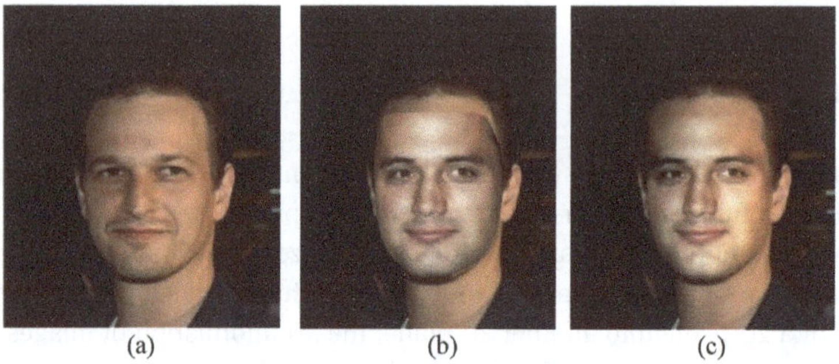


**Figure 15.8:** (a) The destination image; (b) Face swap before blending; (c) Final swapped face image.

Formally, the process involved in replacing the target face Q with the source face P is detailed in the following steps:

i. The coordinates of each triangle of the target face Q are calculated using Equation (15.1):

$$\begin{bmatrix} Q_{a,x} & Q_{b,x} & Q_{c,x} \\ Q_{a,y} & Q_{b,y} & Q_{c,y} \\ 1 & 1 & 1 \end{bmatrix} \begin{bmatrix} \alpha \\ \beta \\ \gamma \end{bmatrix} = \begin{bmatrix} x \\ y \\ 1 \end{bmatrix} \tag{15.1}$$

Here, the matrix $[\alpha \ \ \beta \ \ \gamma]^T$ gives the coordinates. The inverse of the $3 \times 3$ matrix of each triangle that helps to compute the coordinates is given by the following Equation (15.2):

$$\begin{bmatrix} \alpha \\ \beta \\ \gamma \end{bmatrix} = Q_{\Delta}^{-1} \begin{bmatrix} x \\ y \\ 1 \end{bmatrix} \tag{15.2}$$

After the computation of the values $\alpha$, $\beta$ and $\gamma$, the point is inside triangle if $\alpha \in [0, 1]$, $\beta \in [0, 1]$, $\gamma \in [0, 1]$ and $\alpha + \beta + \gamma \in [0, 1]$.

ii. Once the coordinates of the point are obtained, the indices of the pixel of face image $P$ are calculated using Equation (15.3):

$$\begin{bmatrix} x_P \\ y_P \\ z_P \end{bmatrix} = P_{\Delta} \begin{bmatrix} \alpha \\ \beta \\ \gamma \end{bmatrix} \tag{15.3}$$

where

$$P_{\Delta} = \begin{bmatrix} P_{a,x} & P_{b,x} & P_{c,x} \\ P_{a,y} & P_{b,y} & P_{c,y} \\ 1 & 1 & 1 \end{bmatrix} \tag{15.4}$$

Then, the matrix $[x_P \ \ y_P \ \ z_P]^T$ is computed and it is converted into homogeneous coordinates as given in Equation (15.5):

$$x_P = \frac{x_P}{z_P} \text{ and } y_P = \frac{y_P}{z_P} \tag{15.5}$$

iii. Now, the pixel value at location $(x_P, y_P)$ is copied back from the source face image $P$ to the target face image $Q$.

5. Finally, to make the swapped face images appear more natural, seamless cloning is applied as shown in Figure 15.8(c).

### 15.3.3 Face swap attack detection using WLMP and SVM

We propose FSAD which is based on WLMP features [16] and SVM [17] to distinguish the swapped face images from the original. Smoothing and blending are two basic digital image editing techniques for removing irregularities in fake face photos. The texture surfaces are sometimes more than 90% uniform as a result of these procedures. When face images are exchanged or swapped, WLMP-based features that encode the differences obtained by a center pixel with its neighbour pixels are effective in highlighting the most affected areas. The pixel that is closest to the center is given more weight than the other pixel values. That is, rather than binarizing the weight, it delivers it reciprocally in proportion to the difference values of the pixels from the center pixel. These face features also appear to attenuate low-frequency information while keeping high-frequency information.

The procedure for the proposed FSAD is as shown in Figure 15.9. The steps are detailed as follows:

1. The input face image is split into numerous $3 \times 3$ blocks.
2. The differences between the center pixel and its neighbours are determined for each block. The absolute differences between them are also computed.
3. There are eight pixel differences since there are eight neighbor pixels. To give closer pixel values more weight, the differences of pixel values are sorted in non-decreasing order and each value is multiplied by $2^p$, where $p = 0, 1, \ldots, 7$.

   The final result is then converted to a value in the range of 0 to 255. If the final value is more than 255, it is set to 255.
4. The histogram feature vector is then computed.
5. Finally, it provides the SVM classifier with the extracted feature vectors from the training dataset for the detection of swapped face images.

## 15.4 Experimental results

A real-world face attributes dataset is utilized to evaluate the effectiveness of face swapping using the extracted augmented facial features and the performance of the proposed FSAD. The proposed approach's face swap results are compared to the 68-facial landmarks approach's face swap outcomes. Furthermore, various types of SVM classifiers are employed to demonstrate the proposed FSAD's accuracy.

### 15.4.1 Dataset

We trained and tested the proposed system on CelebFaces Attributes Dataset (CelebA) [36]. It is a large-scale face attributes collection with 2,02,599 celebrity face

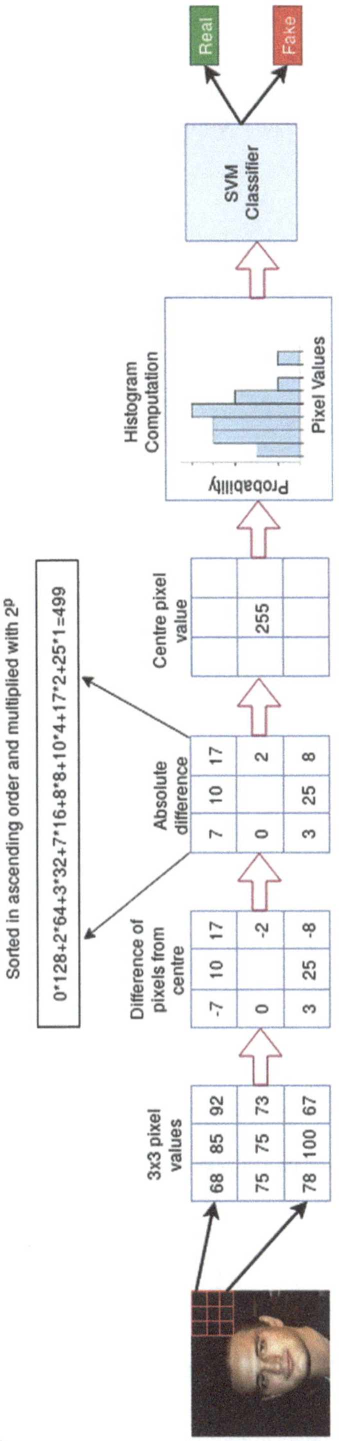


**Figure 15.9:** The procedure of the proposed FSAD approach.

images. Each image contains 40 attribute annotations. It includes face images of people with background clutter and a wide range of poses. It also includes facial images with detailed annotations and large diversities, including 10,177 identities. Figure 15.10 shows the examples of the face swapping results obtained.

**Figure 15.10:** Source image, destination image and final swapped face image (from left to right in each row).

## 15.4.2 Face swap comparison

The effectiveness of face swapping is assessed using [14] 68 facial landmark points as well as the proposed augmented 81 facial landmarks. The 68 facial landmarks and the extracted augmented 81 facial landmarks of a face image are shown in Figure 15.11.

Face swap findings reveal that our method with augmented 81 facial landmarks produces entire face swaps, whereas face swap with 68-landmarks excludes the forehead, resulting in unnatural images. That is, when the skin tones of the source and destination images differ, our technique outperforms the 68-landmark face swap. Source images, destination images, face swap results for 68 landmarks and face swap results for the augmented 81 facial landmarks are shown in Figure 15.12. The face swap findings show that, despite differences in skin tone and lighting conditions, the proposed method preserves facial features such as pose and expression. The proposed augmented facial landmarks also encompass the forehead, making the face swap results appear more natural and unfiltered. Our method, however, fails to effectively swap face photos in the following situations: (i) When there are significant differences in feature size of face images, (ii) When there are major variations in colour of face images, and (iii) When the seam crosses non-skin regions. The inaccurate face swap results are shown in Figure 15.13.

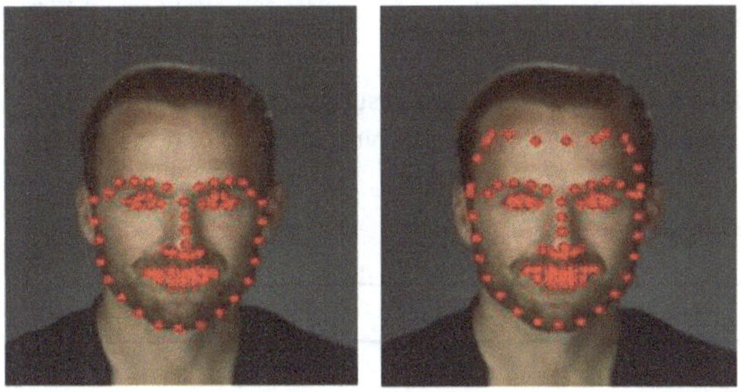

**Figure 15.11:** The 68 vs augmented 81-facial landmark points of a face image.

### 15.4.3 Performance of the proposed FSAD

The proposed FSAD is trained and tested on a real-world dataset. Its performance is demonstrated with different types of SVM classifiers such as linear SVM, polynomial SVM, Gaussian SVM and Sigmoid SVM. The experimental results in Table 15.1 show that the classifiers linear SVM and polynomial SVM detect the swapped face images with 95% accuracy and achieve 96% precision and recall value of 94%. On the other hand, the classifiers Gaussian SVM and Sigmoid SVM detect with an accuracy of 74% and precision and recall values are the same for both swapped and original face images. Therefore, the classifiers linear SVM and polynomial SVM outperform Sigmoid SVM and Gaussian SVM classifiers in terms of all evaluating parameters.

**Figure 15.12:** Shows source image, destination image, face swap results with 68-facial landmark points and face swap results with augmented 81-facial landmark points in each row respectively.

**Table 15.1:** The detection statistics of the proposed FSAD.

| S.No | Classifier | Precision | Recall | F1-Score | Accuracy |
|---|---|---|---|---|---|
| 1 | Linear SVM | 0.96 | 0.94 | 0.95 | **95** |
| 2 | Polynomial SVM | 0.96 | 0.94 | 0.95 | **95** |
| 3 | Sigmoid SVM | 0.78 | 0.68 | 0.73 | 74 |
| 4 | Gaussian SVM | 0.78 | 0.68 | 0.73 | 74 |

## 15.5 Conclusions

A new method for face swapping based on the augmented 81 facial landmarks and its detection is proposed. It extracts the 81-facial landmarks and effectively swaps two face images based on the extracted facial landmarks despite the face images being under different variations in pose, lighting and color. We use the encoded feature descriptors from the face images and SVM to detect the swapped face images. These feature descriptors effectively differentiate the low-frequency and high-frequency information. The performance of the face swapping shows that our method effectively swaps face images and outperforms face swapping with the 68-landmarks while the face images under different variations in skin-tone, pose and lighting conditions. The face swapping results obtained appear more realistic and unedited. The

**Figure 15.13:** Incorrect face swaps: Source image, destination image and their face swap results in each column.

proposed FSAD is evaluated with different types of SVM classifiers. The detection results demonstrate that both classifiers linear SVM and polynomial SVM outperform the remaining two classifiers with 95% accuracy. We will focus on extending the proposed system to effectively swap face images under large variations in properties such as pose, style and incorporate deep learning model to improve the detection accuracy in future work.

## Bibliography

[1] Koestinger, M., Wohlhart, P., Roth, P.M., Bischof, H. (2011). Annotated facial landmarks in the wild: A large-scale, real-world database for facial landmark localization, in 2011 IEEE international conference on computer vision workshops (ICCV workshops). IEEE, 2144–2151.
[2] Thies, J., Zollhofer, M., Stamminger, M., Theobalt, C., Nießner, M. (2016). Face2face: Real-time face capture and reenactment of rgb videos, in Proceedings of the IEEE Conference on Computer Vision and Pattern Recognition, 2387–2395.
[3] Cao, C., Weng, Y., Lin, S., Zhou, K. (2013). 3d shape regression for realtime facial animation. ACM Transactions on Graphics (TOG), 32(4), 41.
[4] Wu, Y., Ji, Q. (2019). Facial landmark detection: A literature survey. International Journal of Computer Vision, 127(2), 115–142.

[5] Oikawa, M.A., Dias, Z., de Rezende Rocha, A., Goldenstein, S. (2015). Manifoldlearning and spectral clustering for image phylogeny forests. IEEE Transactions on Information Forensics and Security, 11(1), 5–18.

[6] Levi, G., Hassner, T. (2015). Emotion recognition in the wild via convolutional neural networks and mapped binary patterns, in Proceedings of the 2015 ACM on international conference on multimodal interaction, 503–510.

[7] Mosaddegh, S., Simon, L., Jurie, F. (2014). Photorealistic Face De-identification by Aggregating Donors Face Components, in Asian Conference on Computer Vision. Springer, 159–174.

[8] Wolf, L., Freund, Z., Avidan, S. (2010). An eye for an eye: A single camera gaze-replacement method. in 2010 IEEE Computer Society Conference on Computer Vision and Pattern Recognition. IEEE, 817–824.

[9] Goodfellow, I., Pouget-Abadie, J., Mirza, M., Xu, B., Warde-Farley, D., Ozair, S., Courville, A., Bengio, Y. (2014). Generative adversarial nets. In Advances in Neural Information Processing Systems, 2672–2680.

[10] Blanz, V., Scherbaum, K., Vetter, T., Seidel, H.-P. (2004). Exchanging faces in images. Computer Graphics Forum, 23(3), Wiley Online Library, 669–676.

[11] Bitouk, D., Kumar, N., Dhillon, S., Belhumeur, P., Nayar, S.K. (2008). Face swapping: Automatically replacing faces in photographs, in ACM SIGGRAPH 2008 papers, 1–8.

[12] Korshunova, I., Shi, W., Dambre, J., Theis, L. (2017). Fast face-swap using convolutional neural networks, in Proceedings of the IEEE International Conference on Computer Vision, 3677–3685.

[13] Sadu, C., Das, P.K. (2020). Swapping face images based on augmented facial landmarks and its detection, in 2020 IEEE region 10 Conference (TENCON). IEEE, 456–461.

[14] Kazemi, V., Sullivan, J. (2014). One millisecond face alignment with an ensemble of regression trees, 2014 IEEE Conference on ComputerVision and Pattern Recognition, 1867–1874.

[15] Huber, P., Hu, G., Tena, R., Mortazavian, P., Koppen, P., Christmas, W.J., Ratsch, M., Kittler, J. (2016). A multiresolution 3d morphable face model and fitting framework, in Proceedings of the 11th International Joint Conference on Computer Vision, Imaging and Computer Graphics Theory and Applications.

[16] Agarwal, A., Singh, R., Vatsa, M., Noore, A. (2017). Swapped! digital face presentation attack detection via weighted local magnitude pattern, in 2017 IEEE International Joint Conference on Biometrics (IJCB). IEEE, 659–665.

[17] Vapnik, V. (2013). The Nature of Statistical Learning Theory. Springer science & business media.

[18] Ding, L., Martinez, A.M. (2008). Precise detailed detection of faces and facial features, in 2008 IEEE Conference on Computer Vision and Pattern Recognition. IEEE, 1–7.

[19] Cootes, T.F., Edwards, G.J., Taylor, C.J. (2001). Active appearance models. IEEE Transactions on Pattern Analysis and Machine Intelligence, 23(6), 681–685.

[20] Sauer, P., Cootes, T.F., Taylor, C.J. (2011). Accurate Regression Procedures for active appearance models. In BMVC, 1–11.

[21] Tzimiropoulos, G., Pantic, M. (2013). Optimization problems for fast aam fitting in-the-wild, in Proceedings of the IEEE international conference on computer vision, 593–600.

[22] Saragih, J.M., Lucey, S., Cohn, J.F. (2013). Deformable model fitting by regularized landmark mean-shift. International Journal of Computer Vision, 91(2), 200–215, 2011.on computer vision, 593–600.

[23] Asthana, A., Zafeiriou, S., Cheng, S., Pantic, M. (2013). Robust discriminative response map fitting with constrained local models, in Proceedings of the IEEE conference on computer vision and pattern recognition, 3444–3451.
[24] Martinez, B., Valstar, M.F., Binefa, X., Pantic, M. (2012). Local evidence aggregation for regression-based facial point detection. IEEE Transactions on Pattern Analysis and Machine Intelligence, 35(5), 1149–1163.
[25] Yang, H., Patras, I. (2013). Privileged information-based conditional regression forest for facial feature detection, in 2013 10th IEEE International Conference and Workshops on Automatic Face and Gesture Recognition (FG). IEEE, 1–6.
[26] Le, V., Brandt, J., Lin, Z., Bourdev, L., Huang, T.S. (2012). Interactive Facial Feature Localization, in European Conference on Computer Vision. Springer, 679–692.
[27] Belhumeur, P.N., Jacobs, D.W., Kriegman, D.J., Kumar, N. (2013). Localizing parts of faces using a consensus of exemplars. IEEE Transactions on Pattern Analysis and Machine Intelligence, 35(12), 2930–2940.
[28] Mahajan, S., Chen, L.-J., Tsai, T.-C. (2017). Swapitup: A face swap application for privacy protection, in 2017 IEEE 31st International Conference on Advanced Information Networking and Applications (AINA). IEEE, 46–50.
[29] Chen, D., Chen, Q., Wu, J., Yu, X., Jia, T. (2019). Face swapping: Realistic image synthesis based on facial landmarks alignment,. Mathematical Problems in Engineering, 2019.
[30] Nirkin, Y., Keller, Y., Hassner, T. (2019). Fsgan: Subject agnostic face swapping and reenactment, in Proceedings of the IEEE International Conference on Computer Vision, 7184–7193.
[31] Zhang, Y., Zheng, L., Thing, V.L. (2017). Automated face swapping and its detection, in 2017 IEEE 2nd International Conference on Signal and Image Processing (ICSIP). IEEE, 15–19.
[32] Khodabakhsh, A., Ramachandra, R., Raja, K., Wasnik, P., Busch, C. (2018). Fake face detection methods: Can they be generalized? in 2018 International Conference of the Biometrics Special Interest Group (BIOSIG). IEEE, 1–6.
[33] Korshunov, P., Marcel, S. (2018). Deepfakes: A new threat to face recognition? assessment and detection, arXiv preprint arXiv:1812.08685.
[34] Ding, X., Raziei, Z., Larson, E.C., Olinick, E.V., Krueger, P., Hahsler, M. (2020). Swapped face detection using deep learning and subjective assessment. EURASIP Journal on Information Security, 2020, 1–12.
[35] Dalal, N., Triggs, B. (2005). Histograms of oriented gradients for human detection, in 2005 IEEE computer society conference on computer vision and pattern recognition (CVPR'05), vol. 1. IEEE, 886–893.
[36] Liu, Z., Luo, P., Wang, X., Tang, X. (2015 December). Deep learning face attributes in the wild, in Proceedings of International Conference on Computer Vision (ICCV).
[37] Natsume, R., Yatagawa, T., Morishima, S. (2018a). Rsgan: Face Swapping and Editing Using Face and Hair Representation in Latent Spaces. arXiv preprint arXiv:1804.03447.
[38] Natsume, R., Yatagawa, T., Morishima, S. (2018b). Fsnet: An Identity-aware Generative Model for Image-based Face Swapping, in Asian Conference on Computer Vision. Springer, 117–132.

Pavan Kumar Asundi, Deepa Radhakrishna, Kavitha Veeraswamy

# 16 Factors affecting the accuracy of a laser scanner

**Abstract:** Optical 3D scanners have become widely used metrological tool for rapidly capturing dense 3D models of objects. Among the optical 3D data acquisition techniques, the use of structured lighting plays a major role, as it is the structured lighting which assists a camera to capture the 3D detail of any given object. This paper presents the empirical study of various factors affecting the accuracy of a 3D scanner with laser-based structured light source. These factors have been identified based on our work on development of a handheld laser scanner. However, most of these factors can also affect the accuracy of a static laser scanner. The factors identified are contributed mostly by the specification of the hardware elements used, usage of the hardware elements within the scanner, the environmental factors and the image processing module within the scanning application.

**Keywords:** optical 3D scanner, structured light scanner, handheld laser scanner, self-referenced sensors, 3D machine vision

## 16.1 Introduction

Advancement in machine tool and manufacturing industry has necessitated the emergence of technologies such as reverse engineering and rapid prototyping, which aids in swift digitization and product modeling. To realize these technologies; capturing of accurate three-dimensional data, which is spatially dense and well-defined, is imperative.

For accurate measurement data acquisition, contact-based approach such as a coordinate measuring machine (CMM) is conventionally used. A CMM works with touch probe, which moves along *XYZ* axes and makes contact with the part to extract its 3D coordinates. A CMM is a highly accurate instrument capable of even delivering accuracy to the tune of 1 µm under controlled environmental conditions. However, the point cloud generated by a CMM is sparse and cannot be used for reverse engineering applications. Also, the high cost, low portability and high operation time are

**Acknowledgment:** This research was funded by Department of Science and Technology, Government of India, under the Advanced Manufacturing Technology (AMT) Program.

**Pavan Kumar Asundi, Deepa Radhakrishna, Kavitha Veeraswamy,** Central Manufacturing Technology Institute, Bengaluru, India

https://doi.org/10.1515/9783110756722-016

some of the constraints of a CMM which limits its ease of use and rapidity of 3D measurements.

Camera-based non-contact measurement systems have existed from at least five decades. They have evolved from single camera systems capable of measuring simple 2D dimensions to complex systems which can perform 3D measurements from macro to nano scale. 3D imaging techniques which have evolved for macro to micro scale measurements include stereo vision, time of flight systems, laser scanners, digital fringe projection systems and photogrammetry systems. Stereo vision systems can provide dense 3D data in scenarios where there is textural difference observed in the images. In situations where there is no texture in the images, which is mostly the case when industrial components are modeled, structured light approaches like laser scanning or digital fringe projection is used.

Digital fringe projection systems are area scan systems which can provide a one-shot point cloud of the view with high accuracy. They give good depth and lateral resolution and denser point cloud data. However, the main drawback comes when the surface to be scanned is highly reflective or totally absorbing like a dark surface. In such cases, the resultant point cloud will be sparse. To alleviate this problem, contrast supporting surface is created artificially using chalk spray. However, this method may not be preferable where the surface and the inspection rate are critical.

Laser scanners support different types of surfaces. They capture data using single or multiple laser line profiles. For highly reflective components, the laser power can be adjusted to reduce speckle effects. Laser scanners are preferable where surface reflectivity and speed of measurements are of concern in fringe projection systems. Since the laser scanners work on projecting line profiles, generally the test object is moved to capture the complete surface. However, moving the test object requires additional handling mechanism which introduces cost and complexity. There may be some areas which are hard to capture due to occlusion and also moving the test object might not be feasible in cases of heavy or large components. To overcome these drawbacks, handheld laser scanners have evolved. These scanners can be held in hand and maneuvered over the object to capture the complete surface. These systems consist of one or two cameras with a laser source.

## 16.2 Related work

The laser scanner system proposed in [1] comprises a camera and laser source attached with the LED markers in a specific geometry. The LEDs mounted on the laser projector are tracked by the video camera to estimate the pose of the laser projector. The laser plane coordinates are estimated in the marker coordinate system. The projector calibration is performed for every captured frame. The system has got the advantage of flexibility as the calibration is performed in every frame. However, this also affects the accuracy of the system.

In [2], the proposed handheld laser system comprises a static camera and moving laser source and designed to work on noisy laser detections as well. The laser source project two laser planes which are orthogonal to each other and the light plane is calibrated considering the orthogonality and co-planarity constraints. Using the established relation between the light plane and camera plane, 3D reconstruction is achieved. The pose estimation and point cloud registration are based on the approach proposed in [3].

In [4], the proposed system consists of a self-calibrating handheld laser plane. The laser ray from the laser projector intersects both the test surface and a known background. The intersection with the known background is used to extract the pose of the laser plane. Markers are used on the known background in order to calibrate the fixed camera. The 3D data is captured using the information of intersection between the camera's line of sight and the laser plane. Having a background with marker could be a limitation in many situations when the system is to be employed in a factory floor. Also, the authors have reported an RMS error of 1.03 mm which could be insufficient for measurements of most industrial components.

The author of [5] proposed a handheld laser range scanner with two cameras and a laser cross hair. The proposed system is a self-referencing system. Cameras are calibrated using the stereo principles and the laser plane is estimated with respect to the stereo coordinate. The 3D points are computed considering the triangulation of camera plane and the laser plane. A set of fixed points are projected to track the position of the sensor.

In [6], authors proposed a laser calibration method which uses a camera and the laser module at fixed relative positions. A set of images are acquired for different known positions of the target. At each position, the laser line is imaged and the transformation relation between the camera and the laser is established. Since camera and the laser are at fixed positions, iterative calibration is avoided. However, for precision calibration, this method would require precise positioning systems.

On similar approaches [6, 7] proposed an optical laser scanning module mounted on a robot arm for large working distances. The proposed module has a single camera and the two laser sources among which one laser source is at fixed position and the other is mobile. The laser sources projects profiles orthogonal to each other. Conventional calibration of a camera and intersection between the fixed laser plane and the mobile laser plane are used to acquire the 3D data.

Accurate laser line center detection is one of the fundamental challenges faced in laser scanning. There will be differences in contrast of laser line, false edges of laser lines and thickness of laser line caused due to reflectivity of object surface resulting in speckle and also laser source positioning. Considering these aspects, the laser line center detection algorithm should be robust enough to handle such variations.

In [5] laser line detection based on the linear differential operator is briefed and the laser line center points are detected by finding the peak point at zero-crossings.

The author of [8, 9] proposed center detection technique to detect curvilinear structures and their widths from images. The proposed method can be used on lines with arbitrary width and the width of the line for each line point can be determined. For each pixel in the image, the second-order Taylor polynomial is computed with derivatives of a Gaussian smoothing kernel. The center points are detected considering the gradient and the curvature direction of the line. The line center is detected with sub-pixel accuracy.

A scanner which is in motion will require a method to integrate the scans captured in different views, into a single reference coordinate system. This can be done by either by using optical line-of-sight techniques such as optical tracking or passive marker-based registration methods or by using non-optical techniques such as inertial measurement unit (IMU) devices [10].

IMU sensors have been widely used in various applications like in-pose estimation for robotics, autonomous vehicles and human motion tracking. IMU provide high accuracies over short periods. However, they have the issue of occlusions and drift when used for longer periods of time [11]. Hence, precision 3D scanners are usually equipped with optical line of sight techniques such as optical tracking or marker-based registration methods.

Marker-based registration methods have been adopted in [5], where a laser point source projects reference points across the frames that are used for pose estimation and in [1] where LED markers mounted on laser projector is used for tracking.

In [12], authors proposed a self-referencing algorithm for the handheld 3D scanners. The system defined comprises a stereo vision system and the reference targets are placed across the sensor's view. These reference targets are traced while the sensor is being traversed and its local position is found using the epipolar geometry and triangulation principle. Thereafter the transformation between the current frame and the global reference frame of the sensor is estimated.

Marker-based registration is also discussed in [13] where a stereo vision system is combined with a fringe projection scanner. The stereo vision sensor observes the fiducial markers, which are pasted on the object and the successive marker positions are compared to estimate the pose change of the sensor module.

In the due course of development of a self-referencing handheld laser scanner system we found that most of the literature emphasizes the role of calibration, laser line center detection and registration aspects on the accuracy of the scanner. Based on our experimentation, this paper brings out the other factors which would affect the accuracy of the handheld laser scanner system such as depth of field, laser depth of focus, standoff distance, illumination effects, marker quality, exposure time and laser line width. Most of these factors can also affect the accuracy of a static laser scanner.

The paper is organized such that Section 16.3 presents the hardware architecture, Section 16.4 discusses the role of different parameters on the accuracy of the system and Section 16.5 briefs about the experimentation carried out followed by conclusions in Section 16.6.

## 16.3 Related work hardware architecture

A laser scanner working with the principle of triangulation has a construction similar to a binocular camera system, where one of the cameras is replaced with the laser source. The light source projects a laser plane which is imaged in the form of a line by the camera. 3D reconstruction is performed by finding the intersection of the line of sight of camera and the laser plane.

The hardware structure used in this paper consists of two cameras and two orthogonal laser planes based on the approach proposed in [5]. The hardware structure is as shown in Figure 16.1.

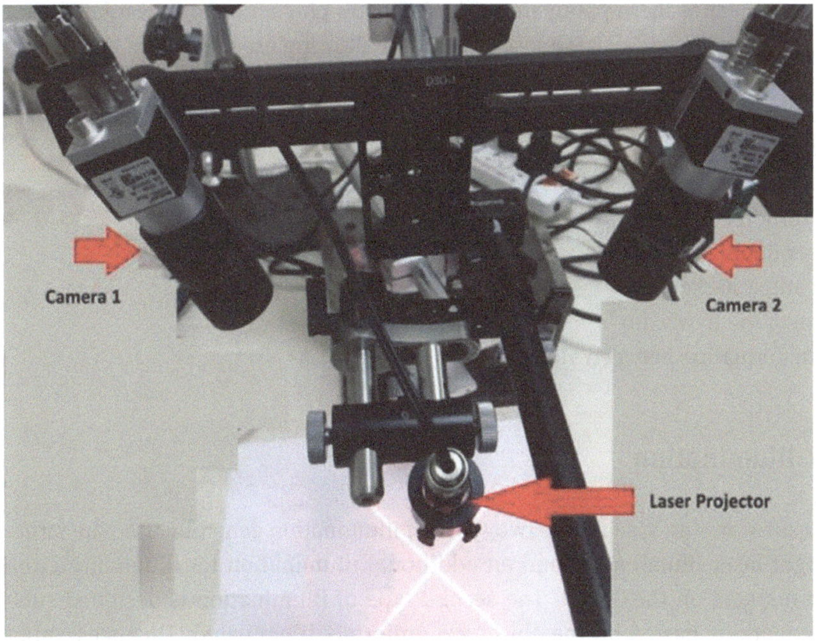


**Figure 16.1:** Construction of 3D laser scanner. Construction of 3D laser scanner with two cameras and a laser projector projecting two orthogonal laser planes.

The internal and external parameters of the cameras and relative positions between the cameras are determined using the conventional binocular camera calibration method [14]. The laser plane pose parameters are estimated with respect to the reference world coordinate system.

Once the relation between the cameras and laser is known, the laser lines extracted from images can be converted to 3D lines. The scanner is used in handheld mode and combining the 3D laser lines acquired from different consecutive views is done by pasting passive markers on the object. As the laser scanner is traversed across the object, the markers visible in the image assist in combining the 3D line data acquired from consecutive views. The corresponding markers detected in consecutive views are used to estimate the transformation matrix of the pose relation between the views. This transformation matrix is also used to transform the 3D line data into a common coordinate system [5].

## 16.4 Factors affecting the accuracy

This section discusses various parameters which were observed to have influence on the accuracy of the developed laser scanner, and the steps taken to address and reduce the effect of these parameters. These factors play an important role when it comes to designing and using a laser scanner. Following are the factors considered and explained further in this paper:

- Illumination
- Laser line detection
- Calibration
- Distortion correction
- Laser depth of focus
- Locating markers
- Pose of the hardware
- Transformation between views

### 16.4.1 Illumination

In the laser scanning system, the two types of illumination generally used are structured light laser illumination and an additional illumination for illuminating the passive markers on the object. The second type of illumination is designed such that the markers pasted on the object are uniformly illuminated. This is essential for ensuring reliable marker detection and hence the correspondence. The camera-lens assembly is also equipped with appropriate filters which restrict unwanted reflections from the object and ambient lighting variations.

Right illumination technique is one which results in high contrast for features of interest and suppression of other features. In our case, the feature of interest is the passive markers that are pasted on the object. Improper illumination would result in increase of complexities in processing of the image.

An efficient synchronization mechanism between the camera and light source is essential for a well-illuminated image acquisition for an object in motion [15]. Since in the case of a laser scanner, either the object or the sensing device is in movement, it is also essential to accurately synchronize the triggering of the light source with the camera acquisition and also synchronize the acquisition of both cameras. Synchronization of both cameras is extremely important in the present context as failure to synchronize between cameras could result in both cameras capturing at different instances of time thereby causing false correspondence and inaccurate 3D point estimation. Figure 16.2 shows the synchronization circuitry. The CamOut1 signal is generated from the camera when the grab is initiated. The CamOut1 signals from both cameras are connected to AN74LS08 AND gate. The output from AND gate is used to externally trigger the cameras. The external triggering produces CamOut2 signal which triggers the light source. This circuitry ensures that both cameras are triggered simultaneously as well as the light sources are synchronously triggered.

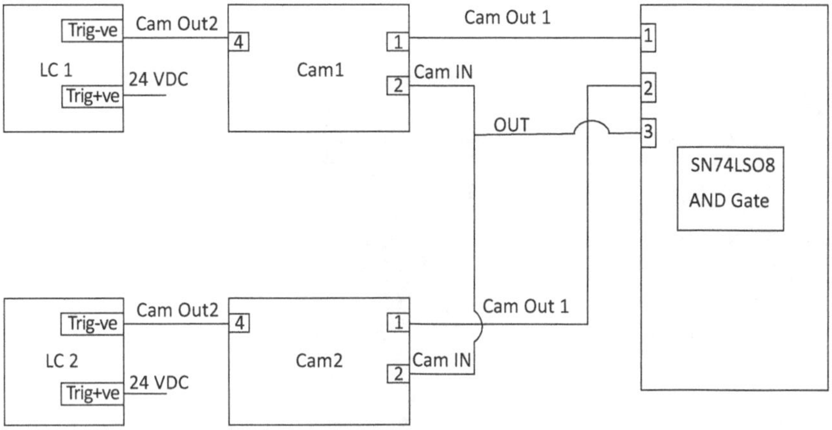


**Figure 16.2:** Synchronization circuit. Synchronization circuitry for synchronizing the cameras and light source.

Figures 16.3 and 16.4 depict the resultant image with uniform illumination and non-uniform illumination.

In certain situations, the illumination from the area light source could interfere with laser line detection and cause discontinuity in the laser line as shown in Figure 16.5. This would result in longer acquisition times, as significant portion of the component scanning could be missed. The combination between exposure setting of the camera, choice of appropriate filters, laser intensity and focus should be appropriately adjusted to address the issue. Figure 16.6 shows a laser line detected completely throughout the test object.

**Figure 16.3:** Image with uniform illumination. Laser line and the markers are distinguishable with good contrast variation.

**Figure 16.4:** Image with non-uniform illumination. Non-uniform illumination resulting from synchronization mismatch between light source and camera.

Varying the exposure time of the camera also results in change in illumination. Since the sensing device is in constant movement, the exposure setting has to be carefully set such that the exposure time is faster than the motion speed in order to avoid blurred images.

Figure 16.5: Discontinuous laser line detection.

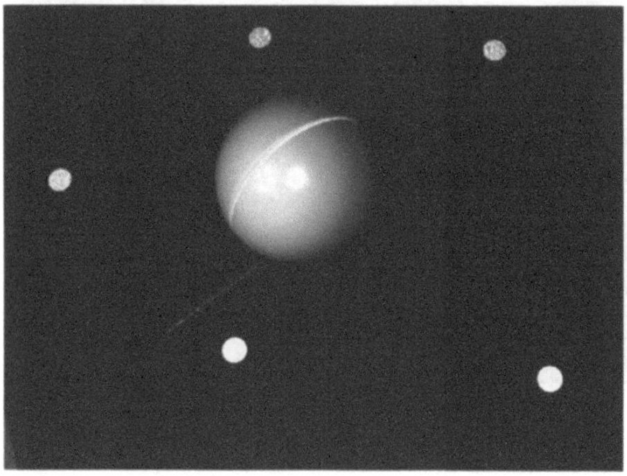

Figure 16.6: Continuous laser line detection.

Too high or too low exposure setting also results in discontinuities in laser line detection, as shown in Figure 16.7.

In order to validate synchronous image capture between cameras, the epipolar constraint can be used. Ideally, for a well-synchronized and well-calibrated stereo system, the difference between row coordinates will be less than a pixel [16] as shown in Figures 16.8 and 16.9.

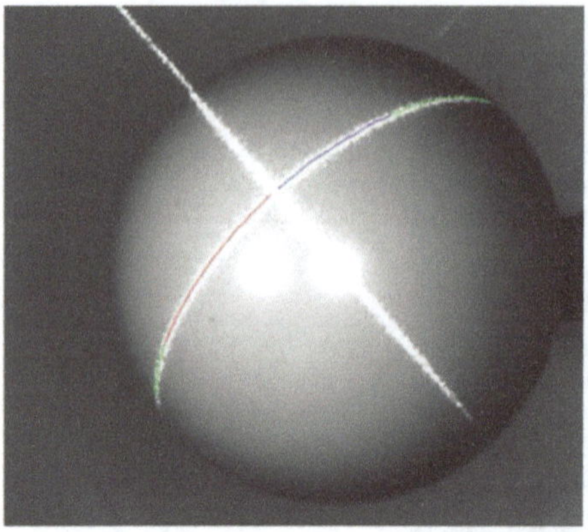

**Figure 16.7:** Over-exposed image resulting in fragmented laser lines.

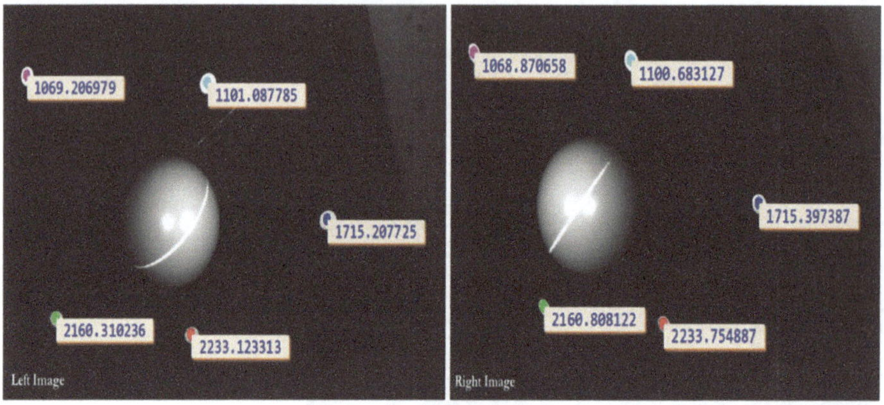


**Figure 16.8:** Epipolar error for synchronized system. Epipolar error for a well-synchronized and well-calibrated stereo system is less than a pixel.

## 16.4.2 Laser line detection

Accurate laser line center detection plays a very crucial role in ensuring the accuracy of measurement. The Steger method [8, 9], which is very widely used for line center detection due to its high robustness and accuracy [17], is used in this paper. In this method, the extraction is done by using partial derivatives of a Gaussian smoothing kernel to determine the parameters of a quadratic polynomial in $x$ and $y$ for each point of the image. The standard deviation of the Gaussian kernel is computed according to equation (16.1),

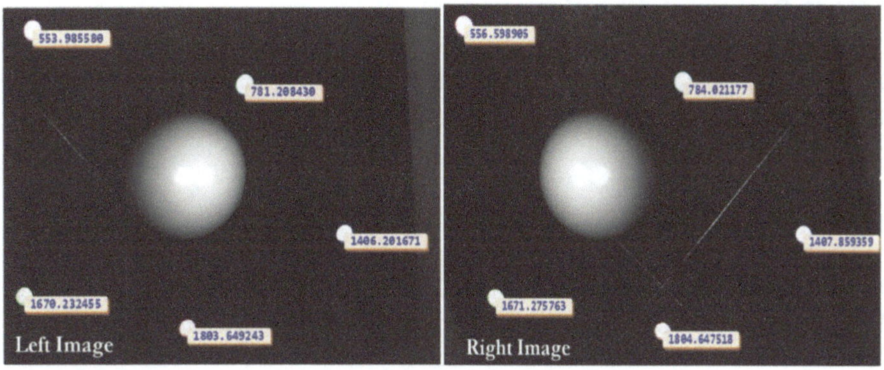


**Figure 16.9:** Epipolar error for asynchronous system. When the left and right cameras are not synchronized some of the markers exhibit more than one pixel error.

$$\sigma \geq \frac{w}{\sqrt{3}} \tag{16.1}$$

where $w$ is the width of the line.

Pixels which exhibit a local maximum in the second directional derivative perpendicular to the line direction are marked as line points. These line points are linked together depending on the upper threshold and lower threshold for the absolute value of the second derivative.

Since the detection is based primarily on line width and contrast parameter, if there are variations in these parameters, it could affect the detection, as shown in Figure 10. The optimal contrast and the line width should be determined during the start of the scan operation.

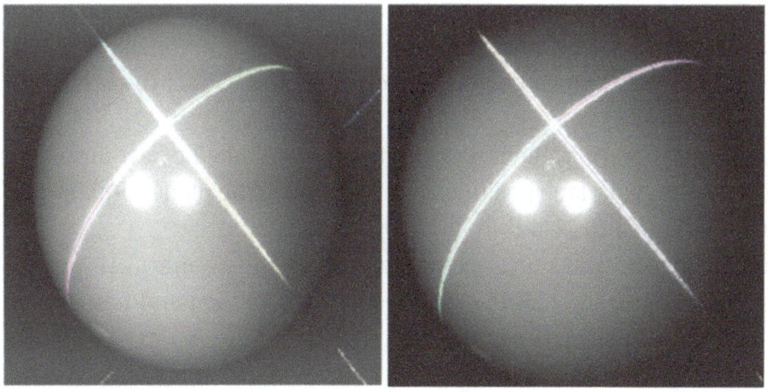

**Figure 16.10(a) and (b):** (a) Discontinuities in line center detection due to higher line width and low contrast. (b) Line center detected completely.

### 16.4.3 Scanner calibration

During calibration of the scanner, we extract the internal and external parameters of both cameras, relative position between the cameras and the laser plane pose.

**Camera calibration**

The calibration artifact may be either a planar target or a 3D target consisting of known features. In this work, a planar 2D calibration target with circular features has been used. The target features are made to an accuracy of 2 µm. Accuracy of the features on the target is an important factor to achieve the desired measurement accuracy.

Table 16.1 shows the calibration error with two types of targets. One with feature accuracy of 100 µm and the other with feature accuracy of 2 µm.

**Table 16.1:** Calibration error with two types of targets.

| Parameter | Target of 100 µm accuracy | Target of 2 µm feature accuracy |
|---|---|---|
| Calibration Error | 0.53 pixels | 0.13 pixels |

The calibration target has to be captured with adequate contrast and focus levels. To validate the results of the calibration, a plane was fit to the observed control points by positioning the calibration plate at different height planes and estimating the plane-to-plane distance error, as shown in Figure 16.11.

**Laser calibration**

Laser calibration extracts the pose of the laser light plane. This can be done by projecting the laser at planes of known heights and combined to extract the laser light plane pose [14].

Similar to camera calibration, the laser calibration is validated by acquiring the laser lines at different known plane positions and measuring the plane-to-plane distance, as depicted in Figures 16.12 and 16.13.

The deviations between the control point plane shown in Figure 16.11 and the laser planes shown in Figure 16.13, at the same height was verified and found to be maximum around 3 µm, as shown in Figure 16.14.

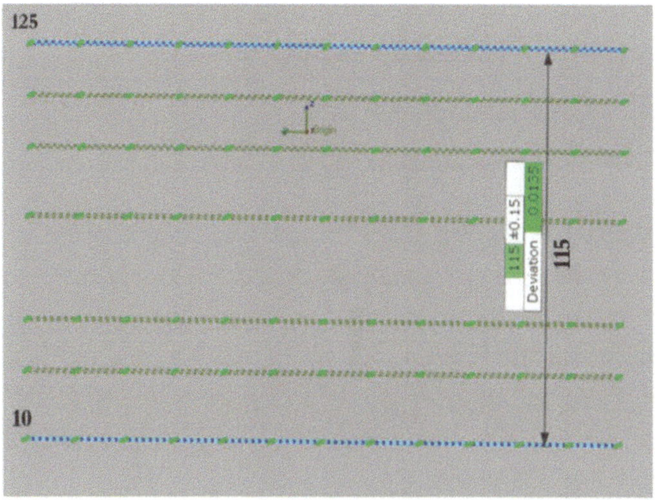


**Figure 16.11:** Plane-to-plane distance measurement using the control points on calibration target.

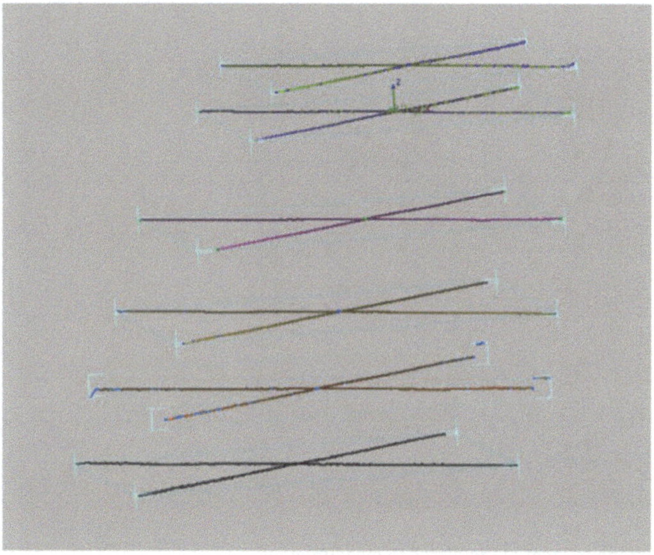

**Figure 16.12:** Construction of laser plane at different heights, to validate laser calibration.

## 16.4.4 Distortion correction

Lens distortion can significantly affect the measurement accuracy. Hence, it needs to be corrected. The distortion effect results in a 3D point being projected to a different pixel position in the imaging sensor, as shown in Figure 16.15.

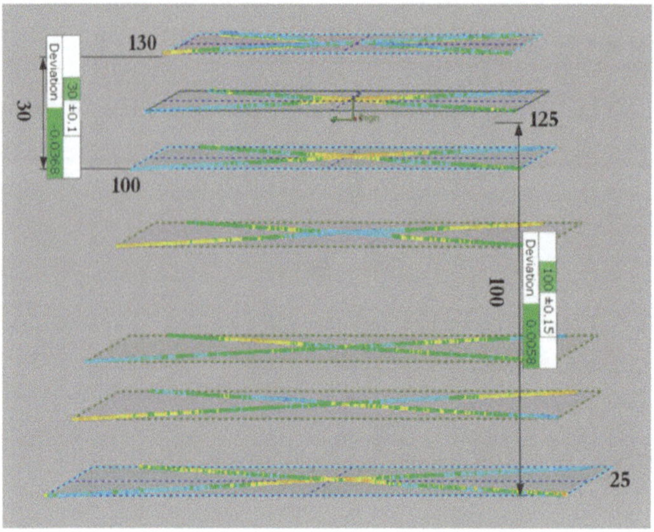


**Figure 16.13:** Laser plane to plane distance measurement.

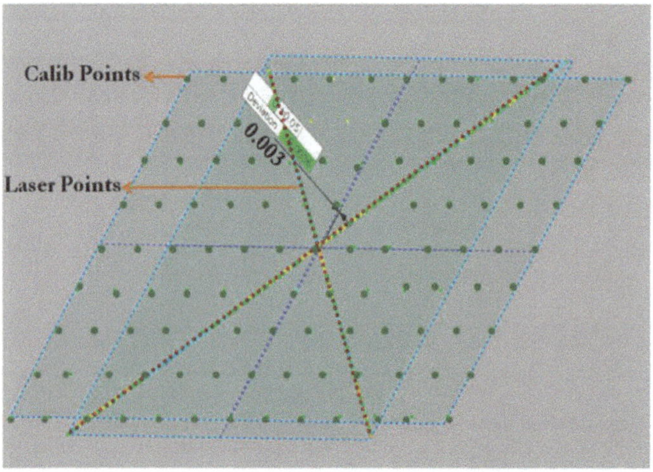


**Figure 16.14:** Control point plane to laser plane distance measurement.

The distortion can be modeled using two types of models namely, the division model and polynomial model. The lens distortion can be modeled in the form of a transformation matrix in the image plane [14]. The division model uses one parameter K to model the radial distortions and the polynomial model uses five parameters namely three parameter K1, K2 and K3 to model radial distortions and two parameters P1 and P2 to model tangential distortions. Radial distortions occur when the

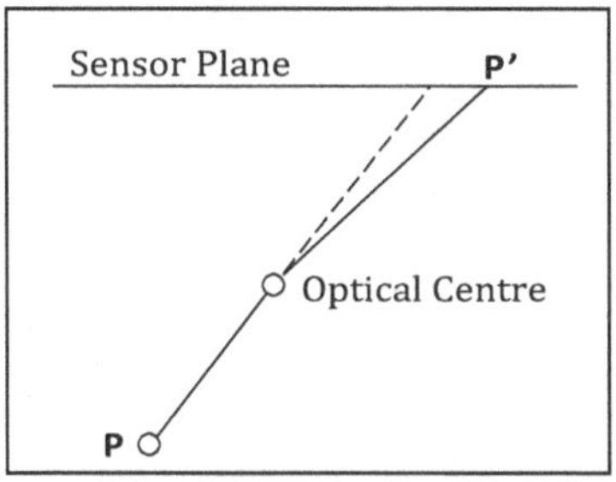


**Figure 16.15:** Schematic illustrating the projection error caused by lens distortions.

light rays bend more near the edges of a lens than they do at the center. Tangential distortions occur when the lens and image plane are not parallel.

The radial distortion model can be expressed by equations (16.2) and (16.3) as

$$u = \frac{\breve{u}}{1 + k(\breve{u}^2 + \breve{v}^2)} \tag{16.2}$$

$$v = \frac{\breve{v}}{1 + k(\breve{u}^2 + \breve{v}^2)} \tag{16.3}$$

The polynomial distortion model can be expressed by equations (16.4) and (16.5) as

$$u = \breve{u} + \breve{u}(K_1 r^2 + K_2 r^4 + K_3 r^6) + P_1(r^2 + 2\breve{u}^2) + 2P_2 \breve{u}\,\breve{v} \tag{16.4}$$

$$v = \breve{v} + \breve{v}(K_1 r^2 + K_2 r^4 + K_3 r^6) + 2P_1 \breve{u}\,\breve{v} + P_2\left(r^2 + 2\breve{v}^2\right) \tag{16.5}$$

where $(u, v)$ are undistorted coordinates while $(\breve{u}, \breve{v})$ are distorted coordinates.

The polynomial model of distortion correction, where radial as well as tangential distortions are estimated, was adopted in this paper. The camera calibration error using equations (16.2) and (16.3) was 0.25 pixels, while the camera calibration error using equations (16.4) and (16.5) was 0.13 pixels.

It is essential to apply distortion correction in the image to ensure accuracy of measurement.

To assess the effect of distortion correction in images, on measurements, the height difference between two planes is measured in 3D, using the stereo setup within the scanner. The two planes are constructed using two length gauges placed at different positions in the field-of-view as shown in Figure 16.16. Two circular targets are pasted on the surface of the length gauges.

Marker center on length gauge are detected. A plane is fit to the points of same elevation and the distance between planes is measured.

Table 16.2 depicts the effect of distortion correction on the plane-to-plane distance measurements.

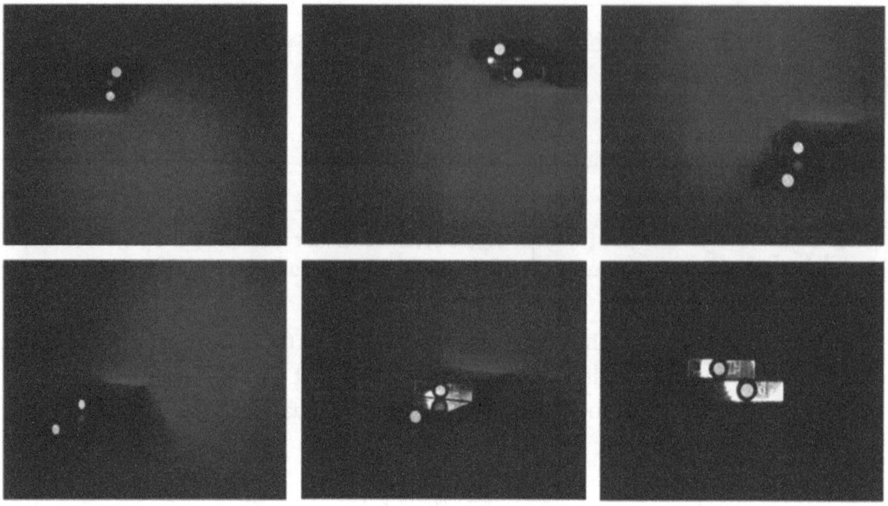

Figure 16.16: Experimentation setup to study the effect of distortion correction.

Table 16.2: Effect of distortion correction on plane-to-plane distance measurements.

| Length gauge combination | Measurements without distortion correction | | Measurements with distortion correction | |
|---|---|---|---|---|
| | Measurements (mm) | Deviation (μm) | Measurements (mm) | Deviation (μm) |
| 130 mm and 90 mm | 40.0512 | 51.2 | 40.0134 | 13.4 |
| 130 mm and 50 mm | 80.053 | 53 | 80.0174 | 17.4 |

## 16.4.5 Laser depth of focus

The importance of the depth of focus of a laser is often not discussed in literature on laser scanning. A defocused laser line affects the accuracy of the system, as the laser center point detection would be affected.

The depth of focus of camera-lens systems can be achieved by choosing suitable lens and their depth of focus achievable is usually higher than the depth of focus of a laser source. However, in case of laser sources, it requires special optics design in order to achieve depth of focus of higher order. This will reduce the effective working volume of the laser scanner.

In order to accurately determine the laser depth of focus, the laser line was projected on a plane at different heights of 5 mm step starting from 115 mm below reference plane to 15 mm above reference plane.

The 3D point cloud of the laser line was extracted and the plane-to-plane distance was measured from every plane to every other plane. It is observed that the laser plane-to-plane distance at too far below the reference plane and too far above the reference plane is giving higher deviation from actual value. In these areas the camera systems are in focus whereas the laser line was observed to be thick compared to other planes indicating a laser focus issue.

As shown in Figure 16.17, the plane-to-plane distance from two planes, namely the reference plane and 90 mm plane to four other planes are depicted. The reference is at 115 mm. The four other planes to which the distances are measured are 20, 25, 125 and 130 mm planes. The maximum deviation value allowed is considered to be ±50 μm. As it can be noted, the distance of reference plane as well as 90 mm plane to 20 mm plane and 130 mm plane is outside the tolerance of ±50 μm. Hence, the effective working volume is restricted between 25 and 125 mm plane.

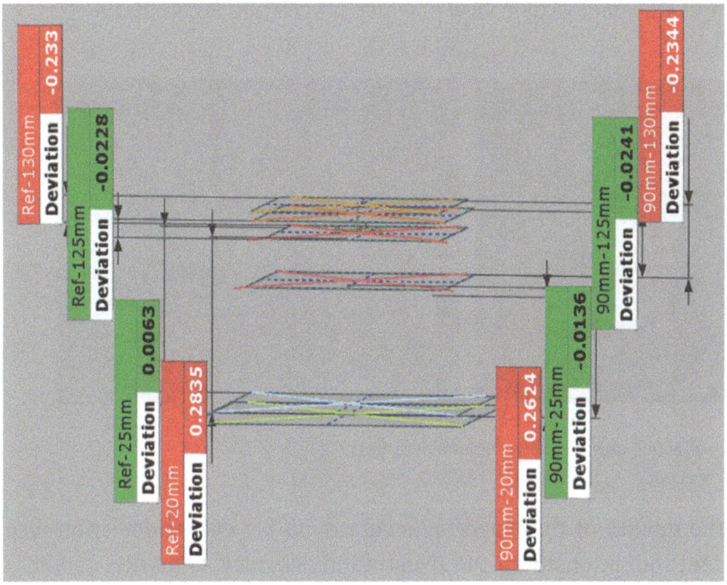


**Figure 16.17:** Laser depth of field determination.

## 16.4.6 Locating markers

As described in Section 16.3, we use markers pasted on the object to assist in self positioning of the handheld device. This can be done either using active markers or passive markers. Active markers generate light which is tracked in the images to assist in the registration process, example LEDs used in [1]. As the test object size increases, more such markers would be required. In such cases, use of active markers increases the cost and complexity. On the other hand, passive markers emit the light that is incident on them. Passive targets have the advantage of ease of use as they have to be just pasted on the test object. Passive targets can be manufactured using retro-reflective material or non-retro-reflective material. The difference in appearance of retro-reflective and non-retro-reflective markers is show in Figure 16.18. Under proper lighting conditions, both types of markers would give similar results. However, in many situations, especially in case of handheld systems, it is quite difficult to maintain a uniform illumination, resulting in failure of marker detection. Hence, for tracking in augmented reality or virtual reality applications, retro-reflective targets are used as they are readily available and these materials reflect even a small amount of light incident on them [18].

Hence, in this study, passive targets with retro-reflective material have been used. The choice of the material would decide the level of retro-reflectivity. The retro-reflective material used in this paper is made up of spherical glass beads embedded in an adhesive layer, the microscopic view of which is shown in Figure 16.19.

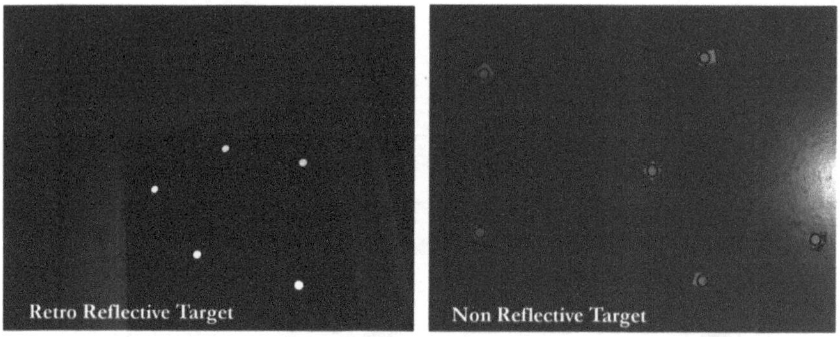


**Figure 16.18:** Retro-reflective and non-retro-reflective targets.

In addition to the quality of the markers, accurate marker center detection also plays an important role in ensuring the registration accuracy as marker center is the parameter which is used to arrive at the transformation from the current view to the previous view.

### 16.4.7 Pose of the hardware

Pose of the scanner which indicates the position and orientation of the scanner device is an important aspect, especially in scenarios where the device is moving, like in the case of a handheld scanner or a robot mounted scanner. The device has to be positioned in a way that the standoff distance is maintained and the cameras and the laser source remain well within their focus levels.

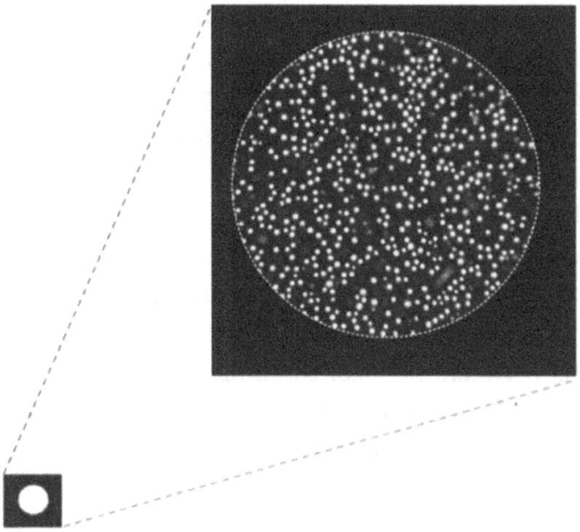

**Figure 16.19:** Microscopic view of the retro-reflective material used in the targets.

To achieve good accuracy, the device is expected to be maintained at a constant working distance from the test object and the line of sight must be approximately perpendicular to the object plane. The pose of the device can be estimated using the targets pasted on the object.

In the case of handheld scanning, ergonomics of the device plays an important role in maintaining the requisite pose of the hardware. Failure to maintain the working distance and orientation of the device can result in issues like loss of accuracy and longer scan time due to loss of frames.

### 16.4.8 Transformation between views

To capture the complete 3D point cloud data, the scanning device is traversed across the object. In every position of the device, the point cloud data is acquired with respect to the device's local coordinate system. The 3D data acquired from successive views are transferred to a common global coordinate system, in order to combine the views. This process is called as registration.

Registration can be done in several ways. Positioning devices like IMU is used to acquire the absolute position of the device, which in turn is used for registering the views. However, IMU sensors are accurate over short periods and suffer from occlusions and drift over longer periods of time [11].

The scanner can be mounted on a robot arm and registration can be carried out using the robot coordinates. However, the positioning accuracy of robots is usually coarse, of the order of 600 μm–1 mm. Hence, for precision applications, this method may not be suitable.

Feature-based registration uses distinct shape based features on object for registering. This method may fail on objects which do not have lot of features on them. Marker-based registration uses targets pasted on the object to merge the views. This method is accurate, however, in some applications pasting targets all over the object may not be a viable task. Optical tracking is another method which tracks the scanner position using retro-reflective spheres mounted on the scanner body. This method is also very accurate; however, optical tracker is a separate system with additional hardware which can increase the cost of the overall system [10].

In this paper, the marker-based registration method was adopted. The accuracy of estimation of transformation matrix between the local coordinate system and the global coordinate system plays a crucial role in ensuring the accuracy of registration. In this method, the markers in the view are converted to 3D coordinates in the local coordinate system using the binocular stereo setup in the scanner. This requires 2D correspondence to be carried out between the left and right camera views. The 3D coordinates of markers are compared with the 3D coordinates of markers in the previous view to estimate the transformation matrix which transforms the markers from local coordinate system to global coordinate system. The same transformation matrix is also used to transform the 3D laser line data. The process is shown in Figure 16.20.

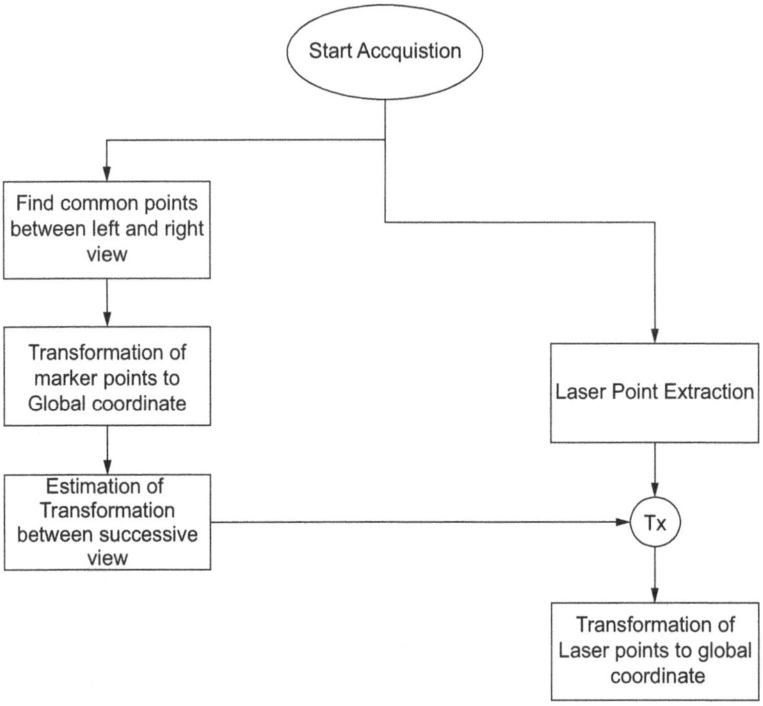


**Figure 16.20:** Marker-based registration technique.

In this process, the various factors that contribute to accurate registration are accurate marker center detection, 2D correspondence and 3D correspondence. Figures 16.21 and 16.22 illustrate the 2D correspondence and 3D correspondence respectively.

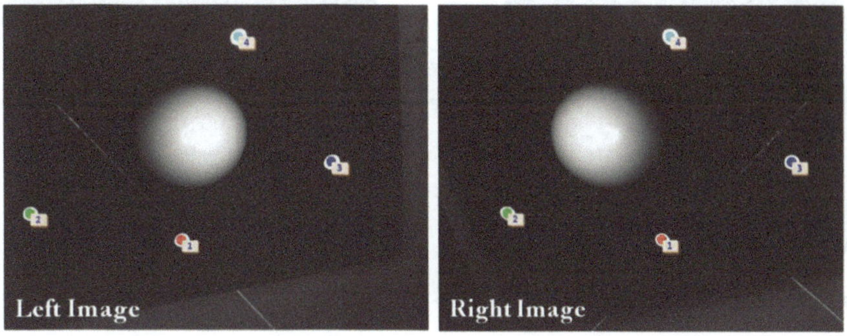


**Figure 16.21:** 2D correspondence between marker points in left and right image.

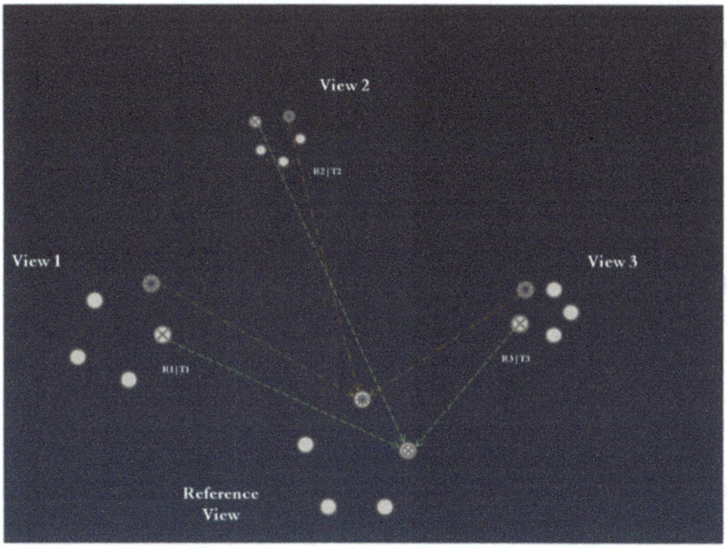


**Figure 16.22:** 3D correspondence. Finding corresponding 3D points between successive views and estimation of transformation for global registration.

Figure 16.23 depicts the point cloud registration, where the point cloud of the laser line generated is transformed to a common global coordinate system.

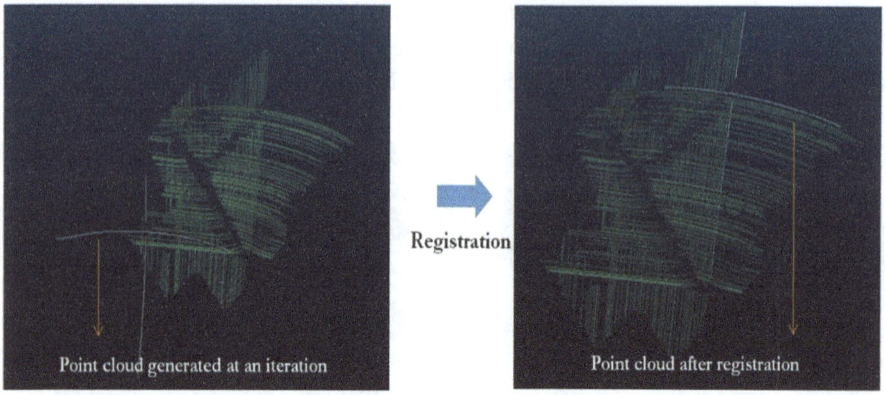


**Figure 16.23:** Point cloud registration.

## 16.5 Experimentation

Considering the various factors discussed in Section 16.4, trials were conducted to evaluate the point cloud generation and accuracy aspects of the generated point cloud.

In order to evaluate the effect of calibration accuracy on measurements, plane-to-plane distance measurements using both binocular stereo setup as well as laser scanning setup were carried out by fixing the scanning device on a tripod.

The planes at different heights were created using length gauges. For evaluating the binocular stereo setup, a target with circular features was placed on different planes of length gauges. The centers of the circles were computed in 3D using the binocular stereo vision system. A plane was fit to the 3D circle centers and plane-to-plane distance was measured. For evaluating the laser scanning, the laser line projected on different planes of length gauges are converted to 3D. A plane was fit to the 3D laser points and plane-to-plane distance was measured.

Tables 16.3 and 16.4 depict the deviation analysis of plane to plane measurements of the stereo vision and the laser scanning, where the camera calibration error is 0.17 pixels.

Tables 16.5 and 16.6 depict the deviation analysis of plane-to-plane measurements of the stereo vision and the laser scanning, where the camera calibration error is 0.13 pixels.

From Tables 16.3 to 16.6, it is evident that improvement in calibration improves the measurements.

A calibrated matt finish ceramic sphere of diameter 38.1013 mm was measured. The sphere was placed on a granite slab prepared with retro-reflective markers for the purpose of handheld scanning.

**Table 16.3:** Plane-to-plane distance measurement of circle center points using stereo vision system. Camera calibration error is 0.17 pixels.

| Reference gauge height (mm) | Varying gauge height (mm) | Actual plane-to-plane distance (mm) | Measured plane-to-plane distance (mm) | Deviation (mm) |
|---|---|---|---|---|
| 45 | 25 | 20 | 20.0380 | 0.0380 |
| | 60 | 15 | 14.9515 | 0.0485 |
| | 90 | 45 | 44.9796 | 0.0204 |
| | 110 | 65 | 64.9390 | 0.0610 |
| | 125 | 80 | 80.0025 | 0.0025 |

**Table 16.4:** Plane-to-plane distance measurement of laser points. Camera calibration error is 0.17 pixels.

| Reference gauge height (mm) | Varying gauge height (mm) | Actual plane-to-plane distance (mm) | Measured plane-to-plane distance (mm) | Deviation (mm) |
|---|---|---|---|---|
| 45 | 25 | 20 | 19.9398 | 0.0602 |
| | 60 | 15 | 14.8535 | 0.1462 |
| | 90 | 45 | 44.8377 | 0.1623 |
| | 110 | 65 | 64.8562 | 0.1438 |
| | 125 | 80 | 79.8354 | 0.1646 |

**Table 16.5:** Plane-to-plane distance measurement of circle center points using stereo vision system. Camera calibration error is 0.13 pixels.

| Reference gauge height (mm) | Varying gauge height (mm) | Actual plane-to-plane distance (mm) | Measured plane-to-plane distance (mm) | Deviation (mm) |
|---|---|---|---|---|
| 45 | 25 | 20 | 20.0434 | 0.0434 |
| | 60 | 15 | 14.9817 | 0.0183 |
| | 90 | 45 | 45.0081 | 0.0081 |
| | 110 | 65 | 65.0191 | 0.0191 |
| | 125 | 80 | 80.0128 | 0.0128 |

Laser line width variation can occur due to varying exposure time settings of camera and varying working distance of both camera and laser resulting in focus issues. The laser line width variation causes discontinuities in laser-line center detection resulting in unfilled areas in the point cloud and error in measurement, as illustrated in Figure 16.24. Figure 16.25 shows a sphere with uniform line width and less deviation compared to Figure 16.24.

**Table 16.6:** Plane-to-plane distance measurement of laser points. Camera calibration error is 0.13 pixels.

| Reference gauge height (mm) | Varying gauge height (mm) | Actual plane-to-plane distance (mm) | Measured plane-to-plane distance (mm) | Deviation (mm) |
|---|---|---|---|---|
| 45 | 25 | 20 | 19.9971 | 0.0029 |
| | 60 | 15 | 14.9934 | 0.0066 |
| | 90 | 45 | 45.0398 | 0.0398 |
| | 110 | 65 | 65.0352 | 0.0352 |
| | 125 | 80 | 80.0519 | 0.0519 |

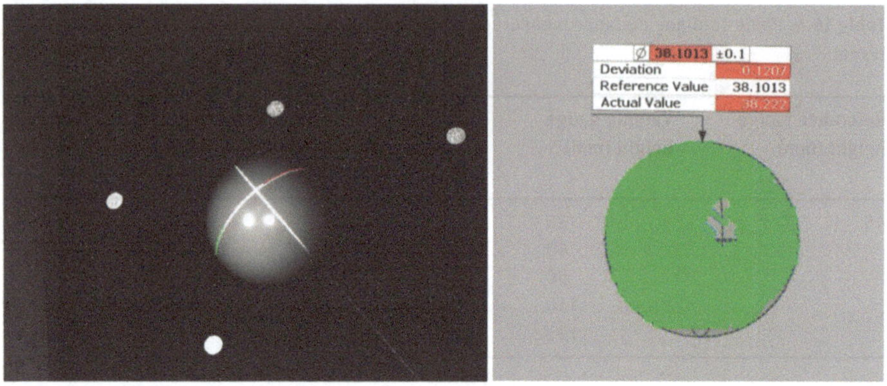


**Figure 16.24:** Image with non-uniform laser line width and its corresponding point cloud.

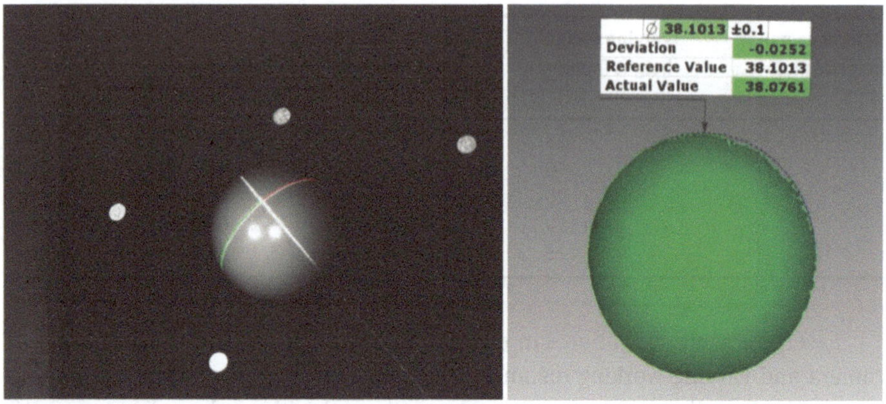


**Figure 16.25:** Image with uniform laser line width and its corresponding point cloud.

In the case of handheld scanning systems, the position and orientation of the scanner keeps varying as it is traversed. Positioning the system too close or too far and too much of orientation can result in error as the appearance of the laser line as well as markers are distorted, as shown in Figure 16.26. This can cause error in marker center detection.

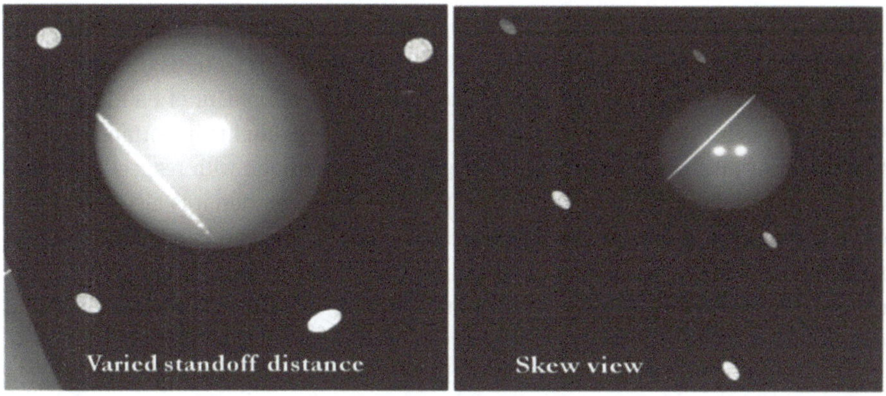


**Figure 16.26:** Images of varied standoff distance orientation.

The error in transformation estimation can result in inaccuracies as shown in Figure 16.27, where it can be seen that the position of the line after transformation is away from the best fit sphere. In the present case, it results in variation of the measured diameter of the sphere.

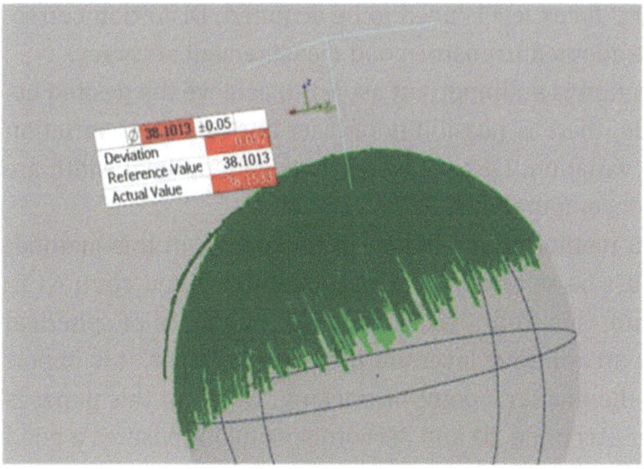


**Figure 16.27:** Position of the line after transformation is away from the best fit sphere.

Thus the scanning module should acquire data at a calibrated standoff distance with approximately perpendicular orientations.

## 16.6 Conclusion

The use of sections to divide the text of the paper is optional and left as a decision for the author. Where the author wishes to divide the paper into sections the formatting shown in Table 16.2 should be used. In the course of development of handheld laser scanning system, we found that most of the literature highlights importance of calibration, laser line center detection and registration technique with respect to the accuracy aspects. During our empirical study and experimentation it is also found that other parameters such as depth of field, laser depth of field, standoff distance, illumination effects, marker quality, exposure time and laser line width also play a significant role in determining the accuracy of the system. The paper has highlighted the importance of these parameters along with conventionally discussed factors.

- Hardware selection and assembly of the scanner structure play a significant role in accuracy aspects. Field of view should be illuminated uniformly. The camera and laser source need to be in focus uniformly throughout the field-of-view. The cameras and light sources need to be synchronized for simultaneous acquisition.
- Exposure time of the camera needs to be optimized such that the laser line width and the image defocus during movement of the scanner can be kept in control.
- Calibration of the sensor module should be carried out to get minimal calibration error. To ensure this a precision calibration target and calibration images with adequate contrast and focus levels need to be acquired. Distortion correction techniques should be adopted to ensure good measurement accuracy.
- The laser line center detection is an important aspect to achieve the desired accuracy. In handheld mode, there is possibility of laser-line thickness variation due to stand-off distance variation. Laser-line segments with optimal width are only considered for further computations.
- Marker-based registration method was adopted in this paper. In this method, the referencing target must be selected such that it exhibits good contrast in the field of view. To ensure this, retro-reflective targets made up of spherical glass beads embedded in an adhesive layer are used in this paper. It is important to accurately detect the marker's point of tracking, which in this paper is the circular marker center. Accurate 2D and 3D correspondence ensures a good registration.
- The position and orientation of the handheld scanning system has to be such that the stand-off distance is maintained and the device is approximately perpendicular to the object surface.

**Funding:** This research was funded by Department of Science and Technology, Government of India, under the Advanced Manufacturing Technology (AMT) Program. The grant number is DST/TDT/AMT/2017/42(G).

**Conflicts of interest:** The authors declare that they have no known competing financial interests or personal relationships that could have appeared to influence the work reported in this paper.

# References

[1] Furukawa, R., Kawasaki, H. (2003). Interactive shape acquisition using marker attached laser projector, Fourth International Conference on 3-D Digital Imaging and Modeling. 3DIM 2003. Proceedings, Banff, AB, Canada, 2003, 491–498, https://doi.org/10.1109/IM.2003.1240286.

[2] Bleier, M., Nuchter, A. (2017). Towards robust self-calibration for handheld 3d line laser scanning. ISPRS – International Archives of the Photogrammetry, Remote Sensing and Spatial Information Sciences, vol. 42, 31–36. https://doi.org/10.5194/isprs-archives-XLII-2-W8-31-2017.

[3] Furukawa, R., Kawasaki, H. (2005). Uncalibrated multiple image stereo system with arbitrarily movable camera and projector for wide range scanning, Fifth International Conference on 3-D Digital Imaging and Modeling (3DIM'05), Ottawa, ON, Canada, 302–309, https://doi.org/10.1109/3DIM.2005.80.

[4] Winkelbach, S., Molkenstruck, S., Wahl, F.M. (2006). Low-cost laser range scanner and fast surface registration approach. In: Franke, K., Müller, K.R., Nickolay, B., Schäfer, R., eds. Pattern Recognition. DAGM 2006. Lecture Notes in Computer Science, vol. 4174. Springer, Berlin, Heidelberg. https://doi.org/10.1007/11861898_72.

[5] Hebert, P. (2001). A self-referenced handheld range sensor, Proceedings Third International Conference on 3-D Digital Imaging and Modeling, Quebec City, QC, Canada, 5–12, https://doi.org/10.1109/IM.2001.924380.

[6] Niola, V., Rossi, C., Savino, S., Strano, S. (2011 April). A method for the calibration of a 3-D laser scanner. Robotics and Computer-integrated Manufacturing. https://doi.org/10.1016/j.rcim.2010.06.030.

[7] Idrobo-Pizo, G.A., Motta, J.M.S.T., Sampaio, R.C. (2019). A calibration method for a laser triangulation scanner mounted on a robot arm for surface mapping. Sensors 19. MDPI AG: 1783. https://doi.org/10.3390/s19081783.

[8] Steger, C. (1996). Extracting curvilinear structures: A differential geometric approach. In: Buxton, B., Cipolla, R., eds. Computer Vision – ECCV '96. ECCV 1996. Lecture Notes in Computer Science, vol. 1064. Springer, Berlin, Heidelberg. https://doi.org/10.1007/BFb0015573.

[9] Steger, C. (1996). Extraction of curved lines from images, Proceedings of 13th International Conference on Pattern Recognition, Vienna, Austria, 251–255 vol.2, https://doi.org/10.1109/ICPR.1996.546827.

[10] Hebert, P., Rioux, M. (1998 March 6). Toward a handheld laser range scanner: Integrating observation-based motion compensation, Proc. SPIE 3313, Three-Dimensional Image Capture and Applications. https://doi.org/10.1117/12.302439

[11] Damgrave, R., Johannes, G., Lutters, D. The drift of the Xsens MoveN motion capturing suit during common movements in a working environment. In Proceedings of the 19th CIRP Design Conference-Competitive Design, Cranfield, UK, 30–31 March 2009.
[12] Khoury, R. (2006). An enhanced positioning algorithm for a self-referencing handheld 3D sensor, The 3rd Canadian Conference on Computer and Robot Vision (CRV'06), Quebec, Canada, 44–44, https://doi.org/10.1109/CRV.2006.10.
[13] Barone, S., Paoli, A., Razionale, A.V. (2012). Three-dimensional point cloud alignment detecting fiducial markers by structured light stereo imaging. Machine Vision and Applications, 23, 217–229. https://doi.org/10.1007/s00138-011-0340-1.
[14] Steger, C., Ulrich, M., Wiedemann, C. (2008). Machine Vision Algorithms and Applications, Second edn. Wiley Publications.
[15] Cyganek, B., Siebert, J.P. (2009). An Introduction to 3D Computer Vision Techniques and Algorithms, Third edn. John Wiley & Sons Ltd.
[16] Hartley, R., Zisserman, A. (2004 March). Multiple View Geometry in Computer Vision, Second edn. Cambridge University Press.
[17] Wei, Z., Zhou, F., Zhang, G. (2005). 3D coordinates measurement based on structured light sensor. Sensors and Actuators A: Physical, 120, 527–535. 10.1016/j.sna.2004.12.007.
[18] Burgess, G., Shortis, M.R., Scott, P. (2011 August). Photographic assessment of retro reflective film properties. Isprs Journal of Photogrammetry and Remote Sensing, 743–750. https://doi.org/10.1016/j.isprsjprs.2011.07.002.

# Index



https://doi.org/10.1515/9783110756722-017

# De Gruyter Frontiers in Computational Intelligence

## Already published in the series

**Volume 14: Artificial Intelligence for Virtual Reality**
Jude Hemanth, Madhulika Bhatia, Isabel De La Torre Diez (Eds.)
ISBN 978-3-11-071374-9, e-ISBN (PDF) 978-3-11-071381-7, e-ISBN (EPUB) 978-3-11-071385-5

**Volume 13: Computational Intelligence in Software Modeling**
Vishal Jain, Jyotir Moy Chatterjee, Ankita Bansal, Utku Kose, Abha Jain (Eds.)
ISBN 978-3-11-070543-0, e-ISBN (PDF) 978-3-11-070924-7,
e-ISBN (EPUB) 978-3-11-070934-6

**Volume 12: Artificial Intelligence and Internet of Things for Renewable Energy Systems**
Neeraj Priyadarshi, Sanjeevikumar Padmanaban, Kamal Kant Hiran, Jens Bo Holm-Nielson, Ramesh C. Bansal (Eds.)
ISBN 978-3-11-071379-4, e-ISBN (PDF) 978-3-11-071404-3,
e-ISBN (EPUB) 978-3-11-071415-9

**Volume 11: Cyber Crime and Forensic Computing. Modern Principles, Practices, and Algorithms**
Gulshan Shrivastava, Deepak Gupta, Kavita Sharma (Eds.)
ISBN 978-3-11-067737-9, e-ISBN (PDF) 978-3-11-067747-8,
e-ISBN (EPUB) 978-3-11-067754-6

**Volume 10: Blockchain 3.0 for Sustainable Development**
Deepak Khazanchi, Ajay Kumar Vyas, Kamal Kant Hiran, Sanjeevikumar Padmanaban (Eds.)
ISBN 978-3-11-070245-3, e-ISBN (PDF) 978-3-11-070250-7,
e-ISBN (EPUB) 978-3-11-070257-6

**Volume 9: Machine Learning for Sustainable Development**
Kamal Kant Hiran, Deepak Khazanchi, Ajay Kumar Vyas, Sanjeevikumar Padmanaban (Eds.)
ISBN 978-3-11-070248-4, e-ISBN (PDF) 978-3-11-070251-4,
e-ISBN (EPUB) 978-3-11-070258-3

**Volume 8: Internet of Things and Machine Learning in Agriculture. Technological Impacts and Challenges**
Jyotir Moy Chatterjee, Abhishek Kumar, Pramod Singh Rathore, Vishal Jain (Eds.)
ISBN 978-3-11-069122-1, e-ISBN (PDF) 978-3-11-069127-6,
e-ISBN (EPUB) 978-3-11-069128-3

**Volume 7: Deep Learning. Research and Applications**
Siddhartha Bhattacharyya, Vaclav Snasel, Aboul Ella Hassanien, Satadal Saha, B. K. Tripathy (Eds.)
ISBN 978-3-11-067079-0, e-ISBN (PDF) 978-3-11-067090-5,
e-ISBN (EPUB) 978-3-11-067092-9

**Volume 6: Quantum Machine Learning**
Siddhartha Bhattacharyya, Indrajit Pan, Ashish Mani, Sourav De, Elizabeth Behrman, Susanta Chakraborti (Eds.)
ISBN 978-3-11-067064-6, e-ISBN (PDF) 978-3-11-067070-7,
e-ISBN (EPUB) 978-3-11-067072-1

**Volume 5: Machine Learning Applications. Emerging Trends**
Rik Das, Siddhartha Bhattacharyya, Sudarshan Nandy (Eds.)
ISBN 978-3-11-060853-3, e-ISBN (PDF) 978-3-11-061098-7,
e-ISBN (EPUB) 978-3-11-060866-3

**Volume 4: Intelligent Decision Support Systems. Applications in Signal Processing**
Surekha Borra, Nilanjan Dey, Siddhartha Bhattacharyya, Mohamed Salim Bouhlel (Eds.)
ISBN 978-3-11-061868-6, e-ISBN (PDF) 978-3-11-062110-5,
e-ISBN (EPUB) 978-3-11-061871-6

**Volume 3: Big Data Security**
Shibakali Gupta, Indradip Banerjee, Siddhartha Bhattacharyya (Eds.)
ISBN 978-3-11-060588-4, e-ISBN (PDF) 978-3-11-060605-8,
e-ISBN (EPUB) 978-3-11-060596-9

**Volume 2: Intelligent Multimedia Data Analysis**
Siddhartha Bhattacharyya, Indrajit Pan, Abhijit Das, Shibakali Gupta (Eds.)
ISBN 978-3-11-055031-3, e-ISBN (PDF) 978-3-11-055207-2,
e-ISBN (EPUB) 978-3-11-055033-7

**Volume 1: Machine Learning for Big Data Analysis**
Siddhartha Bhattacharyya, Hrishikesh Bhaumik, Anirban Mukherjee, Sourav De (Eds.)
ISBN 978-3-11-055032-0, e-ISBN (PDF) 978-3-11-055143-3,
e-ISBN (EPUB) 978-3-11-055077-1